Antonio Arnau (Ed.)

Piezoelectric Transducers and Applications

Springer

Berlin
Heidelberg
New York
Hong Kong
London
Milan
Paris
Tokyo

Antonio Arnau (Ed.)

Piezoelectric Transducers and Applications

With 112 Figures

Springer

Prof. Dr. Antonio Arnau Vives
Departamento de Ingeniería Electrónica
Universidad Politécnica de Valencia
Camino de Vera s/n
E - 46022 Valencia
Spain
aarnau@eln.upv.es

ISBN 3-540-20998-0 Springer-Verlag Berlin Heidelberg New York

Library of Congress Cataloging-in-Publication-Data
A catalog record for this book is available from the Library of Congress.
Bibliographic information published by Die Deutsche Bibliothek.
Die Deutsche Bibliothek lists this publication in the Deutsche Nationalbibliographie; detailed bibliographic data is available in the Internet at http://dnb.ddb.de

Springer-Verlag is a part of Springer Science+Business Media
springeronline.com

Typesetting: Data delivered by editor
Cover design: medio Technologies AG, Berlin
Printed on acid free paper 62/3020/M - 5 4 3 2 1 0

Foreword

One of the most valuable aids to those beginning to investigate a new area of study is a source which will guide them from beginning principles, through detailed implementation and applications. Even for seasoned investigators, it is useful to have a reasonably detailed discussion of closely related topics in a single volume to which one can refer. This is often difficult for many emerging areas of studies because they are so multidisciplinary. The subject matter of the principles, techniques and applications of piezoelectric transducers certainly fits into this category. The host of emerging new uses of piezoelectric devices that are being commercialized as well as the growing number of potential applications ensures that this field will encompass more and more disciplines with passing time. It is extremely fortunate and timely that this volume becomes available to the student at this time.

Piezoelectricity is a classical discipline traced to the original work of Jacques and Pierre Curie around 1880. This phenomenon describes the relations between mechanical strains on a solid and its resulting electrical behavior resulting from changes in the electric polarization. One can create an electrical output from a solid resulting from mechanical strains, or can create a mechanical distortion resulting from the application of an electrical perturbation. In the former case, the unit acts as a receiver of mechanical variations, converting it into electrical output, as in the case of a microphone. In the latter case, the unit can act as a transmitter converting the electrical signal into a mechanical wave. The piezoelectric units can be used both in narrow-band or resonant modes, and under broad-band regimes for detection and imaging applications. One of the remarkable properties of these devices is the ability to use them in a viscous medium, such as a liquid. When excited sinusoidally, these devices can generate waves in the immersing medium. Typically, as a result of the physical size of these devices, the waves are in the ultrasonic regime. From this classical discipline, an astounding number of applications are developing. From its use as a frequency generating standard in the earlier part of the 20th century, additional uses have seen these devices used as highly sensitive mass balances for use both in vacuum deposition and in electrochemical applications, as well as chemical specific sensors, as Doppler devices for fluid velocity measurements and for ultrasonic imagery. There are many other emerging applications in the bio-sciences for example. The number of applications is astounding.

It is clear that the discipline is inter-disciplinary. The authors of the contents of this book are a select group who has all been challenged by the intellectual diversity of the field. To successfully pass on such diverse information, intellectual competence is only a beginning. A devotion to, and love of clear communication is also required. These authors are members of the PETRA organization, (Piezoelectric Transducers and Applications) sponsored by the European Union, devoted to the collection and dissemination of knowledge and skills in the piezoelectric arts to students among the participating universities in Europe and Latin America. I have personally observed many of the authors interacting with students and have been very impressed by their care and mentoring. Contributions from such dedicated and seasoned teachers are now available to the student in this volume. This book fills a real need for a unified source for information on piezoelectric devices, ranging from broadband applications to resonant applications and will serve both experienced researchers and beginning students well.

Kay Keiji Kanazawa
Technical Director, Emeritus
CPIMA
Stanford University
Stanford, CA 94305

Preface

This book is one of the results from the PETRA network (PiezoElectric TRansducers and Applications) following the execution of the project PETRA I co-financed by the European Union in the framework of the ALFA Program (America Latina Formación Académica). It is not the objective of the book to give a complete and detailed treatment of all the facets related to piezoelectricity, piezoelectric transducers and their applications, which would be completely impossible in a single-volume work, but to provide a guide for doctoral degree students and for researchers going into this complex and multidisciplinary issue, at least in those topics in which the members of the network carry out their research.

The book covers, in 16 chapters and two appendices, aspects related to piezoelectric transducers and their applications. Some of the chapters focus on the more basic concepts of the different disciplines involved and are presented in a didactic way, while others go deeper into complex aspects of the specific field of research that they treat, following a similar structure to that of a scientific article. Chapters 1, 11 and Appendices A and B present, respectively, basic concepts of piezoelectricity, electrochemistry, electrostatics and physical properties of crystals. Chapters 4, 6, 13 and 16 delve more deeply into resonant sensors, especially bulk acoustic wave thickness shear mode resonators and their applications as quartz microbalance sensors, their fundamentals and models (Chap. 4), electronic interfaces and associated problems (Chap. 6), the problems associated with the analysis and interpretation of the experimental data (Chap. 16), and complementary techniques (Chap. 13). Chapters 3, 5 and 14 deal with broadband ultrasonic systems, introducing basic aspects and different models of piezoelectric transducers (Chap. 3), electronic interfaces used in broadband configurations (Chap. 5) and ultrasonic schemes and electronic implementations for non-destructive testing industrial applications (Chap. 14). Chapters 8 and 15 introduce specific applications of ultrasonic waves in different disciplines - sonoelectrochemistry (Chap. 8) and hyperthermia (Chap. 15).

Finally, Chapters 2, 7, 9, 10 and 12 present overviews concerning specific or related aspects of resonant sensors and their applications. Chapter 2 is an overview of acoustic sensors, and Chapter 7 provides a guide to the important subject of modification of piezoelectric surfaces in piezoelectric transducers for sensor applications. Chapter 9 provides an overview of chemical sensors, which is of great interest for establishing the differences between chemical sensors based on piezoelectric transducers and those based on other techniques, such as electrochemical, optical, calorimetric and magnetic techniques, and in consequence to facilitate the interpretation of the different data. Chapter 10 treats the specific topic of biosensors from a biological point of view; this treatment is especially useful to understand the mechanism of biological recognition and its potential use for the development of biosensors and, specifically, for piezoelectric biosensors, which is a field of much current interest. Finally, Chapter 12 introduces the concept of viscoelasticity and describes in depth the physical properties of polymers, an aspect of great importance in resonant sensor applications.

The present volume is the result of the effort of a work group that began to be formed nearly 4 years ago and that I have had the honour of coordinating. I will always be in debt to my colleagues, eminent professors and now dear friends that gave me, at that time, the opportunity and the responsibility of creating something, maybe different, and therefore giving sense to life, since, as Gregorio Marañón, a famous Spanish doctor in medicine and writer said: *to live is not only to exist, but to exist and to create, to know how to enjoy and suffer and not to sleep without dreaming, to rest is to begin to die.* Any effort would have been useless without the work and dedication of all of those that carried out the important labour of writing each of the chapters and who reviewed and patiently corrected the different versions of the chapters. Theirs is all the merit, the defects only belong to me, but if the work that is presented can facilitate the study of and the access to this multidisciplinary topic, it will have been worth it. They say that the bad things come without being called while the good things have to be carried out with great effort. I hope this effort has provided something good that helps others. Very recently, the project PETRA II, the second phase of PETRA I, was approved by the European Union for development during the next three years. The project will include 28 research training projects and the creation of a joint course of multidisciplinary training by the 14 members of the network. Undoubtedly, it will require much effort, but it will allow us to create something more, new and different that will help us to continue giving sense to our lives.

Antonio Arnau Vives

December 2003

Contents

Associated Editors and Contributors

Arnau, A.
Departamento de Ingeniería Electrónica, Universidad Politécnica de Valencia
Camino de Vera s/n, Valencia E-46022
Spain

Brett, C.
Departamento de Química, Faculdade de Ciências e Tecnologia, Universidade de Coimbra
Rua Larga, Coimbra-3004-535
Portugal

Calvo, E.
Facultad de Ciencias Exactas y Naturales, Universidad de Buenos Aires
Pabellon 2, Ciudad Universitaria, AR-1428
Argentina

Ferrari, V.
Dipartamento di Electrónica per l'Automazione, Universita' degli Studi di Brescia
Via Branze 38, Brescia I-25123
Italy

González, G.
Departamento de Ingeniería Eléctrica, Centro de Investigación y Estudios Avanzados
Avda. Instituto Politécnico Nacional Nº 2508, San Pedro Zacatenco, Mexico, D.F. 07360
Mexico, D.F.

Jiménez, Y.
Departamento de Ingeniería Electrónica, Universidad Politécnica de Valencia.
Camino de Vera s/n, Valencia E-46022
Spain

Kanazawa, K.
Center of Polymer Interfaces and Macromolecular Assemblies, Stanford University
Stanford University, North-South Mall 381, Stanford, CA 94305-5025
USA

Leija, L.
Departamento de Ingeniería Eléctrica, Centro de Investigación y Estudios Avanzados
Avda. Instituto Politécnico Nacional Nº 2508, San Pedro Zacatenco, Mexico, D.F. 07360
Mexico, D.F.

Lucklum, R.
Institute for Micro and Sensor Systems, Otto-Von-Güericke Universität Magdeburg
Universitätsplatz 2, Magdeburg D-39016
Germany

Muñoz, R.
Departamento de Ingeniería Eléctrica, Centro de Investigación y Estudios Avanzados
Avda. Instituto Politécnico Nacional nº 2508, San Pedro Zacatenco, Mexico, D.F. 07360
Mexico, D.F.

Otero, M.
Facultad de Ciencias Exactas y Naturales, Universidad de Buenos Aires.
Pabellon 2, Ciudad Universitaria, AR-1428
Argentina

Perrot, H.
Laboratoire Interface et Systèmes Electrochimiques, Université P. et M. Curie
Place Jussieu 4, Paris 75252
France

Ramos, A.
Departamento de Señales, Sistemas y Tecnologías Ultrasónicas. Instituto de Acústica. Consejo Superior de Investigaciones Científicas.
Serrano 144, Madrid 28006
Spain

San Emeterio, J.L.
Departamento de Señales, Sistemas y Tecnologías Ultrasónicas. Instituto de Acústica. Consejo Superior de Investigaciones Científicas.
Serrano 144, Madrid 28006
Spain

Soares, D.
Instituto de Física, Departamento de Física Aplicada, Universidade Estadual de Campinas
Caixa Postal 6165, Campinas 13083-970
Brasil

Sogorb, T.
Departamento de Ingeniería Electrónica, Universidad Politécnica de Valencia
Camino de Vera s/n, Valencia E-46022
Spain

Stipek, S.
Institute of Medical Biochemistry, First Faculty of Medicine, Charles University in Prague
Katerinská 32. Prague CZ-121 08
Czech Republic

1 Fundamentals on Piezoelectricity

Antonio Arnau[1] and David Soares[2]

[1] Departamento de Ingeniería Electrónica, Universidad Politécnica de Valencia
[2] Institute de Fisica, Universidade de Campinas

1.1 Introduction

The topic of the following chapter is relatively difficult and includes different areas of knowledge. The piezoelectric phenomenon is a complex one and covers concepts of electronics as well as most of the areas of classical physics such as: mechanics, elasticity and strength of materials, thermodynamics, acoustics, wave's propagation, optics, electrostatics, fluids dynamics, circuit theory, crystallography etc. Probably, only a few disciplines of engineering and science need to be so familiar to so many fields of physics. Current bibliography on this subject is vast though dispersed in research publications, and few of the books on this topic are usually compilations of the authors' research works. Therefore, they are not thought for didactic purposes and are difficult to understand, even for postgraduates. The objective of this chapter is to help understand the studies and research on piezoelectric sensors and transducers, and their applications. Considering the multidisciplinary nature, this tutorial's readers can belong to very different disciplines. They can even lack the necessary basic knowledge to understand the concepts of this chapter. This is why the chapter starts providing an overview of the piezoelectric phenomenon, doing consciously initial simplifications, so that the main concepts, which will be progressively introduced, prevail over the accessories. The issues covered in this chapter must be understood without the help of additional texts, which are typically included as references and are necessary to study in depth specific topics. Finally, the quartz crystal is introduced as a microgravimetric sensor to present the reader an application of the piezoelectric phenomenon, which will be dealt with along the following chapters.

1.2 The Piezoelectric Effect

The word *Piezoelectricity* comes from Greek and means “electricity by pressure” (*Piezo* means pressure in Greek). This name was proposed by Hankel [1] in 1881 to name the phenomenon discovered a year before by the Pierre and Jacques Curie brothers [2]. They observed that positive and negative charges appeared on several parts of the crystal surfaces when comprising the crystal in different directions, previously analysed according to its symmetry.

Figure 1.1a shows a simple molecular model; it explains the generating of an electric charge as the result of a force exerted on the material. Before subjecting the material to some external stress, the gravity centres of the negative and positive charges of each molecule coincide. Therefore, the external effects of the negative and positive charges are reciprocally cancelled. As a result, an electrically neutral molecule appears. When exerting some pressure on the material, its internal reticular structure can be deformed, causing the separation of the positive and negative gravity centres of the molecules and generating little dipoles (Fig. 1.1b). The facing poles inside the material are mutually cancelled and a distribution of a linked charge appears in the material’s surfaces (Fig. 1.1c). That is to say, the material is polarized. This polarization generates an electric field and can be used to transform the mechanical energy used in the material’s deformation into electrical energy.

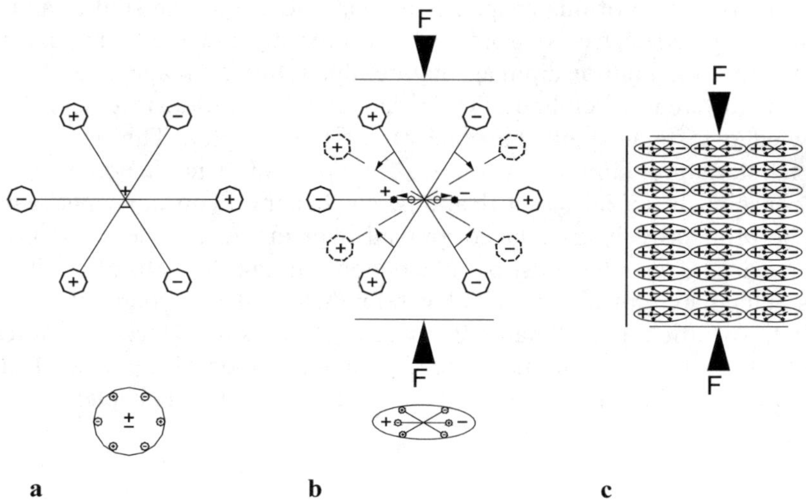

Fig. 1.1. Simple molecular model for explaining the piezoelectric effect: **a** unperturbed molecule; ***b*** molecule subjected to an external force, and ***c*** polarizing effect on the material surfaces

Figure 1.2a shows the piezoelectric material on which a pressure is applied. Two metal plates used as electrodes are deposited on the surfaces where the linked charges of opposite sign appear. Let us suppose that those electrodes are externally short circuited through a wire to which a galvanometer has been connected. When exerting some pressure on the piezoelectric material, a linked charge density appears on the surfaces of the crystal in contact with the electrodes. This polarization generates an electric field which causes the flow of the free charges existing in the conductor. Depending on their sign, the free charges will move towards the ends where the linked charge generated by the crystal's polarization is of opposite sign. This flow of free charges will remain until the free charge neutralizes the polarization effect (Fig. 1.2a). When the pressure on the crystal stops, the polarization will disappear, and the flow of free charges will be reversed, coming back to the initial standstill condition (Fig. 1.2b). This process would be displayed in the galvanometer, which would have marked two opposite sign current peaks. If a resistance is connected instead of a short-circuiting, and a variable pressure is applied, a current would flow through the resistance, and the mechanical energy would be transformed into electrical energy.

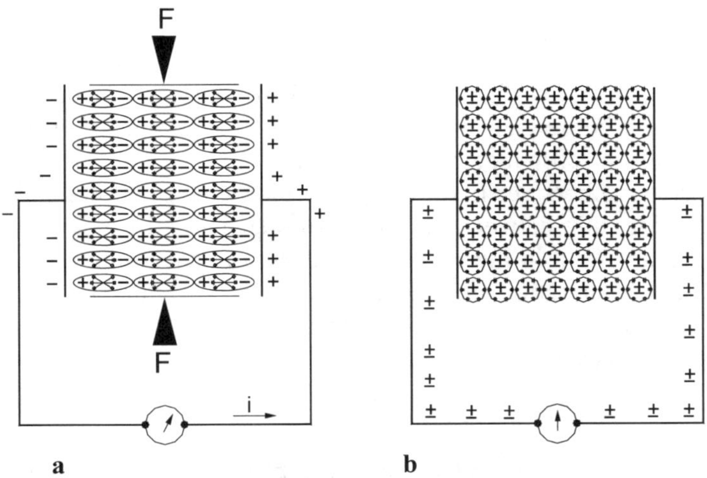

Fig. 1.2. Piezoelectric phenomenon: **a** neutralizing current flowing through the short-circuiting established on a piezoelectric material subjected to an external force; **b** absence of current through the short-circuited material in an unperturbed state

The Curie brothers verified, the year after their discovery, the existence of the reverse process, predicted by Lippmann (1881) [3]. That is, if one arbitrarily names *direct piezoelectric effect*, to the generation of an electric charge, and hence of an electric field, in certain materials and under certain laws due to a stress, there would also exist a *reverse piezoelectric effect* by which the application of an electric field, under similar circumstances, would cause deformation in those materials.

In this sense, a mechanical deformation would be produced in a piezoelectric material when a voltage is applied between the electrodes of the piezoelectric material, as shown in Fig.1.2. This strain could be used, for example, to displace a coupled mechanical load, transforming the electrical energy into mechanical energy.

1.3 Mathematical Formulation of the Piezoelectric Effect. A First Approach

In a first approach, the experiments performed by the Curie brothers demonstrated that the surface density of the generated linked charge was proportional to the pressure exerted, and would disappear with it. This relationship can be formulated in a simple way as follows:

$$P_p = d\,T \tag{1.1}$$

where P_p is the piezoelectric polarization vector, whose magnitude is equal to the linked charge surface density by piezoelectric effect in the considered surface, d is the piezoelectric strain coefficient and T is the stress to which the piezoelectric material is subjected.

The Curie brothers verified the reverse piezoelectric effect and demonstrated that the ratio between the strain produced and the magnitude of the applied electric field in the reverse effect, was equal to the ratio between the produced polarization and the magnitude of the applied stress in the direct effect. Consistently, the reverse piezoelectric effect can be formulated in a simple way, as a first approach, as follows:

$$S_p = d\,E \tag{1.2}$$

where S_p is the strain produced by the piezoelectric effect and E is the magnitude of the applied electric field.

The direct and reverse piezoelectric effects can be alternatively formulated, considering the elastic properties of the material, as follows:

$$P_p = d\,T = d\,c\,S = e\,S \tag{1.3}$$

$$T_p = c\,S_p = c\,d\,E = e\,E \tag{1.4}$$

where c is the elastic constant, which relates the stress generated by the application of a strain ($T = c\,S$), s is the compliance coefficient which relates the deformation produced by the application of a stress ($S = s\,T$), and e is the piezoelectric stress constant. (Note that the polarizations, stresses, and strains caused by the piezoelectric effect have been specified with the p subscript, while those externally applied do not have subscript. Although unnecessary, it will be advantageous later on.

1.4 Piezoelectric Contribution to Elastic Constants

The piezoelectric phenomenon causes an increase of the material's stiffness. To understand this effect, let us suppose that the piezoelectric material is subjected to a strain S. This strain will have two effects. On the one hand, it will generate an elastic stress T_e which will be proportional to the mechanical strain $T_e = c\,S$; on the other hand, it will generate a piezoelectric polarization $P_p = e\,S$ according to Eq. (1.3). This polarization will create an internal electric field in the material E_p given by (see Appendix A):

$$E_p = \frac{P_p}{\varepsilon} = \frac{eS}{\varepsilon} \tag{1.5}$$

where ε is the dielectric constant of the material.

This electric field, of piezoelectric origin, produces a force against the deformation of the material's electric structure, creating a stress $T_p = e\,E_p$. This stress, as well as that of elastic origin, is against the material's deformation. Consistently, the stress generated as a consequence of the strain S will be:

$$T = T_e + T_p = cS + \frac{e^2}{\varepsilon}S = \left(c + \frac{e^2}{\varepsilon}\right)S = \overline{c}\,S \tag{1.6}$$

Therefore, the constant $\overline{c}$ is the piezoelectrically stiffened constant, which includes the increase in the value of the elastic constant due to the piezoelectric effect. This coefficient will appear later on.

1.5 Piezoelectric Contribution to Dielectric Constants

When an external electric field E is applied between two electrodes where a material of dielectric constant ε exists, an electric displacement is created towards those electrodes, generating a surface charge density $\sigma = \sigma_o + \sigma_p$

which magnitude is $D = \varepsilon E$ [1]. If that material is piezoelectric, the electric field E produces a strain given by: $S_p = d\,E$. This strain of piezoelectric origin increases the surface charge density due to the material's polarization in an amount given by: $P_p = e\,S_p = e\,d\,E$ (Fig. 1.3). Because the electric field is maintained constant, the piezoelectric polarization increases the electric displacement of free charges towards the electrodes in the same magnitude ($\sigma_p = P_p$). Therefore, the total electrical displacement is:

$$D = \varepsilon E + P_p = \varepsilon E + ed\,E = \overline{\varepsilon}\,E \tag{1.7}$$

where $\overline{\varepsilon}$ is the effective dielectric constant which includes the piezoelectric contribution.

1.6 The Electric Displacement and the Internal Stress

As shown in the previous paragraph, the electric displacement produced when an electric field E is applied to a piezoelectric and dielectric material is:

$$D = \varepsilon E + P_p = \varepsilon E + e S_p \tag{1.8}$$

Under the same circumstances we want to obtain the internal stress in the material. The reasoning is the following: the application of an electric field on a piezoelectric material causes a deformation in the material's structure given by: $S_p = d\,E$. This strain produces an elastic stress whose magnitude is $T_e = c\,S_p$. On the other hand, the electric field E exerts a force on the material's internal structure generating a stress given by: $T_p = e\,E$. This stress is, definitely, the one that produces the strain and is of opposite sign to the elastic stress which tends to recover the original structure. Therefore, the internal stress that the material experiences will be the resultant of both. That is:

$$T = cS_p - eE \tag{1.9}$$

---

[1] The free charge density which appears on the electrodes, will be the sum of the charge density which appears in vacuum plus the one that appears induced by the dielectric effect, i.e.:

$$\sigma_o + \sigma_d = \varepsilon_o E + \chi E = (\varepsilon_o + \chi) E = \varepsilon E$$

where ε_o is the vacuum dielectric permittivity and χ is the dielectric susceptibility of the material.

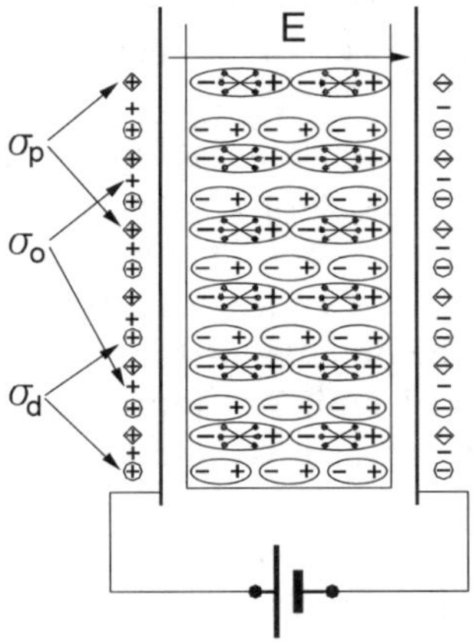

Fig. 1.3. Schematic diagram that explains different electrical displacements associated with a piezoelectric and dielectric material

Eventually, both stresses will be equal leaving the material strained and static. If a variable field is applied, as it is the common practice, the strain will vary as well, producing a dynamic displacement of the material's particles. This electromechanical phenomenon generates a perturbation in the medium in contact with the piezoelectric material. This effect is used in transducers, sensors and actuators, as it will be seen along the following chapters.

1.7 Basic Model of Electric Impedance for a Piezoelectric Material Subjected to a Variable Electric Field

In the previous section, the expressions for the electric displacement and the internal stress produced in a piezoelectric material subjected to an electric field have been obtained.

The electric field is created when a voltage difference is applied between two electrodes deposited on certain surfaces of the material. If the applied voltage difference changes, the electric field as well as the electric displacement change inducing an electric current through the electrodes. The ratio between the applied voltage and the induced electric current is the electric impedance of the piezoelectric component. For example, if it only has dielectric properties, the resulting electric impedance corresponds to a capacitance. Piezoelectric devices are included in electric and electronic circuits to use their electromechanical properties in both direct and reverse applications. Therefore, it is important to obtain an electric model that allows including the piezoelectric component in electric circuits. This will greatly facilitate the analysis of the circuit and the understanding of its operation.

Next, the basic equivalent electric model mentioned will be obtained. In the obtaining of the model some simplifications will be made to minimize the mathematical formulation. These simplifications do not essentially modify the results and let us show the qualitative physical concepts of the model. On the other hand, the obtained expressions for the electric parameters of the model will be very similar to those obtained from a more rigorous mathematical development, as it will be shown in Appendix 1.A.

Figure 1.4 shows the transversal section of a bar of piezoelectric material of thickness *l*. Let us suppose that when applying a field in the direction of the thickness (direction *Y*) by the application of a voltage difference between the electrodes, the material deforms as shown in the left part of Fig. 1.4. When the field is reversed, the strain is reversed as well (right part).

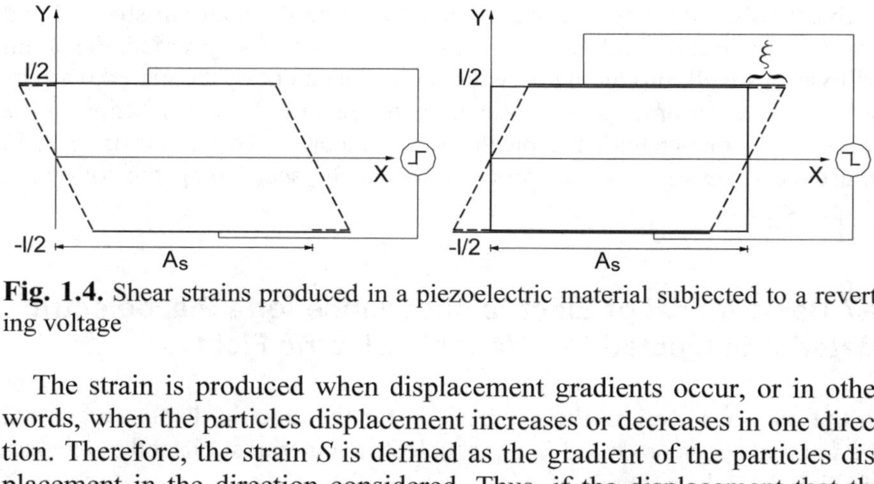

Fig. 1.4. Shear strains produced in a piezoelectric material subjected to a reverting voltage

The strain is produced when displacement gradients occur, or in other words, when the particles displacement increases or decreases in one direction. Therefore, the strain *S* is defined as the gradient of the particles displacement in the direction considered. Thus, if the displacement that the particles experience along a distance *y* is $\xi(y)$, the strain produced along this section will be:

$$S(y) = \frac{\xi(y)}{y} \tag{1.10}$$

Figure 1.4 shows how the particles displacement increases with the coordinate *y*, being null on the abscissas axis[2]. Consistently, the maximum

[2] This type of strain is called in thickness shear mode, and is very common. Precisely, bars of quartz crystal obtained from the AT cut (bars obtained through cut-

strain is produced at $y=l/2$ and is the same in both ends but of opposite sign due to the change of sign in the displacement. Therefore, the strain at $y=l/2$ will be:

$$S(l/2)=\frac{\xi}{l/2}=\frac{2\xi}{l} \tag{1.11}$$

where ξ is the particle displacement at the coordinate $y=l/2$ at a generic instant.

Figure 1.5 shows the forces acting on the material ends when the electric field is applied. This electric field creates a force in the X direction which produces a piezoelectric stress given by Eq. (1.4). An elastic stress $T_e = c\,S_p$ is against the piezoelectric stress and tries to avoid the strain of the material. The internal friction that the particles experience in their displacement is also against the piezoelectric stress since it makes the particles displacement more difficult. The stress due to internal friction is usually considered proportional to the gradient of the particle displacement velocity, as in the case of a viscous phenomenon, that is:

$$T_v = \eta\frac{dv}{dy} = \eta\frac{d^2\xi}{dy\,dt} = \eta\frac{dS}{dt} \tag{1.12}$$

where constant η is named viscosity.

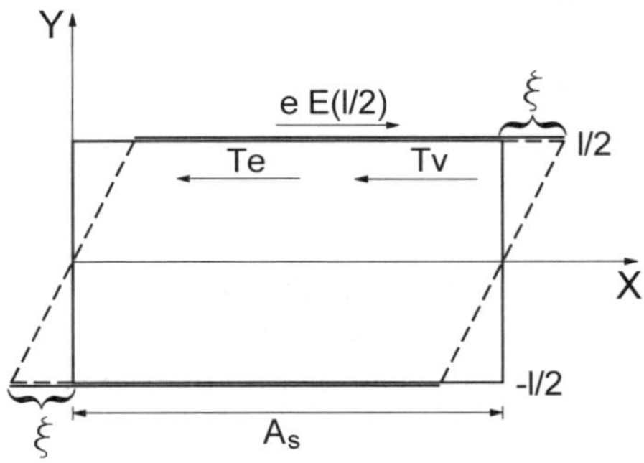

Fig. 1.5. Shear strain and stresses produced at the end of a piezoelectric plate subjected to an electric field

tings done with an angle of 35°15' in relation to the optical axis Z [4]) present a very pure shear vibration mode when an electric field is applied in the direction of the thickness. The anisotropy of the quartz is the responsible for this phenomenon. The anisotropy complicates the mathematical formulation of the elastic, dielectric and piezoelectric effects (see Appendix B). A deeper study of the piezoelectricity considering the anisotropy phenomenon can be found elsewhere [4,5].

The resultant of the forces will be equal to the product of mass by the acceleration of the particles. As stresses are being considered, it will be necessary to take into account the surface mass density ρ_s. Therefore, Newton's first law applied to the material surface at the coordinate $y=l/2$ will be:

$$\sum T = e\,E(l/2) - c\,S(l/2) - \eta \frac{dS(l/2)}{dt} = \rho_s \frac{d^2\xi}{dt^2} \tag{1.13}$$

Considering Eq. (1.11), Eq. (1.13) and that $E(l/2)=V/l$, where V is the voltage difference between the electrodes, the following expression for the voltage V is obtained:

$$V = \frac{2\eta}{e}\frac{d\xi}{dt} + \frac{\rho_s l}{e}\frac{d^2\xi}{dt^2} + \frac{2c}{e}\xi \tag{1.14}$$

On the other hand, the electric displacement on the electrodes is given by Eq. (1.8). The time derivative of the electric displacement provides the density of the induced current J given by:

$$J = \frac{dD(l/2)}{dt} = \varepsilon\frac{dE(l/2)}{dt} + e\frac{dS(l/2)}{dt} = J_d + J_p \tag{1.15}$$

The first term of the second member J_d corresponds to the density of the induced current by the dielectric effect and the second term J_p to the current induced by the piezoelectric effect. Let us analyse the second term, which can be written from Eq. (1.11) as:

$$J_p = \frac{2e}{l}\frac{d\xi}{dt} \tag{1.16}$$

Taking into account that the surface density current $J_p = i_p / A_S$, where i_p is the current induced by piezoelectric effect and A_S is the electrodes surface, the following relationship can be obtained from Eq. (1.16):

$$\frac{d\xi}{dt} = \frac{l}{2eA_S} i_p \tag{1.17}$$

By substituting Eq. (1.17) in Eq. (1.14), it is definitely obtained:

$$V = \frac{\eta\, l}{A_S e^2} i_p + \frac{\rho_s l^2}{2A_S e^2}\frac{di_p}{dt} + \frac{c\, l}{A_S e^2}\int i_p\, dt \tag{1.18}$$

The voltage arising between the ends of a series circuit formed by a resistance R_m, an inductance L_m and a capacitance C_m through which an i_p current flows, has the following expression:

$$V = R_m i_p + L_m \frac{di_p}{dt} + \frac{1}{C_m} \int i_p \, dt \tag{1.19}$$

Therefore, the current induced by the piezoelectric effect, i.e., by the electromechanical effect, in the material is the same as the one that would flow through a series electric circuit formed by a resistor, a coil and a capacitor with the following magnitudes of resistance, inductance and capacitance, respectively:

$$R_m = \frac{\eta l}{Ae^2} = K_R \eta, \quad L_m = \frac{\rho_s l^2}{2Ae^2} = K_L \rho_s, \quad C_m = \frac{Ae^2}{cl} = K_C \frac{1}{c} = K_C s$$

The former expressions make clear the relationships among the electrical parameters and mechanical properties of the material, which are: the resistive electric parameter is proportional to the viscosity and models the physical phenomenon of energy loss due to viscous effects. The inductive parameter is proportional to the surface mass density and models the energy storage by inertial effect, and the capacitive parameter which is proportional to the elastic compliance models the energy storage by elastic effect. These relationships, which settle a clear analogy between the physical properties and the electric parameters, are very useful when evaluating the physical phenomena which take place when the piezoelectric material is used as a micro-gravimetric sensor, at least in simple cases, as it will be seen in Chaps. 4, 12, and 16.

Apart from the i_p component, it is also necessary to consider the component i_d associated with the dielectric effect. In fact, it can be written from Eq. (1.15) as follows:

$$i_d = A_S J_d = A_S \varepsilon \frac{dE(l/2)}{dt} = \varepsilon \frac{A_S}{l} \frac{dV}{dt} \tag{1.20}$$

Equation (1.20) corresponds to the current induced through a capacitor $C_o = \varepsilon A_S / l$ when a variable voltage difference V is applied.

Consistently, the circuit that models the electrical impedance of a piezoelectric and dielectric material subjected to a variable voltage difference is shown in Fig. 1.6. The electric circuit is formed by two parallel branches: one of them is the so-called motional branch formed by a series $R_m L_m C_m$ circuit that models the motional physical phenomenon. The other is the so-called static branch formed by a capacitor C_o which is associated to the electrical capacitance arising from the dielectric material placed between the two electrodes.

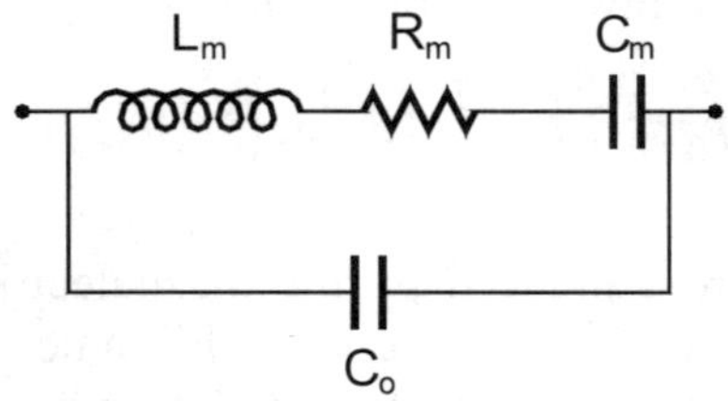

Fig. 1.6. Equivalent electrical model of a piezoelectric material vibrating at frequencies near resonance

The electrical model obtained, even with the corresponding simplifications, represents the real electrical impedance of the component when it vibrates at a frequency near some of its natural vibrating frequencies or resonant frequencies (see next section). In the Appendix 1.1, a more exact calculation of the component's electric admittance (reciprocal to the electric impedance) is developed. The process followed in the appendix is similar to the one made in Chap. 4 (Appendix 4.A) to determine the electrical admittance associated with a piezoelectric sensor in contact with a medium and, in consequence, its reading is recommended.

1.8 Natural Vibrating Frequencies

1.8.1 Natural Vibrating Frequencies Neglecting Losses

In the previous section the concept of natural vibrating frequencies or resonant frequencies has arisen. In this section these concepts will be studied. Let us suppose that a piezoelectric material, of characteristics similar to those presented in the previous section, is subjected to a strain as the one illustrated in Fig. 1.7 (upper part). The stress that the particles present under these conditions is given by Eq. (1.6). At a certain instant, the external force which maintains the strain is removed and the material starts to vibrate freely. Let us analyse that vibration.

Let us consider a slice of material of thickness *dy* located at the coordinate *y*. This slice is subjected to forces at both ends, as shown in Fig.1 7 (central part). The resultant of the forces will be equal to the product of the slice's mass by the acceleration to which the slice is subjected. This can be mathematically written as follows:

$$\frac{\partial F(y,t)}{\partial y}dy = \rho_v\, A_S\, dy\, \frac{\partial^2 \xi(y,t)}{\partial t^2} \tag{1.21}$$

In the former expression, it has been assumed that the force F and the displacement ξ depend on both coordinate y and the time t. Also, the mass has been written as the product of the material's density ρ_v by the slice's differential volume $A_S\, dy$; where A_S is the surface perpendicular to the paper plane.

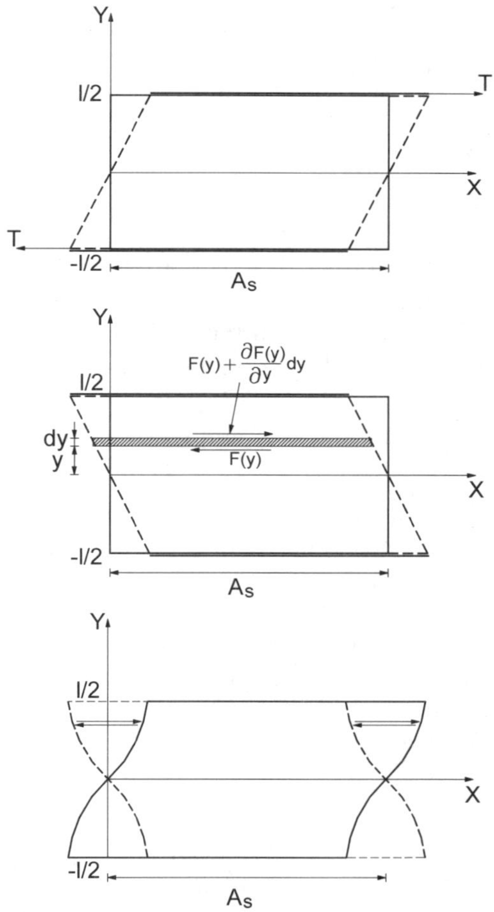

Fig. 1.7. Figures that explain the natural vibration of a piezoelectric resonator: The upper part shows the resonator bar subjected to an external stress, the central part shows the forces that an internal thin slice of a strained piezoelectric material experience, and the lower part shows the displacement profile of a piezoelectric material subjected to a sinusoidal electric field

Equation (1.21) can be written in terms of stress dividing by the surface A_S in both members. Considering Eq. (1.6) and writing $S=\partial\xi/\partial y$, Eq. (1.21) results into:

$$\frac{\partial T}{\partial y}=\bar{c}\frac{\partial^2\xi}{\partial y^2}=\rho_v\frac{\partial^2\xi}{\partial t^2} \tag{1.22}$$

Now, let us assume that the particle displacement near equilibrium has a sinusoidal dependence with time. Thus, the time derivative of particle displacement can be replaced by the product $j\omega$, where j is the complex base $\sqrt{-1}$ and $\omega=2\pi f$ is the oscillating angular frequency of the particles; where f is the frequency. Therefore, Eq. (22) becomes:

$$v_o^{\;2}\frac{\partial^2\xi}{\partial y^2}+\omega^2\xi=0 \tag{1.23}$$

where $v_o=\sqrt{\bar{c}/\rho_v}$.

The resolution of the former differential equation will provide the instantaneous profiles of the particle displacement with regard to the coordinate y. These profiles correspond to those shown in Fig. 1.7 (lower part). The displacement function fulfilling Eq. (1.23) is:

$$\xi=\xi_o\sin(2\pi k y+\sigma)=\xi_o\sin\left(\frac{2\pi}{\lambda}y+\sigma\right) \tag{1.24}$$

where $\lambda=v_o/f$ is the wave length, $k=1/\lambda=f/v_o=\omega/2\pi v_o$ is called the wave number since it corresponds to the number of complete wave lengths in the distance unit; ξ_o is the maximum amplitude of oscillation and σ is a constant to determine consistently with the boundary conditions.

In this case, the particles displacement is null at the coordinate y=0. This condition implies σ=0. Also, the amplitude of oscillation must be a maximum at the ends where $y=\pm l/2$. Therefore, it is necessary that the following condition be fulfilled:

$$2\pi k\frac{l}{2}=n\frac{\pi}{2} \quad\Rightarrow\quad n=1,2,3\ldots \tag{1.25}$$

This condition forces the oscillation at frequencies f_{no} which have to be odd multiples of a frequency f_o according to the following expression:

$$f_{no}=n f_o=n\frac{v_o}{2l} \quad\Rightarrow\quad n=1,2,3\ldots \tag{1.26}$$

Frequency f_o is called natural vibration fundamental frequency or resonant frequency (see following paragraphs) and its multiples are called harmonics of the fundamental frequency. Notice that v_o is the perturbation's propagation speed in the material without losses. Indeed, the speed is the ratio between the distance covered by the perturbation and the time it takes to go through that distance. From the definition of v_o as a function of the wave length and the oscillation frequency we get $v_o = \lambda f$. This equation indicates that the perturbation covers a space corresponding to a wave length in the time corresponding to a period of the oscillation. This is precisely the definition of the propagation speed.

Equation (1.26) also indicates that the frequencies of natural vibration depend solely on the material's physical properties and on its thickness. It also seems to indicate that the only possible vibrating frequencies are the ones that fulfil that condition. In fact, Eq. (1.26) is the result of simplifying the problem to only one dimension. When the lateral dimensions are infinite in comparison with thickness, the vibrating frequencies relative to those directions are null. However, in practice, the portions of material are three-dimensional with finite dimensions. In practice, two of the dimensions are much bigger than the third one, and they can be considered approximately as two–dimensional systems. In these cases, additional possible vibrating modes take place. Most of these vibration modes are not exact multiples of the fundamental mode; that is, they are not harmonically related to the fundamental; therefore they are called inharmonic modes. An important problem in practical applications is that these inharmonic modes can vibrate at frequencies very close to those of natural vibration. The correct application of electrodes on the piezoelectric material cancels some inharmonic modes. Other modes must be cancelled through an adequate design and additional contouring techniques which are not always applicable. Besides the inharmonic modes, the crystal's anisotropy generates the so called coupled vibrating modes, where a determined vibrating mode also excites another one [4].

1.8.2 Natural Vibrating Frequencies with Losses

In the previous study, the vibration caused by the initially created strain is maintained indefinitely. However, the free vibrations of any real physical system disappear with time. The reason is that any vibrating system involves phenomena which dissipate energy and eventually cause the vibration to stop. The incorporation of the loss effects to the previous vibrating system's physical model represents an approach to reality. This can be done by including the loss stress, modelled as a viscous effect, already formulated in Eq. (1.12). This way, the global recovering stress is:

$$T = \overline{c}S + \eta\frac{\partial S}{\partial t} \tag{1.27}$$

Equation (1.22) is transformed now into:

$$\frac{\partial T}{\partial y} = \overline{c}\frac{\partial^2 \xi}{\partial y^2} + \eta\frac{\partial^3 \xi}{\partial t\,\partial^2 y} = \rho_v\frac{\partial^2 \xi}{\partial t^2} \tag{1.28}$$

In order to solve Eq. (1.28), let us assume that the displacement, which is a function of the coordinate *y* and the time *t*, can be written as the product of two functions of separate variables, i.e., $\xi(y,t) = \zeta(t)\varphi(y)$. Additionally, let us suppose that the particles displacement has a sinusoidal profile with coordinate *y*. In fact, this is the profile obtained in the previous case, when neglecting the losses and is the profile the particles very accurately follow when losses are small, as it will be seen in the next section. In this case, each partial derivate in relation to *y* can be replaced in Eq. (1.28) by *(j2πk)* (see Eq. (1.24)), giving in the following expression:

$$\frac{\partial^2 \zeta}{\partial t^2} + \frac{4\pi^2 k^2 \eta}{\rho_v}\frac{\partial \zeta}{\partial t} + \frac{4\pi^2 k^2 \overline{c}}{\rho_v}\zeta = 0 \tag{1.29}$$

In the former equation the following parameters were defined to simplify the mathematical formulation and to facilitate the understanding of the physical phenomena derived from it.

In first place, the attenuation coefficient or losses coefficient α will be defined as:

$$\alpha = \frac{4\pi^2 k^2 \eta}{\rho_v} \tag{1.30}$$

On the other hand, the coefficient associated with the third term of Eq. (1.29), taking into account the propagation speed in Eq. (1.23), results in the squared natural vibrating angular frequency without losses, that is:

$$\omega_o^2 = \frac{4\pi^2 k^2 \overline{c}}{\rho_v} = \frac{\alpha\overline{c}}{\eta} \tag{1.31}$$

In this way Eq. (1.29) becomes:

$$\frac{\partial^2 \zeta}{\partial t^2} + \alpha \frac{\partial \zeta}{\partial t} + \omega_o^2 \zeta = 0 \tag{1.32}$$

Solving Eq. (1.32) for the displacement ξ results in the following time function:

$$\zeta(t) = A\, e^{j(\Omega t + \theta)} \tag{1.33}$$

where the constants Ω and θ will have to be determined according to the boundary conditions.

By substituting Eq. (1.33) in Eq. (1.32) it is obtained:

$$(j\Omega)^2 + \alpha\, j(\Omega) + \omega_o^2 = 0 \tag{1.34}$$

The previous equation makes Ω be a complex number, because α is not null. If Ω is assumed to be $\Omega = \omega_p + j\gamma$, it is obtained by substitution:

$$\omega_p^2 = \omega_o^2 - \frac{\alpha^2}{4}; \quad \gamma = \frac{\alpha}{2} \tag{1.35}$$

Thus, it results from Eq. (1.33), solving for real parts of the complex exponential:

$$\zeta(t) = A\, e^{-\frac{\alpha}{2}t} \cos(\omega_p t + \theta) \tag{1.36}$$

Considering that, at any moment, particle displacement at coordinate y=0 must be null, the following expression for the displacement is obtained:

$$\xi(y,t) = A\, e^{-\frac{\alpha}{2}t} \cos(\omega_p t + \theta) \sin\left(\frac{2\pi}{\lambda} y\right) \tag{1.37}$$

Thus, ω_p defined in Eq. (1.35) is identified as the natural vibrating angular frequency of the damped system. This frequency must be coherent with the boundary conditions which establish that the displacements at $y=l/2$ must always be a maximum. Consistently, that maximum condition is formulated as follows:

$$\left.\frac{\partial \xi}{\partial y}\right|_{y=l/2} = A\, e^{-\frac{\alpha}{2}t} \cos(\omega_p t + \theta) \frac{2\pi}{\lambda} \cos\left(\frac{2\pi}{\lambda}\frac{l}{2}\right) = 0 \tag{1.38}$$

The previous condition is fulfilled if:

$$\frac{2\pi}{\lambda}\frac{l}{2}=2\pi k\frac{l}{2}=\frac{\omega_p}{v_p}\frac{l}{2}=n\frac{\pi}{2} \quad \Rightarrow \quad n=1,2,3\ldots \tag{1.39}$$

where v_p is the speed propagation in the medium with losses.

Therefore, the damping vibrating frequencies f_{np} are odd multiples of the natural vibrating fundamental frequency of the damped system f_p according to the following expression:

$$f_{np}=nf_p=n\frac{v_p}{2l} \quad \Rightarrow \quad n=1,2,3\ldots \tag{1.40}$$

Notice that Eq. (1.40) has the same formulation as Eq. (1.26) where the speed propagation is v_p. We will see that for small losses, what is true in most practical cases, the difference between v_o and v_p is negligible.

On the other hand, the value of θ in Eq. (1.37) must be zero since the displacement in the initial instant (*t*=*0*) at the end (*y*=*l*/2) must be a maximum, i.e., $\xi(l/2,0)=A$. Equation (1.37) can be written as:

$$\xi(y,t)=\frac{A}{2}e^{-\frac{\alpha}{2}t}\left[\sin\frac{2\pi}{\lambda}\left(y-\frac{\omega_p}{2\pi k}t\right)+\sin\frac{2\pi}{\lambda}\left(y+\frac{\omega_p}{2\pi k}t\right)\right] \tag{1.41}$$

This expression corresponds to a damped stationary wave. Stationary waves are generated by superposing two waves. One is called progressive wave (first sine in Eq. (1.41)) which displaces in the positive direction of *Y*, and the other is regressive and displaces in the opposite direction (second sine in Eq. (1.41)). This superposition creates a stationary wave, named so because it seems as it does not displace in space. Its spatial profile is sinusoidal and is formed by zones which do not vibrate and are called nodes (as that slice of material located in the centre of the material) and by zones of maximum amplitude of vibration, (as those slices located at the ends). The particles vibrate around their equilibrium positions according to a sinusoidal relation with time. In the case of Eq. (1.41), the maximum amplitudes of vibration decrease exponentially with time until they disappear. The wave propagation speed corresponds to the term that goes with time *t* in Eq. (1.41). This velocity corresponds to that of the material with losses v_p and has the following value:

$$v_p=\frac{\omega_p}{2\pi k}=\frac{1}{2\pi k}\sqrt{\omega_o^2-\frac{\alpha^2}{4}}=\frac{\omega_o}{2\pi k}\sqrt{\left(1-\frac{\alpha^2}{4\omega_o^2}\right)}=v\sqrt{\left(1-\frac{\alpha^2}{4\omega_o^2}\right)} \tag{1.42}$$

It is evident that the damped vibratory movement is characterized by the two parameters ω_o and α. ω_o is the oscillating angular frequency without losses and α is the time needed for the energy of oscillation to decrease to *1/e* of its initial value. In fact, the expression for the energy of a thin slice of material with a mass *dm* located at the coordinate *y,* as a function of time *t*, associated with its harmonic movement, is given by:

$$W_t = \frac{1}{2} dm\omega^2 A(y)^2 e^{-\alpha t} \tag{1.43}$$

Therefore, in each span of time equivalent to *1/*α, the energy decreases in *1/e* in relation to the one at the end of the previous span.

According to Eq. (1.43) the decrease of energy by time unit will be given by:

$$\frac{dW_t}{dt} = -\alpha \frac{1}{2} dm\omega^2 A(y)^2 e^{-\alpha t} = -\alpha W_t \tag{1.44}$$

Consequently, the energy lost in a cycle corresponding to a period *T* will be:

$$\Delta W_t(T) = -\alpha W_t T \tag{1.45}$$

The two parameters that characterize the damped vibratory movement can be combined into another named *quality factor Q* of the oscillating system. *Q* is defined as the ratio between the energy stored and the one dissipated by the oscillating system during a cycle, multiplied by the factor 2π, it is to say:

$$Q = 2\pi \frac{\text{Energy stored per cycle}}{\text{Energy dissipated per cycle}} = 2\pi \frac{W_t}{\alpha W_t T} = \frac{\omega_o}{\alpha} \tag{1.46}$$

Considering Eq. (1.31) $Q = \overline{c} / \omega_o \eta$.

According to Eq. (1.42) the propagating speed in the medium with losses becomes:

$$v_p = v_o \sqrt{\left(1 - \frac{1}{4Q^2}\right)} \tag{1.47}$$

Let us observe that if the losses are small the system's quality factor is high and, according to Eqs. (1.35) and (1.47), the natural vibrating angular frequency of the damped oscillations ω_p as well as the propagation speed with losses v_p, are very similar to those of the system without losses.

1.8.3 Forced Vibrations with Losses. Resonant Frequencies

The previous analysis provides the natural vibrating frequencies of free systems without losses and also of those which have losses. Therefore, the conclusion is that natural vibrating frequencies must follow some specific relations. In a great number of applications the piezoelectric materials are subjected to a forced vibration of certain frequency. For example when subjected to a variable field of an established frequency. It is also important to study what particle displacement is like when the frequency is different from the previously obtained natural vibrating frequencies. It is also interesting to mention that from Eq. (1.17) the displacements are directly related to the intensity induced by the piezoelectric effect. This characteristic will be of special interest.

Next, the situation in which a material of piezoelectric characteristics as previously described is subjected to an alternative sinusoidal field of angular frequency ω will be analysed. The losses in the material will be taken into account. The equations for the electric displacement and for the internal stress, including the losses in the material, are obtained from Eqs. (1.8) and (1.9) giving:

$$T = cS_p - eE + \eta \frac{\partial S_p}{\partial t} \tag{1.48}$$

$$D = \varepsilon E + eS_p \tag{1.49}$$

The analysis of forces presented in Fig. 1.7 (central part) is still valid. From it, one comes to Eq. (1.21) that, considering Eq. (48), finally becomes:

$$\frac{\partial T}{\partial y} = \rho_v \frac{\partial^2 \xi}{\partial t^2} = c\frac{\partial S_p}{\partial y} - e\frac{\partial E}{\partial y} + \eta \frac{\partial^2 S_p}{\partial y \, \partial t} \tag{1.50}$$

The equation of the electric displacement can be used to establish the relation between the electric field applied and the particle displacement. Indeed, as in the inside of the piezoelectric material it is assumed that there is no free charge, the divergence of the electric displacement vector must be zero. As the piezoelectric polarization only exists in direction Y, the Gauss law for electric displacement gives:

$$\frac{\partial D}{\partial y} = 0 \quad \Rightarrow \quad \varepsilon \frac{\partial E}{\partial y} = -e\frac{\partial S_p}{\partial y} \tag{1.51}$$

Substituting Eq. (1.51) in Eq. (1.50), results in:

$$\bar{c}\frac{\partial^2 \xi}{\partial y^2} + \eta\frac{\partial^3 \xi}{\partial y^2 \partial t} = \rho_v \frac{\partial^2 \xi}{\partial t^2} \tag{1.52}$$

The former expression is the wave equation for the particle displacement. In this movement, the system, after certain transitory time where it will try to vibrate at some of its natural vibrating frequencies, will end up oscillating with a forced vibration at a frequency equal to that imposed by the external field applied. Therefore, the particles' displacement will follow a harmonic movement of the same angular frequency as the one of the applied variable electric field. This sinusoidal variation in time lets us write the particle displacement as follows:

$$\xi(y,t) = \zeta(y)\, e^{j\omega t} \tag{1.53}$$

Consequently, Eq. (1.52) is reduced to:

$$\frac{\partial^2 \zeta}{\partial y^2} = -\frac{\rho_v \omega^2}{(\bar{c} + j\omega\eta)}\zeta \tag{1.53}$$

In the previous section, it was assumed that the particle displacement had a sinusoidal profile with regard to the coordinate y, whenever there were small losses. This assumption, done at that time to simplify the calculations, will be deduced after the following general analysis. The following solution will be tried for ξ:

$$\zeta = Ae^{\gamma y} + Be^{-\gamma y} \tag{1.54}$$

In first place, the condition for a zero-displacement at y=0 implies that B=-A. By substituting Eq. (1.54) in Eq. (1.3) one gets the value for γ, that results in:

$$\gamma^2 = -\frac{\rho_v \omega^2}{\bar{c} + j\omega\eta} = -\frac{\omega^2 \dfrac{\rho_v}{\bar{c}}}{1 + j\dfrac{\omega\eta}{\bar{c}}} = -\frac{\dfrac{\omega^2}{v_o^2}}{1 + j\dfrac{1}{Q}} \tag{1.55}$$

If the quality factor of the piezoelectric material is much greater than unity, as is the usual case, the constant γ can be approximated to:

$$\gamma = j\frac{\omega}{v_o}\left(1 + j\frac{1}{Q}\right)^{-1/2} \approx \frac{\omega}{2Q v_o} + j\frac{\omega}{v_o} = \frac{\alpha}{2v_o} + j\frac{\omega}{v_o} \tag{1.56}$$

On the other hand, considering Eq. (1.2), the strain at coordinate $y= l/2$ will be:

$$S_p\Big|_{y=\frac{l}{2}} = \frac{\partial \xi}{\partial y}\Bigg|_{y=\frac{l}{2}} = d\,E\Big|_{y=\frac{l}{2}} = d\frac{V_m}{l}e^{j\omega t} \tag{1.57}$$

where V_m is the maximum voltage difference applied to the material between the electrodes located at $y=\pm l/2$.

The application of this boundary condition provides the value of the constant A, which results in:

$$A = \frac{dV_m}{\gamma l}\frac{1}{e^{\gamma\frac{l}{2}}+e^{-\gamma\frac{l}{2}}} \tag{1.58}$$

In this way Eq. (1.54) results in:

$$\zeta = \frac{dV_m}{\gamma l}\frac{e^{\gamma y}-e^{-\gamma y}}{e^{\gamma\frac{l}{2}}+e^{-\gamma\frac{l}{2}}} = \frac{dV_m}{\gamma l}\frac{\sinh(\gamma y)}{\cosh\left(\frac{\gamma l}{2}\right)} \tag{1.59}$$

Notice that if the losses are small, that is, that the Q factor is high, the constant $\gamma \approx j\omega/v_o$, and Eq. (1.59) is reduced to:

$$\zeta \approx \frac{dV_m}{j\frac{\omega}{v_o}l}\frac{\sinh\left(j\frac{\omega}{v_o}y\right)}{\cosh\left(j\frac{\omega l}{2v_o}\right)} = \frac{dV_m v_o}{\omega l}\frac{\sin\left(\frac{\omega}{v_o}y\right)}{\cos\left(\frac{\omega l}{2v_o}\right)} \tag{1.60}$$

The previous expression, although approximated for null losses, provides much information. In fact, it is enough for our immediate interest to consider the particles' displacement at the piezoelectric material's ends. Considering Eqs. (1.53) and (1.60), the displacement for $y=l/2$ will be:

$$\xi\left(\frac{l}{2},t\right) = \frac{dV_m v_o}{\omega l}\frac{\sin\left(\frac{\omega l}{2v_o}\right)}{\cos\left(\frac{\omega l}{2v_o}\right)}\sin(\omega t) \tag{1.61}$$

In first place, it can be observed that the approximation to a sinusoidal displacement profile in relation to the coordinate y for small losses, used in the previous section, was completely valid. In second place, it can be noticed that if the excitation frequency applied corresponds to those frequencies equal to odd multiples of the vibrating fundamental frequency, i.e., $\omega_n = n\omega_o = n2\pi f_o = n\pi v_o / l$, where n is odd (see Eq. (1.26)), particle displacement becomes infinite. This effect produces a great displacement, theoretically infinite, even for very small excitations. This phenomenon is known as *resonance* and the frequencies that cause it are called *resonant frequencies*. It is evident that the infinite displacement amplitudes are a consequence of disregarding the losses, but the previous result indicates that a vibration forced into frequencies near those of natural vibration causes very big mechanical displacements.

In third place, it can be noticed that excitations at frequencies which are even multiples of the natural vibrating fundamental frequency, that is, $\omega_{2n} = n\omega_o = n2\pi f_o = n\pi v_o / l$ where n is even, do not cause a displacement at the ends of the piezoelectric material.

The last two observations should be more carefully commented. In fact, the current induced by piezoelectric effect, i.e. the current that flows through the motional branch in the model shown in Fig. 1.6, is proportional to the speed of the particle displacement according to Eq. (1.17). Therefore, the current due to the piezoelectric effect will be a maximum when the excitation frequencies coincide with the natural vibration frequencies without losses. These frequencies must coincide with the series resonance frequencies of the motional branch. The circulating current is a maximum for these frequencies. Consequently, the electric model indicated in Fig. 1.6 must be applied to each frequency of natural vibration. This can be done including the motional branches in parallel as shown in Fig. 1.8.

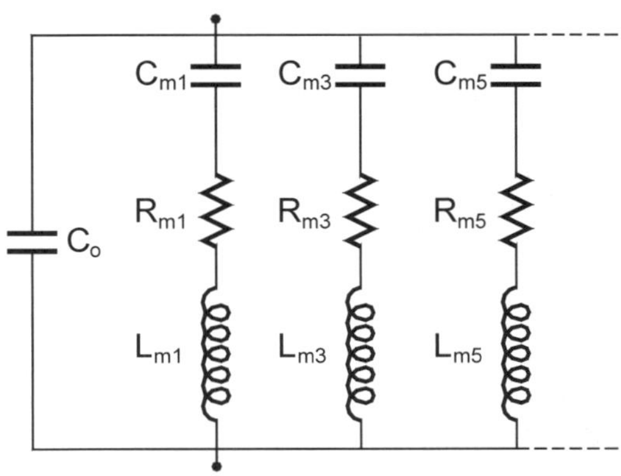

Fig. 1.8. Equivalent electrical model of a piezoelectric resonator vibrating at frequencies near any of its resonant frequencies

It is necessary to indicate that this piezoelectric model is, therefore, an approximation of the electrical impedance response of a piezoelectric material subjected to a variable electric field, whose excitation frequencies are near the natural vibration frequencies of the material.

In a similar way, when the excitation frequencies are even multiples of the natural oscillation fundamental frequency, the displacement speed is null because there is no particle displacement at the ends and, therefore, no current by piezoelectric effect is induced. Under these circumstances, the electric model of the piezoelectric material is reduced to the branch formed by the capacitor representing the current induced by the dielectric effect. This result is of practical importance, as it will be shown in Chap. 6.

As it has already been mentioned, these results were obtained after neglecting the material's losses. However, they are really true for relatively small losses such as those that occur in most practical cases. A mathematical expression which included the loss effects in a more rigorous way can be obtained from Eq. (1.59). In fact, considering the relations between the trigonometric and hyperbolic functions, the equation becomes:

$$\zeta = \frac{dV_m}{\gamma l}\frac{\sinh(\gamma y)}{\cosh\left(\frac{\gamma l}{2}\right)} \approx \frac{dV_m}{l}\left(-j\frac{v_o}{\omega}\right)\frac{\frac{\alpha y}{2v_o}\cos\frac{\omega y}{v_o}+j\sin\frac{\omega y}{v_o}}{\cos\frac{\omega l}{2v_o}+j\frac{\alpha l}{4v_o}\sin\frac{\omega l}{2v_o}} \tag{1.62}$$

where the following approximations have been done:

$$\gamma \approx j\frac{\omega}{v_o},\ \sinh\frac{\alpha y}{2v_o} \approx \frac{\alpha y}{2v_o},\ \sinh\frac{\alpha l}{4v_o} \approx \frac{\alpha l}{4v_o},$$

$$\cosh\frac{\alpha y}{2v_o} \approx 1 \text{ and } \cosh\frac{\alpha l}{4v_o} \approx 1.$$

Operating in Eq. (1.62) and disregarding terms equal to or higher than the second order in term α, one gets to the following expression for the displacement at $y = l/2$:

$$\xi(l/2,t) \approx \frac{dV_m v_o}{2l\omega}\frac{\sqrt{\sin^2\frac{\omega l}{v_o}+\frac{\alpha^2 l^2}{4v_o^2}}}{\cos^2\frac{\omega l}{2v_o}+\frac{\alpha^2 l^2}{16v_o^2}\sin^2\frac{\omega l}{2v_o}}\sin(\omega t+\theta) \tag{1.63}$$

where $\theta = -\arctan\left[(\sin \omega l/v_o)/(\alpha l/2v_o)\right]$ is the out of phase between the electric voltage applied and the displacement produced in the particles at the coordinate $y = l/2$.

As it can be deduced from Eq. (1.63), the losses limit the maximum displacement at resonant frequencies and keep a small displacement for the even harmonics of the natural vibration fundamental frequency without losses. However, it can be proved that the frequencies which have the maximum displacement speed are still those that correspond to the natural vibration frequencies without losses [5]. This last result implies that the frequencies that maximize the current by the series branch of the equivalent electric model shown in Fig. 1.8, that is, those at series resonance of the motional branch, are still the natural vibration frequencies without losses. Next, the different characteristic frequencies of a one-dimensional bar of piezoelectric material, are summarized:

Free natural vibration frequencies without losses:	$f_{no} = nf_o = n\dfrac{v_o}{2l}$
Damped natural vibration frequencies:	$f_{np} = f_{no}\sqrt{1+\dfrac{1}{4Q^2}}$
Forced vibration frequencies with losses for a maximum displacement:	$f_{nf} = f_{no}\sqrt{1+\dfrac{1}{2Q^2}}$
Frequencies for maximum displacement speed in forced vibration with losses:	$f_{nv} = f_{no} = nf_o = n\dfrac{v_o}{2l}$

1.9 Introduction to the Microgravimetric Sensor

An AT cut quartz crystal vibrates in thickness shear mode. This vibration mode is the one chosen to develop the previous sections. However, the results obtained are general when considering a piezoelectric material where only one dimension determines the vibrating state of the bar or section of material. From the natural vibrating fundamental frequency of the material, the physical fundamentals that have permitted to use the piezoelectric crystal as a micro–gravimetric sensor can be understood. Among the piezoelectric crystals, the AT cut quartz is the most commonly used as a sensor for this type of applications.

According to Eq. (1.26), the natural vibrating fundamental frequency of a piezoelectric material is given by $f_o = v_o/2l$. As a result, that frequency depends on the intrinsic properties of the material and on the dimension that determines the vibrating state, in this case the thickness. Therefore, if the physical properties of the material are considered as constant, the frequency is substantially determined by its thickness and can be written as

$f_o = N/l$, where N is the so-called *frequency constant* and depends on the material and the type of cut.

Thus, a change in the thickness will imply a variation in the system's vibration frequency. This variation can be mathematically obtained in a simple way by taking logarithms and deriving the expression for the frequency f_o. Consequently, the following relation is obtained:

$$\frac{\Delta f}{f_o} = -\frac{\Delta l}{l} \tag{1.64}$$

The change in the thickness can be written according to the mass change as:

$$\Delta l = \frac{\Delta m}{\rho_v A_S} \tag{1.65}$$

where A_S is the surface. Considering the relation between the thickness and frequency, the variation of the frequency settled by Eq. (1.64) can be written as:

$$\Delta f = -\frac{f_o^2}{\rho_v N} \frac{\Delta m}{A_S} = -C_f \rho_s \tag{1.66}$$

The previous equation indicates that if the resonant frequency is chosen as parameter, the shift in the resonant frequency provides a measurement of the surface mass density on the sensor. An important factor to know is that it has been assumed that the frequency shift was due to an increase in the material's thickness. The properties of the material have been used to set the relations between the changes in the thickness and in the mass. So that the previous equation is still valid for masses of different materials to that used as sensor; thus, it is necessary to assume that the effect on the vibration frequency is the result of a merely inertial perturbation; i.e., the viscoelastic properties of the material deposited must not affect the resonant frequency. This assumption assumes that the layer of material deposited on the sensor does not deform and is, therefore, an approximation fulfilled under certain conditions [6]. However, it has been proved to be precise in many practical applications.

It is interesting to establish to what extent the measurement of surface mass density is sensitive. For this, a 10 MHz AT cut crystal will be used as piezoelectric material whose properties are shown in Table 1.1. A resolution of 0.1 Hz will be set for the vibration frequency measurement. Under these conditions, the maximum sensitivity for the surface mass density will be:

$$\Delta\rho_s = \frac{\rho_v N}{f_o^2}\Delta f = \frac{\rho_v N}{f_o^2}0.1 \approx 4\text{pg mm}^{-2} \tag{1.67}$$

This great sensitivity, one million times higher than conventional static balance systems, is due to the enormous acceleration that the particles joined rigidly to the quartz surface experience.

To evaluate the particles' acceleration, firstly the vibration amplitude of a quartz crystal with a quality factor Q has to be estimated.

The vibration amplitude of a system with a quality factor Q at resonance can be set according to the static amplitude, for quality factors higher than 5 as [7]:

$$A = Q A_o \tag{1.68}$$

where A_o is the amplitude at zero frequency, i.e, the static amplitude.

The static displacement can be calculated from the thickness and the strain as:

$$A_o = l d E\big|_{y=\frac{l}{2}} = d l \frac{V_m}{l} = d V_m \tag{1.69}$$

where d is the piezoelectric strain coefficient for an AT cut crystal and V_m is the maximum voltage difference to what the quartz plate is subjected between its electrodes.

Consequently, for a 10 MHz AT quartz crystal, with a quality factor of 80.000 (very reasonable value in practice), which has been subjected to a variable voltage difference with a maximum amplitude of V_m=250 mV, the amplitude of resonance given by Eq. (1.68) will be around 330 Å (this result is in agreement with the measurements done by some researchers [8-10] for a vibrating quartz in an unperturbed state).

Table 1.1. Properties of typical 10MHz AT-cut quartz

Quartz Parameter	Value	Description
ε_{22}	3.982×10^{-11} A^2 s^4 Kg^{-1} m^{-3}	Permittivity
η_q	9.27×10^{-3} Pa s	Effective viscosity
$\bar{c}_{66}$	2.947×10^{10} N m^{-2}	Piezoelectrically stiffened shear modulus
e_{26}	9.657×10^{-2} A s m^{-2}	Piezoelectric constant
ρ_q	2651 Kg m^{-3}	Density
A_S	2.92×10^{-5} m^2	Effective electrode surface area
$l=l_q$	166.18×10^{-6} m	Thickness

Therefore, the maximum acceleration of the oscillating system will be:

$$a = \omega^2 A = \omega^2 dV_m Q = 1{,}3 \cdot 10^8 \text{ m} \cdot \text{s}^{-2} \approx 10^7 g \tag{1.70}$$

where g is the acceleration of gravity. In other words, this result means that a mouse of 100 g subjected to this acceleration would weigh one thousand tons.

Appendix 1.A The Butterworth Van-Dyke Model for a Piezoelectric Resonator

1.A.1 Rigorous Obtaining of the Electrical Admittance of a Piezoelectric Resonator. Application to AT Cut Quartz

The electrical admittance of a piezoelectric resonator considered as a one-dimensional system will be obtained in this appendix. An equivalent electrical model at frequencies near the resonant frequencies of the piezoelectric resonator will be derived from the expression of this admittance. An AT-cut quartz crystal will be used to represent the piezoelectric resonator but it will not reduce the generality of result.

A one-dimensional AT cut quartz plate undergoes a strain x_y (see Appendix B) when an electric field in the thickness direction Y is applied. The responsible of this effect is the piezoelectric stress coefficient e_{26} (see Appendix B to understand the meaning of subscripts) which is not null in the quartz [4,5]. Rigorously, a strain z_x corresponding to the CT and DT cuts can arise through the coefficient $d_{25} = -d_{14}$. However, in the high frequency range, where these resonators are used, the fundamental strain is $S_6 = x_y$. Consistently, Fig. 1.A.1 shows a cross section of an AT-cut plate vibrating in thickness shear mode. In this vibrating mode particle displacement is perpendicular to the wave propagation direction, creating a transversal wave propagating in the thickness direction. Thus the particle moves in direction X around its rest position with amplitude which depends on the coordinate y.

When the electric field is variable, so is the strain. Thus, when the applied voltage is sinusoidal, it is assumed that the strain is sinusoidal as well. In this case, the transitory solution corresponding to the free oscillation state vanishes with time. The pursuit of a stationary solution assumes that the particles' displacement is harmonic with the same angular frequency as the external phenomenon which produces the oscillation. Next, the wave equation of the movement is deduced.

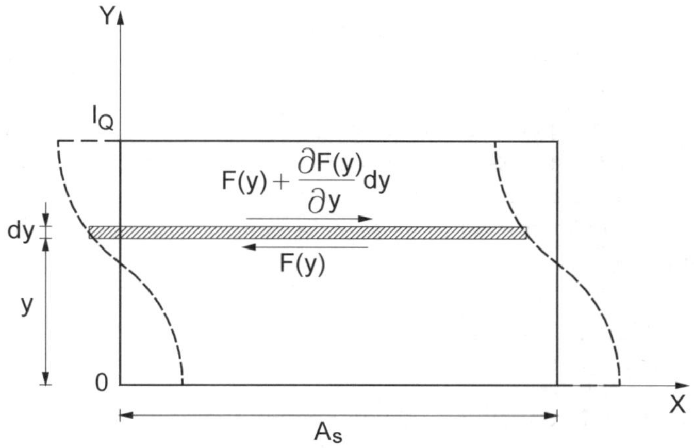

Fig. 1.A.1. Shear strain profile and forces in an internal thin slice of an AT-cut quartz plate subjected to a sinusoidal electric field

The recovery force has an elastic component through the elastic constant c_{66} and a component corresponding to the internal friction in the material. According to Eq. (1.48), the internal stress is:

$$T_6 = c_{66} S_6 - e_{26} E_2 + \eta_Q \frac{\partial S_6}{\partial t} \tag{1.A.1}$$

where the lossless coefficient has been represented with the viscosity of quartz and indicated with the subscript Q.

Considering the following relationships between the electric field and the voltage, between the strain and the particles' displacement and keeping in mind that only displacement in the direction X exist, that is:

$$E_2(y,t) = -\frac{\partial V(y,t)}{\partial y} \tag{1.A.2}$$

$$S_6 = 2S_{12} = \left(\frac{\partial \xi_x}{\partial y} + \frac{\partial \xi_y}{\partial x} \right) = \frac{\partial \xi_x}{\partial y} = \frac{\partial \xi(y,t)}{\partial y} \tag{1.A.3}$$

Equation (1.A.1) becomes:

$$T_6 = c_{66} \frac{\partial \xi(y,t)}{\partial y} + e_{26} \frac{\partial V(y,t)}{\partial y} + \eta_Q \frac{\partial^2 \xi(y,t)}{\partial t \partial y} \tag{1.A.4}$$

On the other hand, the electric displacement in direction Y is:

$$D_2 = \varepsilon_{22} E_2 + e_{26} S_6 = e_{26} \frac{\partial \xi(y,t)}{\partial y} - \varepsilon_{22} \frac{\partial V(y,t)}{\partial y} \tag{1.A.5}$$

Since inside the quartz there is no free charge and the electrical displacements in the X and Z directions are null, the Maxwell equation for the divergence of the electric displacement indicates that the displacement in the Y direction not being null has to be a constant (see Appendix A) and consequently:

$$\frac{\partial D_2}{\partial y} = 0 \quad \Rightarrow \quad \frac{\partial^2 V(y,t)}{\partial y^2} = \frac{e_{26}}{\varepsilon_{22}} \frac{\partial^2 \xi(y,t)}{\partial y^2} \tag{1.A.6}$$

Thus, the partial derivative of the voltage will be

$$\frac{\partial V(y,t)}{\partial y} = \frac{e_{26}}{\varepsilon_{22}} \frac{\partial \xi(y,t)}{\partial y} + C(t) \tag{1.A.7}$$

and Eq. (1.A.4) results into:

$$\begin{aligned} T_6 &= \left(c_{66} + \frac{e_{26}^2}{\varepsilon_{22}} \right) \frac{\partial \xi(y,t)}{\partial y} + \eta_Q \frac{\partial^2 \xi(y,t)}{\partial t \partial y} + e_{26} C(t) = \\ &= \bar{c}_{66} \frac{\partial \xi(y,t)}{\partial y} + \eta_Q \frac{\partial^2 \xi(y,t)}{\partial t \partial y} + e_{26} C(t) \end{aligned} \tag{1.A.8}$$

The equilibrium of forces in a thin slice of material of thickness *dy* shown in Fig. 1.A.1 results in the following equation:

$$\frac{\partial T_6}{\partial y} = \rho \frac{\partial^2 \xi}{\partial t^2} \tag{1.A.9}$$

By substituting Eq. (1.A.8) in Eq. (1.A.9) the following equation is obtained:

$$\rho \frac{\partial^2 \xi(y,t)}{\partial t^2} = \bar{c}_{66} \frac{\partial^2 \xi(y,t)}{\partial y^2} + \eta_Q \frac{\partial^3 \xi(y,t)}{\partial t \partial y^2} \tag{1.A.10}$$

In the forced vibration in stationary state the displacements do not vanish with time. Let us assume that these displacements are sinusoidal and have amplitude dependent on coordinate y. Consequently they can be formulated as follows:

$$\xi(y,t) = (A e^{j\gamma_Q y} + B e^{-j\gamma_Q y}) e^{j\omega t} = \Phi(y) e^{j\omega t} \tag{1.A.11}$$

By substituting Eq. (1.A.11) in Eq. (1.A.10) the following expression is obtained:

$$\gamma_Q = \omega\sqrt{\frac{\rho}{\bar{c}_{66} + j\omega\eta_Q}} = \omega\sqrt{\frac{\rho}{\bar{c}_{66}}}\frac{1}{\sqrt{1 + j\dfrac{\omega\eta_Q}{\bar{c}_{66}}}} = \frac{\omega}{v}\frac{1}{\sqrt{1 + j\dfrac{1}{Q}}} \approx \frac{\omega}{v}\left(1 - j\frac{1}{2Q}\right) \tag{1.A.12}$$

where it has been assumed that the quality factor given by Eq. (1.46) is much greater than unity.

From Eq. (1.A.12) the real and imaginary parts of γ_Q are obtained as follows:

$$\mathrm{Re}(\gamma_Q) \approx \frac{2\pi}{\lambda} \approx \frac{\omega}{v} \tag{1.A.13}$$

$$\mathrm{Im}(\gamma_Q) \approx \frac{\omega}{2Qv} \tag{1.A.14}$$

Parameters A and B in Eq. (1.A.11) are determined with appropriate boundary conditions. Because we are dealing with forced vibrations driven with an alternate voltage applied between the contacting electrodes, it is appropriate to establish the boundary conditions derived from this situation. From Eq. (1.A.7) the voltage as a function of the coordinate y and time t is obtained. Considering Eq. (1.A.11) one obtains:

$$V(y,t) = \left(\frac{e_{26}}{\varepsilon_{22}}\Phi(y) + Cy + D\right)e^{j\omega t} \tag{1.A.15}$$

Consequently the problem is defined with the following equations:

$$\xi(y,t) = (Ae^{j\gamma_Q y} + Be^{-j\gamma_Q y})e^{j\omega t} \tag{1.A.16}$$

$$V(y,t) = \left(\frac{e_{26}}{\varepsilon_{22}}(Ae^{j\gamma_Q y} + Be^{-j\gamma_Q y}) + Cy + D\right)e^{j\omega t} \tag{1.A.17}$$

Next, parameters A, B, C and D will be determined according to the following boundary conditions:

a) $T_6 = 0$, at $y = 0$.
b) $T_6 = 0$, at $y = l_Q$.
c) $V(0,t) = \varphi_o e^{j\omega t}$, voltage boundary condition at $y = 0$.
d) $V(0,t) = -\varphi_o e^{j\omega t}$, voltage boundary condition at $y = l_Q$.

Applying the former conditions to Eq. (1.A.16) and Eq. (1.A.17) gives the following expressions:

$$j\gamma_Q\bar{\bar{c}}_{66}\left[A-B\right]+e_{26}\,C=0 \tag{1.A.18}$$

$$j\gamma_Q\bar{\bar{c}}_{66}\left[Ae^{j\gamma_Q l_Q}-Be^{-j\gamma_Q l_Q}\right]+e_{26}\,C=0 \tag{1.A.19}$$

$$\frac{e_{26}}{\varepsilon_{22}}\left[A+B\right]+D=\varphi_o \tag{1.A.20}$$

$$\frac{e_{26}}{\varepsilon_{22}}\left[Ae^{j\gamma_Q l_Q}+Be^{-j\gamma_Q l_Q}\right]+C\,l_Q+D=-\varphi_o \tag{1.A.21}$$

where $\bar{\bar{c}}_{66}=\bar{c}_{66}+j\omega\eta_Q$

The former expressions can be formulated in a matrix as follows:

$$\begin{pmatrix} j\gamma_Q\bar{\bar{c}}_{66} & -j\gamma_Q\bar{\bar{c}}_{66} & e_{26} & 0 \\ j\gamma_Q\bar{\bar{c}}_{66}e^{j\gamma_Q l_Q} & -j\gamma_Q\bar{\bar{c}}_{66}e^{-j\gamma_Q l_Q} & e_{26} & 0 \\ \dfrac{e_{26}}{\varepsilon_{22}} & \dfrac{e_{26}}{\varepsilon_{22}} & 0 & 1 \\ \dfrac{e_{26}}{\varepsilon_{22}}e^{j\gamma_Q l_Q} & \dfrac{e_{26}}{\varepsilon_{22}}e^{-j\gamma_Q l_Q} & l_Q & 1 \end{pmatrix}\begin{pmatrix} A \\ B \\ C \\ D \end{pmatrix}=\begin{pmatrix} 0 \\ 0 \\ \varphi_o \\ -\varphi_o \end{pmatrix} \tag{1.A.22}$$

Parameters *A*, *B*, *C* and *D* can be determined from the former expression following traditional methods. Their values can be found elsewhere [11]. However, as we will show, to obtain the electrical admittance of the resonator, only the parameter *C* is necessary and its value is:

$$C=\frac{2\,\varphi_o\bar{\bar{c}}_{66}\gamma_Q}{2\dfrac{e_{26}^2}{\varepsilon_{22}}\tan\dfrac{\gamma_Q l_Q}{2}-\bar{\bar{c}}_{66}\gamma_Q l_Q} \tag{1.A.23}$$

From Eq. (1.A.5) and Eq. (1.A.7) the electric displacement D_2 is:

$$D_2=-\varepsilon_{22}\,C\,e^{j\omega t} \tag{1.A.24}$$

Consequently, the current density *J* is:

$$J=\frac{\partial D_2}{\partial t}=-j\omega\varepsilon_{22}C\,e^{j\omega t} \tag{1.A.25}$$

Therefore, if the current density is assumed uniform, the total current will be:

$$I = J\,A_S = -j\omega\varepsilon_{22} A_S C\, e^{j\omega t} \tag{1.A.26}$$

Considering the voltage difference applied through the quartz which is $2\varphi_o e^{j\omega t}$, the electrical admittance Y is:

$$Y = -j\omega\varepsilon_{22} A_S \frac{C}{2\varphi_o} \tag{1.A.27}$$

where it can be notice that only the parameter C appears. Thus the final expression for the electrical admittance of the resonator is:

$$Y = -j\omega\varepsilon_{22}\frac{A_S}{l_Q}\frac{\overline{\overline{c}}_{66}\gamma_Q l_Q}{2\dfrac{e_{26}^2}{\varepsilon_{22}}\tan\dfrac{\gamma_Q l_Q}{2} - \overline{\overline{c}}_{66}\gamma_Q l_Q} \tag{1.A.28}$$

By introducing C_o as the value of the capacitor formed by the quartz as dielectric material between the electrodes given by $C_o = \varepsilon_{22}A_S/l_Q$, Eq. (1.A.28) can be rewritten as follows:

$$Y = j\omega C_o + \frac{1}{Z_m} = -j\omega C_o \frac{\overline{\overline{c}}_{66}\gamma_Q l_Q}{2\dfrac{e_{26}^2}{\varepsilon_{22}}\tan\dfrac{\gamma_Q l_Q}{2} - \overline{\overline{c}}_{66}\gamma_Q l_Q} \tag{1.A.29}$$

where Z_m is:

$$Z_m = \frac{i}{\omega C_o}\left(1 - \frac{\overline{\overline{c}}_{66}\gamma_Q l_Q}{2\dfrac{e_{26}^2}{\varepsilon_{22}}\tan\dfrac{\gamma_Q l_Q}{2}}\right) \tag{1.A.30}$$

The former expression corresponds to the electrical impedance of the resonator as a vibrating system due to the piezoelectric effect. This is the reason why such impedance is called *motional impedance*.

For a better understanding of such impedance, it is necessary to simplify Eq. (1.A.30). For that it will be assumed that the resonator is working near any of its resonant frequencies. At those frequencies, the product $\gamma_Q l_Q$ is approximately $n\pi$ where $n = 1, 3, 5$, etc., being equal to $n\pi$ when the losses are neglected (Eq. (1.A.13)). For these frequencies, the tangent in equation (1.A.30) has a pole. Such a trigonometric function can be expanded through its poles as follows [12].

$$\tan\frac{\gamma_Q l_Q}{2} \approx \frac{4\gamma_Q l_Q}{(n\pi)^2 - (\gamma_Q l_Q)^2} \tag{1.A.31}$$

By substituting the previous expansion in Eq.(1.A.30) and keeping in mind that the results obtained by the application of this expansion are restricted to frequencies near resonance, one obtains:

$$Z_m = \frac{j}{\omega C_o}\left(1 - \frac{(n\pi)^2 - (\gamma_Q l_Q)^2}{8K^2}\right) \tag{1.A.32}$$

where K^2 has been defined as follows:

$$K^2 = \frac{e_{26}^2}{\overline{\overline{c}}_{66}\varepsilon_{22}} = \frac{e_{26}^2}{\overline{c}_{66}\varepsilon_{22}} \frac{1}{1 + j\dfrac{1}{Q}} = K_o^2 \frac{1}{1 + j\dfrac{1}{Q}} \tag{1.A.33}$$

It can be notice that, neglecting the losses, all the terms in Eq. (1.A.33) are reactive and the motional impedance Z_m is null for the following value of the product $\gamma_Q l_Q$ (which is the spatial phase of the propagating mechanical wave):

$$\gamma_Q l_Q = \frac{\omega_s^2 l_Q^2}{v^2} = (n\pi)^2 - 8K_o^2 \tag{1.A.34}$$

For this value, the electrical admittance of the resonator is infinite, if the losses are neglected, and the corresponding frequency is called the *motional series resonant frequency* f_s.

By substituting the complex values of γ_Q and K^2 given by Eqs.(1.A.12) and (1.A.33) in Eq. (1.A.32), the following expression for the motional impedance is obtained:

$$Z_m = \frac{j}{\omega C_o}\left(1 - \frac{(n\pi)^2\left(1 + j\dfrac{1}{Q}\right) - \dfrac{\omega^2 l_Q^2}{v^2}}{8K_o^2}\right) \tag{1.A.35}$$

By substituting the ratio $l_Q^2 / v^2 = \left[(n\pi)^2 - 8K_o^2\right] / \omega_s^2$ obtained from Eq. (1.A.34) in Eq. (1.A.35) and taking into account that the angular frequency ω is very near to ω_s, Eq. (1.A.35) can be written as follows:

$$Z_m = \frac{(n\pi)^2}{8K_o^2 \omega C_o Q} + j\omega \frac{(n\pi)^2}{8K_o^2 \omega_s^2 C_o} + \frac{1}{j\omega} \frac{(n\pi)^2}{8K_o^2 C_o} \quad (1.A.36)$$

It can be easily notice that Eq. (1.A.36) is analogous to the impedance of a series *R*, *L*, *C* circuit whose resonant frequency is ω_s, which cancels the imaginary part of the impedance. Considering the relations of the electrical parameters in Eq. (1.A.36) and the physical magnitudes of the resonator, the equivalent parameters *R*, *L*, *C* of the model can be obtained as follows:

$$R = \frac{(n\pi)^2 \eta_Q l_Q}{8 e_{26}^2 A_S}; \qquad L = \frac{\rho l_Q^3}{8 e_{26}^2 A_S}; \qquad C = \frac{8 e_{26}^2 A_S}{l_Q (n\pi)^2 \overline{c}_{66}} \quad (1.A.37)$$

Thus, the equivalent electrical circuit modelling the impedance of a piezoelectric resonator at frequencies near resonance is a circuit, as shown in Fig. 1.A.2, formed by two parallel branches: one being a capacitor C_o which corresponds to that one arisen due to the dielectric material between the electrodes, and the other one being a series *R*, *L*, *C* branch modelling the motional impedance of the resonator. The values derived for the parameters of the equivalent model match with those obtained by other authorities [4, 5]. In Chapter 4, an equivalent model for a quartz sensor in contact with a viscoelastic medium will be derived following a similar approach.

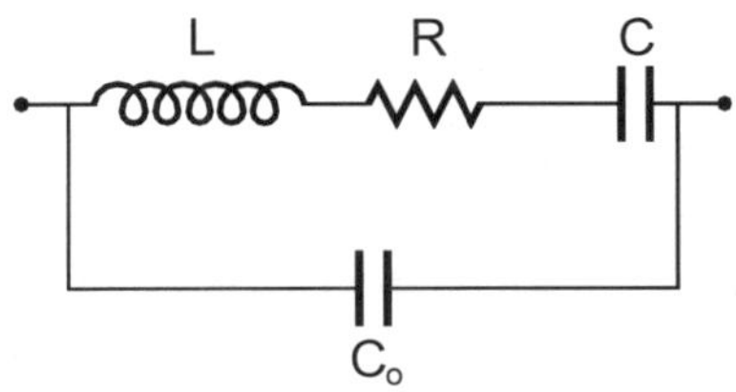

Fig. 1.A.2. Butterworth Van-Dyke model of a piezoelectric resonator vibrating at frequencies near resonance

1.A.2 Expression for the Quality Factor as a Function of Equivalent Electrical Parameters

Next, the quality factor of the piezoelectric resonator will be derived as a function of the equivalent electrical parameters. This expression will be of importance in some of the next chapters.

The quality factor associated to the motional impedance of the equivalent electrical model at the motional series resonant angular frequency ω_s is:

$$Q = \frac{L\omega_s}{R} \tag{1.A.38}$$

Considering the relations given by Eq. (1.A.37), the former equation can be rewritten as follows:

$$Q = \frac{\rho l_Q^2 \omega_s}{(n\pi)^2 \eta_Q} \tag{1.A.39}$$

The motional series resonant frequencies coincide with the natural vibration frequencies without looses given by Eq. (1.26). Consequently, the motional series resonant angular frequency ω_s is given by:

$$\omega_s = n\frac{v_o}{2l_Q} \tag{1.A.40}$$

where v_o is given by Eq. (1.23) which is rewritten as follows:

$$v_o = \sqrt{\frac{\bar{c}_{66}}{\rho}} \tag{1.A.41}$$

By substituting Eqs. (1.A.41) and (1.A.40) in Eq. (1.A.39), the following expression for the quality factor of the resonator is obtained:

$$Q = \frac{\bar{c}_{66}}{\eta_Q \omega_s} \tag{1.A.42}$$

It can be notice that the former expression coincides with Eq. (1.46) derived from a physical point of view and is consistent with it. However, it is necessary to make clear that the previous expression for the quality factor is related to the motional branch and that the electrical contribution of the parallel capacitor is not included. Therefore the expression obtained for the quality factor must be considered an approximate equation for the quality factor of the resonator at frequencies very close to the motional series resonant frequencies.

References

1. W.G. Hankel, Abh. Sächs., 12, 457 (1881).
2. P. & J. Curie, *Comptes Rendus,* 91, 294 (1880).
3. G. Lippmann, *Annales De Physique Et De Chimie, 5th Series,* **24**, 145 (1881).

4. V.E. Bottom, *Introduction to Quartz Crystal Unit Design,* (Van Nostrand, New York, 1982).
5. W.G. Cady, *Piezoelectricity (An Introduction To The Theory And Applications Of Electromechanical Phenomena In Crystals)*, (Dover Publication, Inc., New York, 1964) *2nd Ed 1964 (II Vols).*
6. G. Sauerbrey, Zeitschrift Fuer Physik, **155**, 206 (1959).
7. A.P. French, *Vibraciones y Ondas, Ed. Reverté S.A.*, 1974.
8. G. Sauerbrey, Zeitschrift Fuer Physik, **178**, 457 (1964).
9. L. Wimmer and S. Hertl, Rev. Sci. Instrum., **55**, 605 (1984).
10. V.M. Mecea, Journal Of Physics. E, **22**, 59 (1989).
11. C. Reed, K. Kanazawa and J.H. Kaufman, J. Appl. Phys., **68**, 1993 (1990).
12. J.F. Rosenbaum, *Bulk Acoustic Wave Theory And Devices*, (Artech House, Inc. Boston, 1988).
13. A. Samartin Quiroga, *Curso de Elasticidad*, (Librería Editorial Bellisco, 1990).
14. W.P. Mason, *Electromechanical Transducers And Wave Filters*, (Van Nostrand Company, Inc. (2nd Ed.*)*, 1948).
15. H. Lamb, *The Dinamical Theory Of Sound,* (Dover Publications, Inc. New York,,1960).
16. J.L. Davis, *Wave Propagation In Solids And Fluids*, (Springer-Verlag New York Inc., 1988).
17. J. Zelenka, *Piezoelectric Resonators and their Applications*, (Elsevier, 1986).
18. R.A. Heising, *Quartz Crystals For Electrical Circuits*, (D. Van Nostrand Company, Inc. New York,1946).
19. W.P. Mason, *Piezoelectric Crystals And Their Applications To Ultrasonic*, (D.Van Nostrand Company, Inc. 1ªEd 1950, 2ªEd 1956, 3ªEd 1960, 4ªEd 1964,1964).
20. W.P. Mason, *Piezoelectricity, Its History And Applications*, Journal Of Acoustical Society Of America, **70**, 1561 (1981).
21. H.F. Tiersten, *Linear Piezoelectric Plate Vibrations*, (*Plenum Press, New York,*,1969).
22. D. Belincourt, *Piezoelectric Ceramics: Characteristics And Applications*, Journal Of Acoustical Society Of America, **70**, 1586 (1981).
23. A.H. Love, *Theory Of Elasticity*, (*Cambridge University Press, 4th Edition,*,1934).

2 Overview of Acoustic-Wave Microsensors

Vittorio Ferrari[1] and Ralf Lucklum[2]

[1] Dipartimento di Elettronica per l'Automazione, Università di Brescia
[2] Institute for Micro and Sensor Systems, Otto-von-Güericke-University Magdeburg

2.1 Introduction

The term acoustic-wave microsensor in its widest meaning can be used to indicate a number of significantly different devices. Their common characteristic is the fact that acoustic waves are involved in their operating principles.
Acoustic-wave microsensors can be grouped into the following three classes.

1. Microfabricated, or miniaturized, sensors where acoustic waves, i.e. matter vibrations propagating in elastic media, are involved in the sense that they define the domain of the measurand quantity. Examples of this type of devices are accelerometers, microphones, and acoustic-emission pick-ups. The piezoelectric effect, though often used, is not necessarily required in this class of sensors.
2. Microfabricated, or miniaturized, sensors that emit and receive acoustic waves in a surrounding medium along a distance which is typically longer than several wavelengths, in order to sense the properties of the medium and/or the presence and nature of internal discontinuities. This type of devices essentially includes ultrasound transducers, both single-element and arrays, for acoustic inspection, monitoring, and imaging in air, solids, and liquids. The majority, though not the totality, of these devices base their functioning on the piezoelectric effect, mostly because of its reversibility and efficiency.
3. Microfabricated, or miniaturized, sensors in which acoustic waves propagate and interact with a surrounding medium, in such a way that the degree of interaction or the properties of the medium can be sensed and measured from the characteristics of the acoustic or electro-acoustic field in the sensor itself [1].

Sensors of this latter kind essentially behave as acoustic waveguides which, depending on the configurations, can be made responsive to a wide range of physical quantities, like applied stress, force, pressure, temperature, added surface mass, density and viscosity of surrounding fluids. In addition, sensors can be made responsive to chemical and biological quantities by functionalizing their surface with a coating which, depending on its composition, is (bio)chemically active and works as a "receptor" for the analytes to be detected. The coating film has the role of a (bio)chemical-to-physical transducer element, as it converts signals from the (bio)chemical domain into variations of physical parameters, typically the equivalent mass, stiffness, or damping, that the acoustic sensor can detect and measure (see Chaps. 4, 7, 9, 10).

This class of acoustic-wave sensors makes an extensive use of the piezoelectric effect and comprises a number of device types that differ either in the nature of the acoustic waves involved or in the configurations adopted.

In the following sections, the main characteristics of piezoelectric acoustic-wave microsensors belonging to class 3 will be illustrated.

2.2 General Concepts

The basic principle of operation for a generic acoustic-wave sensor is a traveling wave combined with a confinement structure to produce a standing wave whose frequency is determined jointly by the velocity of the traveling wave and the dimensions of the confinement structure. Consequently, there are two main effects that a measurand can have on an acoustic-wave microsensor: the wave velocity can be perturbed, or the confinement dimensions can be changed. In addition, the measurand can also cause a certain degree of damping of the traveling wave.

An important distinction between sensor types can be made according to the nature of the acoustic waves and vibration modes involved in different devices. The devices usually have the same name as the wave dominant in the device.

In the case of a piezoelectric crystal resonator, the traveling wave is either a bulk acoustic wave propagating through the interior of the substrate, or a surface acoustic wave propagating on the surface of the substrate (see Fig. 2.1).

In the bulk of an ideally infinite unbounded solid, two types of bulk acoustic waves (BAW) can propagate. They are the longitudinal, also called compressional/extensional, waves, and the transverse, also called shear, waves, which respectively identify vibrations where particle motion is parallel and perpendicular to the direction of wave propagation. Longitudinal waves have higher velocity than shear waves.

When a single plane boundary interface is present forming a semi-infinite solid, surface acoustic waves (SAW) can propagate along the boundary. Probably the most common type of SAWs are the Rayleigh waves, which are actually two-dimensional waves given by the combination of longitudinal and transverse waves and are confined at the surface down to a penetration depth of the order of the wavelength. Rayleigh waves are not suited for liquid applications because of radiation losses.

Shear horizontal (SH) particle displacement has only a very low penetration depth into a liquid (see Chap. 4), hence a device with pure or predominant SH modes can operate in liquids without significant radiation losses in the device. By contrast, waves with particle displacement perpendicular to the device surface can be radiated into a liquid and cause significant propagation losses, as in the case of Rayleigh waves. The only exception are devices with wave velocities in the device smaller than in the liquid.

Other surface waves with important applications in acoustic microsensors are Love waves (LW), where the acoustic wave is guided in a foreign layer and surface transverse waves (STW), where wave guiding is realized with so-called gratings.

Plate waves, also called Lamb waves, require two parallel boundary planes. The lowest anti-symmetric mode is the so-called flexural plate wave (FPW).

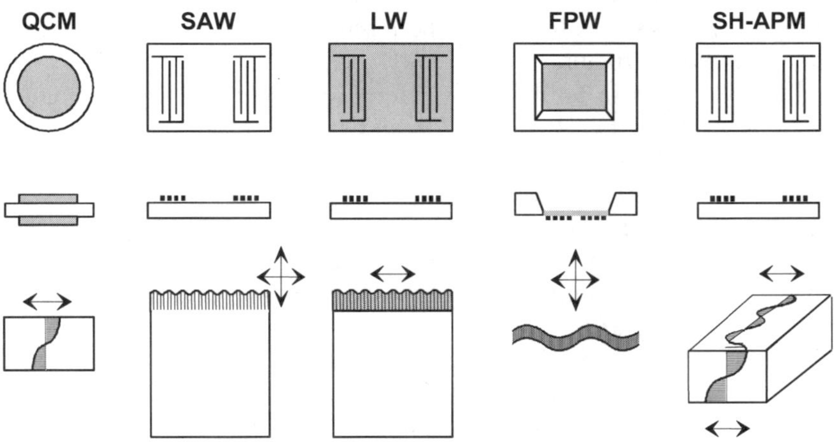

Fig. 2.1. Different types of acoustic-wave sensors

Acoustic plate modes (APM), although generated at the device surface, belong to BAWs.

Devices based on acoustic waves shown in Fig. 2.1 are shortly described in the next section. Other types of waves or devices not described here are pseudo-SAW (or leaky SAW) [2], surface skimming bulk waves [3], Bleustein-Gulyaev-waves [4] as well as magneto-SAWs [5] and the magneto-acoustic-resonator sensor (MARS) [6].

2.3 Sensor Types

2.3.1 Quartz Crystal Thickness Shear Mode Sensors

The oldest application of quartz crystal resonators (QCR) as sensors is the quartz crystal microbalance (QCM or QMB). These sensors typically consist of a thin AT-cut quartz plate with circular electrodes on both parallel main surfaces of the crystal. BAWs are generated by applying an electrical high-frequency (HF) signal to the electrodes. QCMs are operated as resonators in an almost pure thickness-shear mode, hence the sensors are also called TSM sensors.

The sensor resonant frequencies are inversely proportional to the crystal thickness. For the fundamental mode, resonance frequencies of 5 to 30 MHz are typical. For higher frequencies the crystals can be operated at overtones. Nowadays high-frequency QCRs with fundamental frequencies up to 150 MHz are available. The required crystal thickness down to 1 μm is prepared by chemical milling and, for mechanical stability reasons, the etching of the crystal is limited to the region of the electrode area, leading to inverted-mesa structures.

After their first use as frequency-reference elements in time-keeping applications in 1921 by W. Cady [7] and as a microbalance in 1959 by G. Sauerbrey [8], quartz crystals have become probably the most common acoustic-wave sensors, finding application in the measurement of several other quantities and, in turn, opening the way to the development of newer and more specialized sensors. The typical configuration is as single-element sensors, but multisensor arrays on the same crystal have been recently proposed [9, 10].

The basic effect, common to the whole class of acoustic-wave microsensors, is the decrease in the resonant frequency caused by an added surface mass in the form of film. This gravimetric effect motivates the denomination of quartz-crystal microbalance (QCM) and is exploited, for instance, in thin-film deposition monitors and in sorption gas and vapor sensors using a well-selected coating material as the chemically-active interface [11, 12].

Within a certain range, the frequency shift Δf is sufficiently linear with the added loading mass Δm regardless of the film material properties, and the sensitivity $\Delta f/\Delta m$ is proportional to f^2 [8]. For higher loading, the sensor departs from the gravimetric regime and the frequency shift becomes a function of the mass as well as of the viscoelastic properties of the film [13] (see Chaps. 4, 16).

TSM quartz sensors can also operate in liquid, due to the predominant thickness-shear mode. In this case, the frequency shift is a function of liquid density and viscosity [14], which makes it possible to use TSM quartz resonators as sensors for fluid properties [15]. In addition, the mass sensitivity and in-liquid operation can be advantageously combined, and TSM sensors coated with (bio)chemically-active films can be used for in-solution (bio)chemical analysis, for instance in the chemical, biomedical and environmental fields [16] (see Chaps. 9, 10).

Mass sensitivity and liquid density-viscosity sensitivity are two special cases of the more general sensitivity of all acoustic-wave microsensors to the so-called surface acoustic load impedance, which is discussed in Chap. 4. Because of its importance and simplicity we further limit the discussion here to mass sensitivity and applicability of the devices in a liquid environment.

2.3.2 Thin-Film Thickness Longitudinal Mode Sensors

These are BAW sensors based on thickness-mode waves that, as opposed to TSM quartz crystals, are of the longitudinal type. They are made by electroded thin films of piezoelectric materials, such as AlN, in the form of diaphragms photolithographically defined and etched from a silicon substrate. In this way, a very low thickness can be obtained that causes a high resonant frequency, up to 900-1000 MHz [17], and, in turn, a high mass sensitivity in gravimetric applications.

As opposed to homogeneous resonators, composite resonators can also be used where the piezoelectric film is deposed on a non piezoelectric substrate, such as silicon, with possible intermediate matching layers to decrease the effective thickness and improve the thermal stability.

Composite resonators can also be made by resonant piezo-layers (RPL) of lead zirconate titanate (PZT) films screen printed on alumina substrate [18].

Thickness-longitudinal-mode sensors have many analogies with TSM quartz sensors. One important difference is that, in the former ones, the vibrations normal to the sensor surface irradiate energy in a surrounding liquid, which makes thin-film thickness longitudinal mode sensors generally unsuitable for (bio)chemical applications in solutions.

2.3.3 Surface Acoustic Wave Sensors

Surface acoustic wave (SAW) sensors are made by a thick plate of piezoelectric material, typically ST-cut quartz, lithium niobate or lithium tantalate, where predominantly Rayleigh waves propagate along the upper surface [19].

Surface wave generation is efficiently accomplished by a particular electrode configuration named interdigital transducer (IDT) (Fig. 2.2a). An IDT, in its simple version, is formed by two identical comb-like structures whose respective fingers are arranged on the surface in an interleaved alternating pattern. The IDT period length d, or pitch, is the spacing between the center of two consecutive fingers of the same comb. When an AC voltage is applied to the IDT, acoustic waves are generated which propagate along the axis perpendicular to the fingers in both directions. The maximum wave amplitude is obtained when constructive interference among the fingers occurs. This happens at the characteristic or synchronous frequency $f_o = v/d$, where v is the SAW velocity in the material. Typical SAW characteristic frequencies are 30-500 MHz. Two basic configurations are possible: one-port SAW resonators with a single IDT, and two-port SAW delay lines with two IDTs separated by a distance L.

Similarly to what happens with BAW devices, SAWs can be used as high-frequency reference elements in filters and oscillators, but they can also be made responsive to a variety of quantities to have them work as sensors [20].

The primary interaction mechanisms are those that affect the frequency by changing the wave velocity, the IDT spacing or distance, or both. Temperature, strain, pressure, force, and properties of added surface materials are examples of measurand quantities. In particular, the accumulated surface mass produces a decrease in frequency.

Compared to QCMs, the higher values of the unperturbed frequency and the fact that vibrations are localized near the surface, becoming more affected by surface interactions, determines a higher sensitivity of SAWs in gravimetric applications.

This fact is advantageously exploited in sorption gas and vapor sensors where SAWs coated with chemically-active films (Fig. 2.2b) can achieve significantly low detection limits [21].

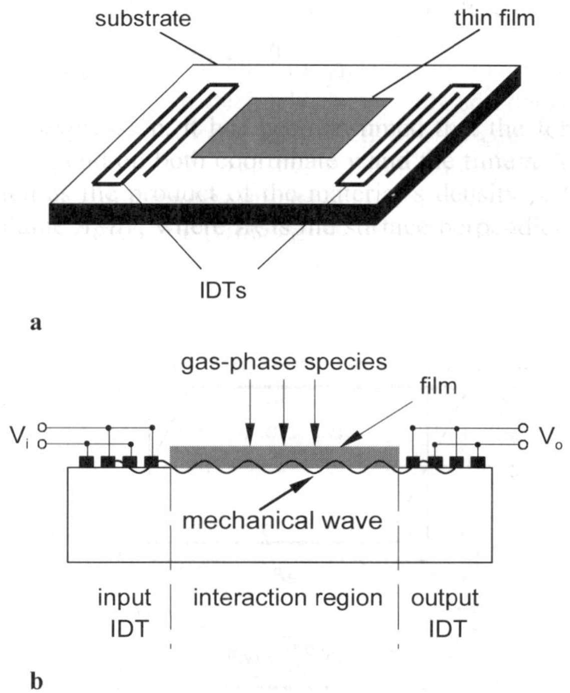

Fig. 2.2. a Interdigital transducer configuration as used in SAW sensors; **b** structure of a SAW sensor

Due to the configuration of the IDT electrodes, SAW sensors are also responsive to the electric properties of the coating film or the surrounding medium by means of the acoustoelectric coupling.

The improvement over quartz crystal TSM sensors offered by SAW sensors in air cannot be extended in liquids because of the vibration component normal to the surface involved in Rayleigh waves, which causes acoustic energy radiation into the liquid with a consequent excess of damping.

In principle, IDTs can generate a spectrum of transversal horizontally and vertically polarized waves as well as longitudinal waves, which propagate on the surface or into the volume of the piezoelectric material [7]. Material properties, crystal cut, and sensor geometry are responsible for which modes appear and in what extent. A whole family of SAW-like devices has been developed. The most important ones are further described.

2.3.4 Shear-Horizontal Acoustic Plate Mode Sensors

Shear-horizontal acoustic plate mode (SH-APM) sensors are quartz plates of a few wavelengths thickness, where shear-horizontal (SH) waves are generated by means of two IDTs positioned on one surface of the plate [22] (Fig. 2.3). Therefore, they are two-port devices.

SH waves have particle displacement predominantly parallel to the plate surface and perpendicular to the propagation direction along the separation path between the two IDTs and hence are suited for operation in contact with liquid. Typical operation frequencies of SH-APM sensors are 20-200 MHz. APMs are a series of plate modes with slightly different frequencies. The difference between these frequencies decreases with decreasing plate thickness. To select a dominant SH mode, material and crystal cut, IDT design and oscillator electronics must be optimized. APMs have antinodes on both device surfaces so that each of them can be used as a sensing surface. In particular, the electrodeless face can be made (bio)chemically active and analysis in solution can be performed with a complete separation between the electric side and the liquid side.

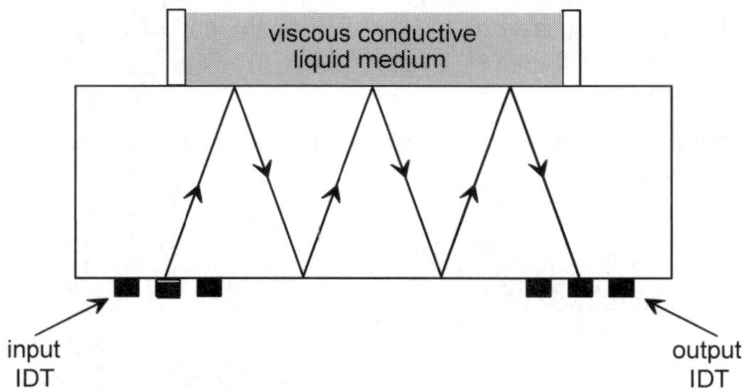

Fig. 2.3. Structure of an APM sensor

2.3.5 Surface Transverse Wave Sensors

Surface transverse wave (STW) sensors are devices in which shear vibrations are confined in a thin surface area on the face where the IDTs are placed. This wave confinement is obtained by inserting a metallic grating between the IDTs that introduces a periodic perturbation in the wave path and lowers the wave velocity at the surface [1, 23].

Since the vibration energy density is concentrated on a thin layer near the surface, the device is very responsive to surface perturbations and, in particular, it provides a high mass sensitivity. As shear vibrations are predominant, STW sensors (also called SH-SAW) are indicated for in-liquid applications and are mainly used with chemically-modified surfaces for analysis in solutions.

2.3.6 Love Wave Sensors

Love wave (LW) sensors are rather similar to STW sensors in that they involve shear vibrations confined in the upper surface.

The wave confinement is in this case obtained by depositing a thin layer of a material with low acoustic-wave velocity over a quartz plate where two IDTs are realized. Such an added overlayer, typically of silicon dioxide or polymethylmethacrylate (PMMA), works as a waveguide and keeps most of the vibration energy localized close to the surface, regardless of the plate thickness. This has the same positive effect on the mass sensitivity as the gratings in STW sensors and, once again, in-liquid operation is permitted by the shear-mode vibrations [24, 25].

Love-mode sensors are mainly used in (bio)chemical analysis in solutions.

A generalized Love-wave theory considers APMs and Love waves as the two solutions of the dispersion equation of a substrate with finite thickness [26].

2.3.7 Flexural Plate Wave Sensors

In thin plates, i.e. diaphragms with thickness smaller than the wavelength, a series of symmetric and antisymmetric plate modes can be generated. These so-called Lamb waves have a particle displacement similar to Rayleigh waves [27, 28], i.e. particle motions describe a retrograde ellipsis with the major and minor axes normal and parallel to the surface, respectively. The wave velocity depends on the plate material and the plate thickness. The advantage of the lowest antisymmetric mode, the so-called flexural plate wave (FPW) mode, is a wave velocity smaller than that of SAW devices. It decreases with decreasing plate thickness and becomes lower than the wave velocity of liquids.

This determines a couple of unique features that makes FPW sensors very attractive. The first is that, for a given wavelength, the corresponding frequency is comparatively low, in the range of 5-20 MHz, which alleviates the requirements on the associated electronics.

The second is that FPW sensors are best suited to the measurement of fluid properties, such as liquid viscosity, and gravimetric (bio)chemical analysis in solutions. In this latter application, the plate being very thin and significantly affected by surface perturbations, the achievable mass sensitivity can be extremely high [29]. Typically, the plate is a few-micron thick rectangular silicon-nitride diaphragm with a piezoelectric overlayer, such as zinc oxide, in which the waves are generated by means of IDTs (Fig. 2.4). Unfortunately, those FPW sensors are still fragile and the fabrication process must be further optimized. Another version of excitation involves a magnetic field [30].

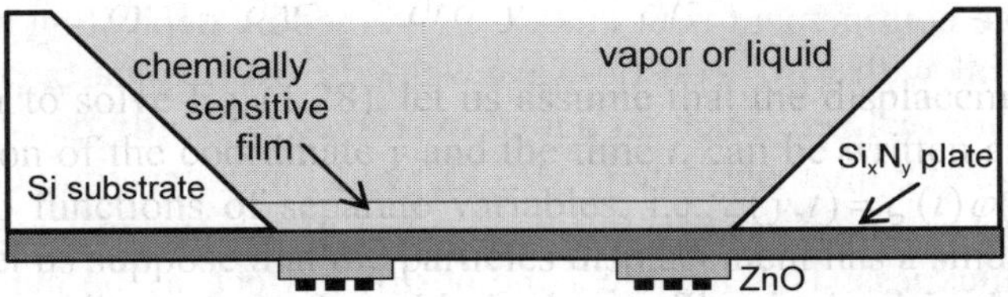

Fig. 2.4. Structure of a FPW sensor

2.4 Operating Modes

Piezoelectric acoustic-wave sensors invariably have an electrical port where a driving AC signal is applied that generates vibrations via the converse piezoelectric effect (induced strain proportional to applied voltage). Such vibrations propagate through the sensor interacting with the measurand quantity and are transduced back to the electrical domain via the direct piezoelectric effect (induced charge proportional to applied stress). Depending on the way the electrical output signal is exploited, two categories of sensors can be distinguished.

In one-port sensors the electrical output can be thought as generated across the same port, i.e. the same couple of electrodes, where the input is applied. In two-port sensors the electrical output is physically available at a second port, distinct from the input one, realized by a dedicated pair of electrodes.

For both one-port and two-port sensors, the effect of the measurand quantity produced on the wave propagation can be measured in two different methods.

In the first method, called the open-loop, or passive, or non-resonant method, an excitation signal coming from an external generator is applied to the sensor input and the corresponding response signal at the output is detected.

Usually, the measurement is performed by a network analyzer which provides the excitation signal as a fixed-amplitude sine wave swept over a frequency range, detects the output, and directly visualizes the output/input ratio as a complex function of frequency, i.e., taking both amplitude and phase into account.

The open-loop operation mode has the advantage of providing the maximum of information on the electrical behavior of the sensor and further on, via the acoustic behavior of the sensor, on the measurand/sensor interaction. The limitations are that extracting such information is not always straightforward, since it implies a certain knowledge of sensor operation and modeling. Moreover, network/impedance analyzers are typically costly instruments.

An alternative to swept-frequency analysis is the use of transient analysis, in which a sinusoidal excitation at the resonant frequency is applied at the sensor input and suddenly removed, and the resulting output oscillatory damped response is examined. This method is mostly used with quartz crystal TSM sensors [31] (see Chap. 6).

In the second method, called the closed-loop, or active, or resonant method, the sensor is configured as the feedback element of an electronic amplifier. In practice, the connection schemes are different for one-port and two-port sensors, but the principle in both cases is actually the same.

By a proper choice of the amplifier it is possible to establish positive feedback around the loop and make the sensor/amplifier combination work as an oscillator, which continuously sustains and tracks oscillations in the sensor at one of its resonant frequencies (see Chap. 6). In one-port devices, like quartz crystal TSM sensors, the sensor behaves like a mechanical resonator. Conversely, SAW, FPW and APM and LW configured as two-port devices behave as acoustic delay lines.

The closed-loop configuration has the advantages that it provides a continuous reading of the resonant frequency, allowing to follow the evolution of the experiment in real time without the need for repeated measurements of the sensor open-loop response.

For comparatively low-frequency sensors, oscillator circuits can be relatively simple and inexpensive, while for higher-frequency sensors the design becomes less straightforward.

A fundamental point to keep in mind with oscillators, that can also become their main limitation in high-accuracy applications, is that, in general, the sensor resonant frequency and the output frequency of the oscillator circuit are not exactly equal under every load conditions. This is due to the combination of the sensor and amplifier phase responses that determine an oscillation condition in the loop at a frequency which, in some cases, can be appreciably different from the sensor resonance (see Chap. 6).

In particular, great care must be taken with oscillators when the sensor is heavily loaded either acoustically, due to a thick viscoelastic coating, or dielectrically, due to immersion in liquid, or both. In such cases, the oscillation frequency of the oscillator can become significantly different from the resonant frequency of the sensor, causing errors in the interpretation of the results. As a limiting case, oscillations can even stop in the circuit, though the sensor resonant frequency of course still exists, with the negative consequence of restricting the operating range.

Special oscillator designs developed for heavy-load conditions should be adopted in these cases.

A further limitation of oscillators is that they usually provide the measurement of a single parameter of the sensor response, namely its resonant frequency. There are oscillators that incorporate circuitry for the simultaneous measurement of the vibration amplitude in addition to its frequency, therefore providing information also on the amount of damping undergone by the sensor. Concepts on electronics are further discussed in Chap. 6.

2.5 Sensitivity

The parameter of acoustic-wave sensors that is primarily employed for measurement is the fundamental resonant frequency f. From theory, in the case of quartz crystal TSM sensors, the series resonant frequency f_s where the real part of admittance has a maximum must be measured to be in accordance with the theoretically predicted values. Additional parameters ranging from damping and phase shift, to the complete spectrum provide an increasing degree of further information.

Limiting to the resonant frequency f, it can be generally expressed as:

$$f = \frac{v}{2l} = \frac{1}{2l}\sqrt{\frac{c}{\rho}} = \frac{1}{2\pi}\sqrt{\frac{K}{M}} \tag{2.1}$$

where l is the frequency determining dimension (e.g. the crystal thickness in a QCM), v is the wave velocity, c is the effective elastic stiffness (e.g. the shear stiffness constant in a QCM), ρ is the mass density, K and M are the lumped equivalent spring and mass associated with the particular vibration mode. Note that M is definitely different from the rest mass of the sensor.

The fractional frequency variation can then be derived as a function of the variations of the individual parameters caused by an external quantity as follows:

$$\frac{df}{f} = \frac{dc}{2c} - \frac{d\rho}{2\rho} - \frac{dl}{l} = \frac{dK}{2K} - \frac{dM}{2M} \tag{2.2}$$

The sensitivity towards a measurand x can be defined as the ratio df/dx.

Despite its simplicity, Eq. (2.2) has a certain general validity in indicating the effect of a measurand on the resonant frequency and in finding the associated sensitivity.

In particular, those measurands that increase the effective stiffness c, or equivalently the spring constant K, cause f to rise. Examples are tensile stress or bending.

On the contrary, those measurands that increase either the effective density ρ, or the length l, or equivalently the equivalent mass M, cause f to decrease. A typical example is mass loading. The Sauerbrey equation for the mass sensitivity of a QCM can be derived from Eq. (2.2) by assuming that the load only changes the thickness l and leaves the average density and stiffness unaltered (see Chap. 1).

In general, the higher the sensor unperturbed frequency f, the greater the frequency shift at parity of measurand value. For instance, SAW sensors in the 100 MHz range have a higher mass sensitivity than TSM sensors in the 10 MHz range.

However, considering sensitivity as a benchmark to compare different sensors can be misleading. In fact, a higher value of the *nominal* sensitivity as apparently granted by a higher resonant frequency does not necessarily imply a higher value of the *usable* sensitivity in a practical device.

For instance, a QCM can be operated with a sensitive coating much thicker than that on a SAW sensor, which results in a higher amount of gas absorbed in the coating. Therefore, it is more appropriate the use of the reduced, or fractional, sensitivity $S=(df/dx)/f$ to normalize for the unperturbed frequency.

The typical fractional mass sensitivities, where the mass is intended for unit surface area, for different sensor types are compared in Table 1 [32, 33, 34].

It should be noted that the sensitivity is only one factor to the ultimate goal of achieving a high resolution, i.e., a discrimination capability of small incremental values of the measurand [35]. High resolution implies good frequency stability.

Short-term frequency stability is mostly determined by the sensor, especially by the coating and the measurement environment, in combination with the oscillator electronics. Sensors with higher values of the quality factor Q for the resonance in question provide a better stability at parity of electronics. Therefore, a significant figure of merit for a sensor is actually the sensitivity-quality factor product SQ [36].

Long-term frequency stability is typically dominated by thermal drift and material aging or degradation, however, these effects must be related to the time scale of sensor signal changes.

To counteract drift effects a differential configuration can be helpful, with one sensor exposed to the measurand and a second identical sensor screened from it. Both sensors are subjected to the influencing quantities, such as temperature. By taking the difference of the signals from the sensor/reference pair, the common-mode perturbing factors can be compensated to some extent.

Table 2.1. Comparison of the characteristics of different acoustic-wave sensors

Sensor type	FRO[a]	S_m [b]	Examples (*)			OL[f]
			FO[c]	FN[d]	S/N[e]	
TSM quartz	5-30	12-70	10	0.2	110	Yes
Thin-film BAW	900-1000	400-700				No
SAW	30-500	100-500	160	2	100	No
SH-APM	20-200	20-40	100	4	5	Yes
STW	100-200	100-200				Yes
LW	100-200	150-500	110	2	125	Yes
FPW	5-20	200-1000	5	1	450	Yes

[a] Frequency Range of Operation [MHz]
[b] Surface mass sensitivity ([Hz $MHz^{-1}\mu g^{-1}cm^2$])
[c] Frequency of Operation [MHz]
[d] Frequency Noise [Hz]
[e] Sensitivity-to-Noise ratio ([$MHz^{-1}\mu g^{-1}cm^2$])
[f] Operation in Liquid
(*) Data taken from: [37-39].

References

1. Ballantine D.S., White R.M., Martin S.J., Ricco A.J., Zellers E.T., Fryre G.C., Wohltjen H., *Acoustic wave sensors*, Academic press, San Diego, 1997.
2. Engan H., Ingebrigtsen K.A., Tonning A., *Appl. Phys. Lett.* 10 (1967), 311-313.
3. Lewis M.F., *Proc. IEEE Ultrasonics Symp.* (1977), 744-752.
4. Bleustein J.L., *Appl. Phys. Lett.* 13 (1968), 412-413. Gulyaev Y.V., *Soviet Phys. JETP Lett.* 9 (1969), 37-38.

5. Chiriac H., Pletea M., Hristoforou E., *Sensors Actuators A* 91 (2001), 107-111.
6. Stevenson A.C., Lowe C.R., *Sensors Actuators A* 72 (1999), 32-37.
7. Peach R., *IEEE Trans. Ultrason., Ferroel., Freq. Contr.* 48 (2001), 1308-1320.
8. Sauerbrey G., *Z. Phys.* 155, (1959), 206-222.
9. Abe T., Esashi M., *Sensors and Actuators A* 82, (2000), 139-143.
10. Boecker P., Horner G., Rösler S., *Sensors and Actuators B* 70, (2000), 37-42.
11. Czanderna A.W., Lu C., in *Applications of Piezoelectric Quartz Crystal Microbalances,* edited by C. Lu and A. W. Czanderna (Elsevier, Amsterdam, 1984), Vol. 7.
12. King W.H., *Anal. Chem.* 36, (1964), 1735-1739.
13. Lucklum R., Behling C., Hauptmann P., *Sensors and Actuators B* 65, (2000), 277-283.
14. Kanazawa K.K., Gordon J.G., *Anal. Chim. Acta* 175, (1985), 99-105.
15. Martin S.J., Frye G.C., Wessendorf K.O., *Sensors and Actuators A* 44, (1994), 209-218.
16. O'Sullivan C.K., Guilbault G.G., *Biosensors & Bioelectronics* 14, (1999), 663-670.
17. O'Toole R.P., Burns S.G., Bastiaans G.J., Porter M.D., *Anal. Chem.* 64, (1992), 1289-1294.
18. Ferrari V., Marioli D., Taroni A., *Sensors and Actuators A* 92, 1-3, (2001), 182-190.
19. White R.M., *Proc. IEEE* 58, (1970), 1238-1276.
20. Nieuwenhuizen M.S., Venema A., *Sensors and Materials* 5 (1989), 261-300.
21. Wohltjen H., Dessy R., *Anal. Chem.* 51, (1979), 1458-1464.
22. Martin S.J., Ricco A.J., Niemczyk T.M., Frye G.C., *Sensors and Actuators* 20, (1989), 253-268.
23. Vellekoop M.J., *Ultrasonics* 36, (1998), 7-14.
24. Kovacs G., Lubking G.W., Vellekoop M.J., Venema A., *Proc. IEEE Ultrasonics Symp.* (1992), 281.
25. Du J., Harding G.L., Ogilvy J.A., Dencher P.R., Lake M., *Sensors and Actuators A* 56, (1996), 211-219.
26. McHale G., Newton M.I., Martin F., *J. Appl. Phys.* 91, (2002), 9701-9710.
27. White R.M., Wicher P.J., Wenzel S.W., Zellers E.D., *IEEE Trans. Ultrason., Ferroel., Freq. Cont.* 34, (1987), 162-171.
28. White R.M., Wenzel S.W., *Appl. Phys.Lett.* 26 (1988), 1653-1655.
29. Wenzel S.W., White R.M., *Sensors and Actuators A*, 21-23, (1990), 700-703.
30. Martin S.J., Buttler M.A., Spates J.J., Schubert W.K., Mitchell M.A., *IEEE Trans. Ultrason., Ferroel., Freq. Cont.* 45 (1998), 1381-1387.
31. Rodahl M., Höök F., Krozer A., Brzezinski P., Kasemo B., *Rev. Sci. Instrum.* 66, (1995), 3924-3930.
32. Wenzel S.W., White R.M., *Appl. Phys. Lett.* 54, (1989), 1976-1978.
33. Wang Z., Cheeke J.D.N., Chen C.K., *Electron. Lett.*, 26, 18, (1990), 1511-1513.

34. Martin S.J., Frye G.C., Spates J.J., Butler M.A., *Proc. IEEE Ultrasonics Symp.* (1996), 423-434.
35. Vig J.R., *IEEE Trans. Ultrason., Ferroel., Freq. Contr.* 38, 3, (1991), 311.
36. Benes E., Gröschl M., Seifert F., Pohl A., *IEEE Trans. Ultrason., Ferroelec., Freq. Cont.* 45, 5, (1998), 1314-1330.
37. Grate J.W., Martin S.J., White R.M., *Anal. Chem.* 65, (1993), 987A-996A.
38. Gizeli E., *Smart Mater. Struct.* 6, (1997), 700-706.
39. Harding G.L., Du J., *Smart Mater. Struct.* 6, (1997), 716-720.

3 Models for Piezoelectric Transducers Used in Broadband Ultrasonic Applications

José Luis San Emeterio and Antonio Ramos

Departamento de Señales, Sistemas y Tecnologías Ultrasónicas. Instituto de Acústica (CSIC).

3.1 Introduction

Piezoelectric transducers are key elements of many broadband ultrasonic systems, either pulse-echo or through-transmission, used for imaging and detection purposes. In ultrasonic broadband applications such as medical imaging, or non-destructive testing, piezoelectric transducers should generate/receive ultrasonic signals with good efficiency over a large frequency range. This implies the use of piezoelectric transducers with high sensitivity, broad bandwidth and short-duration impulse responses. High sensitivity provides large signal amplitudes which determine a good dynamic range for the system and the short duration of the received ultrasonic signal provides a good axial resolution.

The most important and common type of piezoelectric transducer elements used in ultrasonic broadband applications is a thin piezoelectric plate, with lateral dimensions much greater than the thickness, driven in a simple thickness extensional mode of vibration [1-2]. They usually operate in the frequency range 0.5-15 MHz. Different types of piezoelectric materials are used for the active transducer element. Ferroelectric ceramics, like lead zirconate titanate (PZT), lead metaniobate, etc., have a high piezoelectric coupling coefficient. Piezoelectric polymers like polyvinylidene difluoride (PVDF) and copolymers have useful low-acoustic impedances. Piezoelectric composites are mixtures of piezoceramics with nonpiezoelectric polymers.

When designing a broadband piezoelectric transducer or when finding optimal transducer system configurations, it is useful to be able to predict the global response by means of theoretical calculations, bearing in mind that there is a large number of materials and configuration parameters involved in the global system [3, 4]. The aim of this chapter is to summarize the basic modeling approaches describing the electrical and ultrasonic characteristics of broadband multilayer transducers. In the active piezoelectric plates, the length and width to thickness ratios are sufficiently large so that one-dimensional models are good approximations to predict the properties of the transducer [5-10].

Modeling the transducer as a two-port network permits the use of the transfer matrix formalism of the circuit theory. In this chapter, a general methodology for the treatment of all the components of a transducer system, including acoustic matching layers and electric matching components, as a set of cascade networks [11-13], is also described. A computer program for design and optimization of transducer systems can be easily developed [11].

3.2 The Electromechanical Impedance Matrix

Figure 3.1 shows a simple diagram of a broadband piezoelectric transducer. A piezoelectric layer of thickness t, with very thin electrodes of area A at its surfaces, is embedded between an attenuating backing material and the irradiated medium (load). Usually, a high attenuating material ("backing") is bonded to the back face of the transducer element in order to enlarge the emission/reception (two-way) frequency band and therefore to shorten the impulse response (at the expense of a loss on sensitivity and signal amplitude). One or more acoustic matching layers are bonded in the front face in order to optimize the transmission of energy to the explored medium [5-11].

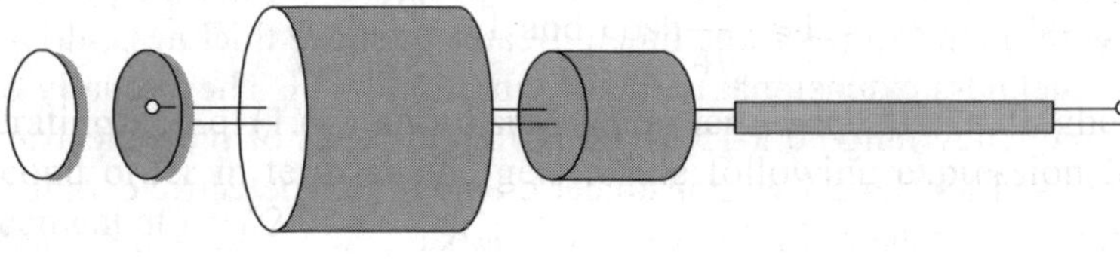

Fig. 3.1. Constructive scheme of a thin piezoelectric plate transducer: ***1*** matching layer; ***2*** piezoelectric active element; ***3*** casing; ***4*** backing; and ***5*** coaxial cable

Piezoelectric transducers convert electrical energy into mechanical energy and vice versa. The direct piezoelectric effect consist of the change in polarization in a material induced by an applied mechanical stress, and is used in the ultrasonic reception stage. The converse piezoelectric effect consists of the dimensional change (mechanical strain) in a material induced by an applied electric field, and is used in the ultrasonic emission stage. The linear piezoelectric constitutive equations define the interrelationships among the electric displacement D, the electric field E, the mechanical stress T, and the elastic strain S. The complexity of the equations involved depends on both the symmetry of the piezoelectric material and the particular geometry (mechanical and electrical boundary conditions) of the transducer element [1, 2, 5-7].

Most of piezoelectric materials used in the fabrication of broadband ultrasonic transducers, (i.e. ferroelectric ceramics, piezoelectric polymers, and piezoelectric composites) become piezoelectric by a process of electrical poling along the thickness direction ($Z \equiv 3$ axis). The poled plates present a symmetry which can be associated with the crystallographic class 6 mm, reducing the number of fundamental material constants to 5 elastic, 3 piezoelectric and 2 dielectric coefficients. This type of symmetry will be considered in the following.

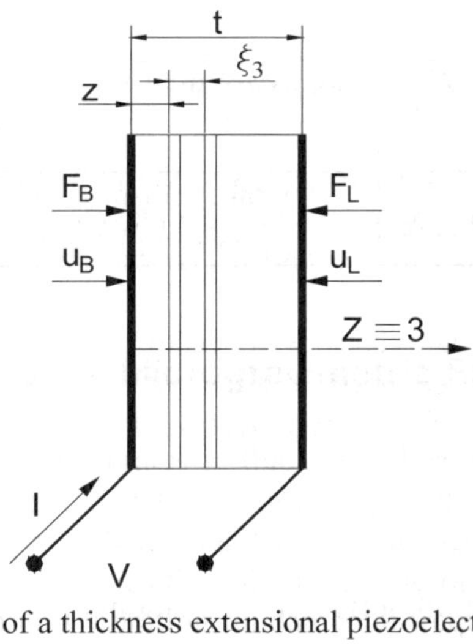

Fig. 3.2. Cross section of a thickness extensional piezoelectric transducer element

Figure 3.2 shows a diagram of the cross section of a thin piezoelectric plate, poled along the thickness t, driven by electrodes on the mayor surfaces. The electrical field is applied along the same $Z \equiv 3$ axis, (thickness excitation (TE)) [1, 2]. This transducer element can be regarded as a three port device, with one electrical port (electrical terminals carrying a voltage V and a current I) and two mechanical ports corresponding to the back (F_B, u_B) and front (F_L, u_L) faces developing forces F and particle velocities u at the faces, being:

V: voltage across the electrodes of the transducer element.
I: current through the electrodes of the transducer element.
F_L: force on the front face (radiating surface).
u_L: particle velocity at the front face.
F_B: force on the back face.
u_B: particle velocity at the back face.

The lateral dimensions of the piezoelectric plate are much larger than thickness t, so that S_3 is the only non-zero strain component and D_3 the only non-zero electric displacement component. According to these electrical and mechanical boundary conditions, the piezoelectric plate will vibrate in a thickness extensional mode of vibration and the pertinent piezoelectric equations are (see Appendix B for understanding the meaning of subscripts):

$$T_3 = c_{33}^D S_3 - h_{33} D_3 \tag{3.1}$$

$$E_3 = - h_{33} S_3 + \frac{D_3}{\varepsilon_{33}^S} \tag{3.2}$$

where the sub index $3 \equiv Z$ corresponds to the poling axis, c_{33}^D is the elastic stiffened constant, $h_{33} = e_{33} / \varepsilon_{33}^S$ is the piezoelectric constant (e_{33} is the piezoelectric stress constant), and ε_{33}^S is the clamped (high frequency) dielectric constant.

The piezoelectric material is assumed to be a perfect insulator and according to the charge equation of electrostatics the displacement component D_3 is constant inside the piezoelectric plate (see Appendix A):

$$\frac{\partial D_3}{\partial z} = 0 \tag{3.3}$$

The stress equation of motion, as a function of the particle displacement ξ_3, being $S_3 = \partial \xi_3 / \partial z$, is

$$\frac{\partial T_3}{\partial z} = \frac{\rho \partial^2 \xi_3}{\partial t^2} \tag{3.4}$$

From Eqs. (3.1) and (3.4), the following plane-wave equation is easily obtained:

$$\frac{\rho\,\partial^2\xi_3}{\partial t^2}=\frac{c_{33}^D\,\partial^2\xi_3}{\partial z^2} \tag{3.5}$$

The solution of this plane wave equation for harmonic excitations [5, 6] is:

$$\xi_3=\left(C_1 e^{-j\beta z}+C_2 e^{j\beta z}\right)e^{j\omega t} \tag{3.6}$$

where the constants C_1 and C_2 have to be determined from the mechanical boundary conditions ($u=\partial\xi_3/\partial t$, and $F=T_3\,A$), at the transducer surfaces ($z=0$, backing; and $z=t$, load) [6]:

$$u_B=\left.\frac{\partial\xi_3}{\partial t}\right|_{z=0} \quad ; \quad u_L=-\left.\frac{\partial\xi_3}{\partial t}\right|_{z=t} \tag{3.7a, b}$$

$$F_B=-\left.AT_3\right|_{z=0} \quad ; \quad F_L=-\left.AT_3\right|_{z=t} \tag{3.7c, d}$$

In addition, the expressions for the current I through the transducer and the voltage V across the transducer terminals have to be introduced:

$$I=j\,\omega\,A\,D_3 \quad ; \quad V_3=\int_0^t E_3\,dz \tag{3.8a, b}$$

From Eqs. (3.1) to (3.8), three equations, which specify the relations among the terminal variables, present at the ports of the transducer, are obtained. This set of equations can be written in matrix form:

$$\begin{pmatrix} F_L \\ F_B \\ V \end{pmatrix}=\begin{pmatrix} Z_0\,A/\,j\tan\beta t & Z_0\,A/\,j\sin\beta t & h_{33}/\,j\omega \\ Z_0\,A/\,j\sin\beta t & Z_0\,A/\,j\tan\beta t & h_{33}/\,j\omega \\ h_{33}/\,j\omega & h_{33}/\,j\omega & 1/\,j\omega C_0^S \end{pmatrix}\begin{pmatrix} u_L \\ u_B \\ I \end{pmatrix} \tag{3.9}$$

where $C_0^S=A\varepsilon_{33}^S/t$ is the clamped (zero strain, high frequency) capacitance of the piezoelectric plate, being A the area of the electrodes; $Z_0=(\rho c_{33}^D)^{1/2}$ is the characteristic stiffened acoustic impedance of the piezoelectric material, being ρ the density, and $\beta=\omega/(c_{33}^D/\rho)^{1/2}$ the propagation constant.

The specific or characteristic impedance gives the ratio of the stress to the particle velocity for acoustic plane waves [7]. The characteristic impedances of the backing and the load are $Z_B=(\rho_B\,c_B)^{1/2}=F_B/(A\,u_B)$ and $Z_L=(\rho_L\,c_L)^{1/2}=F_L/(Au_L)$, respectively, being c_B and c_L the corresponding elastic constants.

3.3 Equivalent Circuits

The electrical and acoustic characteristics of a TE piezoelectric layer can be evaluated from the previous set of Eqs. (3.9). Nevertheless, it is frequently more convenient to devise an electrical equivalent circuit using the fundamental electromechanical analogies. According to the classical analogy, the mechanical force can be considered as the analog of the electrical voltage and the particle velocity as the analog of the electrical current.

The previous 3x3 electromechanical matrix in Eq. (3.9), relates the terminal variables (F, u, V, I) present at the mechanical and electrical ports. From a functional point of view, three components can be distinguished in this matrix: i) The first 2X2 sub-matrix corresponds to the equations of a mechanical transmission line of length t (thickness of the piezoelectric plate), characteristic impedance Z_0 and phase velocity $(c_{33}^D/\rho)^{1/2}$, which could be thought as representing the ultrasonic propagation in the piezoelectric medium; ii) the terms including the piezoelectric constant h_{33} correspond to the electromechanical coupling; iii) the last matrix element is the electrical impedance of the plate capacitance.

Using the electromechanical analogy, different three-port (six terminals) equivalent circuits can be obtained. Figures 3.3, 3.4 and 3.5 show Mason [14], Redwood [15] and KLM [16] equivalent circuits, which have been widely used for piezoelectric transducer simulation, design and optimization.

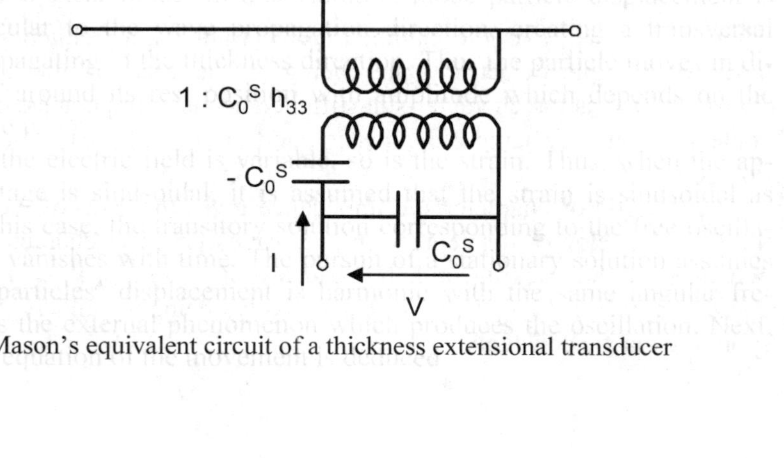

Fig. 3.3. Mason's equivalent circuit of a thickness extensional transducer

Figure 3.3 shows the pioneer Mason equivalent circuit for a thickness extensional transducer in which the electrical port is coupled, through an ideal transformer (transformation ratio $N = C_0^S h_{33} = A e_{33} / t$), to a T-network which represents the acoustic propagation in the piezoelectric material (the first 2x2 sub-matrix). The mechanical ports of the transducer are located at the extremes of the T-network. It should be noted the presence of the negative capacitance $-C_0^S$ close to the electrical terminals. Both, the negative capacitance and the ideal transformer account for the electro-mechanical coupling.

Redwood equivalent circuit shown in Fig. 3.4 is very similar to Mason circuit and can be derived from it by simply substituting the T-network with the equivalent transmission line (2x2 sub-matrix). In this circuit, the ideal transformer is connected to the external shield of the transmission line. This equivalent circuit has been used for SPICE implementations of the transducer model [17-19].

Figure 3.5 shows the KLM equivalent circuit proposed by Krimholtz, Leedom, and Matthaei in 1970 [16], and widely used since then. A mechanical transmission line represents the wave propagation in the piezoelectric material (2x2 sub-matrix). The mechanical ports are located at its extremes. The electrical port is connected through a transformer (with a frequency dependent transformer ratio Φ) to the center of the transmission line. The transformer is connected through a capacitance C_0^S in series with a reactance X (also frequency dependent) to the transducer electrical terminals.

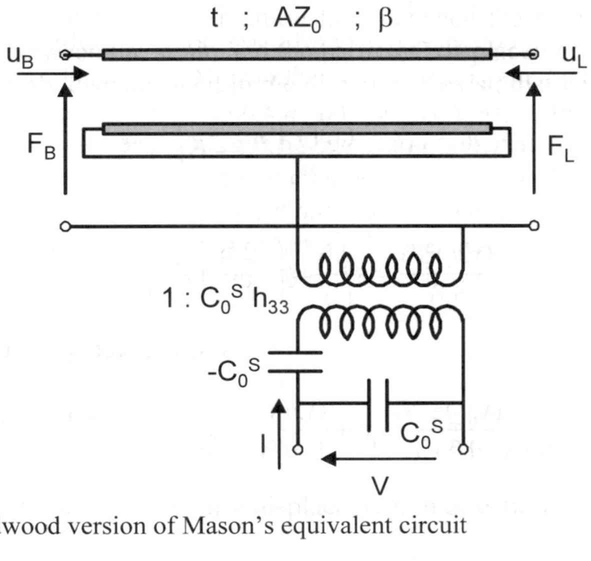

Fig. 3.4. Redwood version of Mason's equivalent circuit

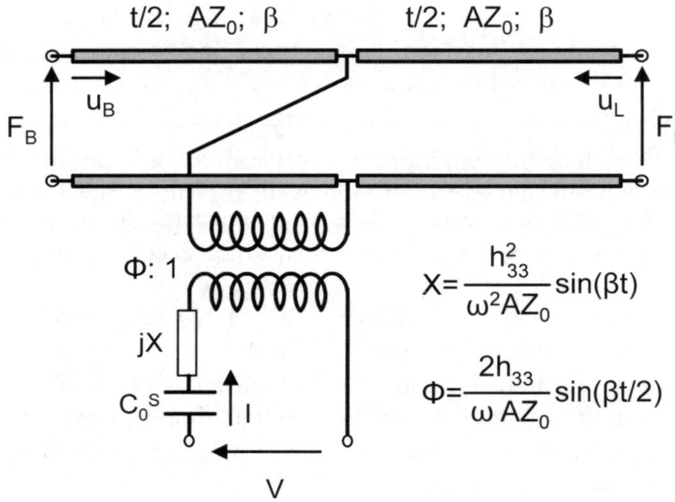

Fig. 3.5. KLM equivalent circuit of a thickness extensional piezoelectric transducer

Mason and Redwood circuits, with the explicit presence of the clamped capacitance in parallel with the transducer electrical terminals, provide a good representation of the electrical port of the transducer. On the other hand, the KLM network, with the explicit transmission line analogy, represents better the mechanical sections of the transducer. In fact this model has been used to design multiple acoustic matching layers, establishing alternative design criteria [3, 10].

Although these equivalent networks are not physically realizable with discrete components (negative capacitance $-C_0^S$, transformer ratio Φ and reactance X with a complex dependence on frequency), and contain no information other than that contained in the 3x3 electromechanical matrix, they are particularly useful for broadband multilayer piezoelectric analysis. Different non-piezoelectric layers (i.e. matching layers) can also be represented by transmission lines and the powerful techniques of network analysis can be used. In addition, the circuit diagram helps to represent the otherwise complex mathematics [2].

All these circuits correspond exactly to the 3x3 electromechanical matrix Eq. (3.9) and can be derived from them. Their complete equivalence can be established from the equality of the open and short circuit input electrical impedance [10]. A closed form expression for the electrical impedance at the transducer terminals $Z_{in}(\omega)$, in the most general case, when the mechanical ports are terminated by arbitrary acoustic impedances Z_B and Z_L, can be derived from any of these exact one-dimensional models:

$$Z_{in}(\omega) = \frac{1}{j\omega C_0^S}\left(1 + \frac{k_t^2}{\beta t}\,\frac{j(Z_L + Z_B)Z_0 \sin\beta t - 2Z_0^2(1-\cos\beta t)}{(Z_0^2 + Z_L Z_B)\sin\beta t - j(Z_L + Z_B)Z_0 \cos\beta t}\right) \tag{3.10}$$

where $k_t = h_{33}(\varepsilon_{33}^S / c_{33}^D)^{1/2}$ is the electromechanical coupling coefficient.

3.4 Broadband Piezoelectric Transducers as Two-Port Networks. Transfer Functions

When the back mechanical port is closed with an absorbing material of characteristic impedance Z_B, the transducer can be represented by a two-port linear and reciprocal network as shown in Fig. 3.6. Any ratio between the four remaining unknowns V, I, F_L, and u_L can be obtained.

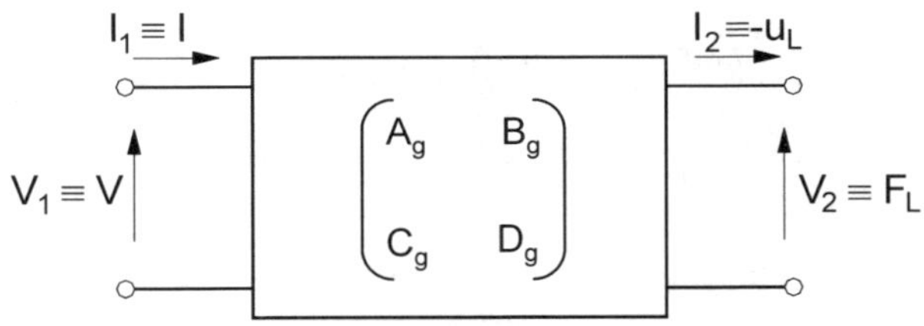

Fig. 3.6. Two-port representation of a broadband transducer system when the back mechanical port is closed with a mechanical impedance ($A\ Z_B$). The electrical and mechanical variables at the terminals are related by the ABCD matrix

The transfer matrix formalism of the circuit theory (with the transfer matrix ABCD relating the parameters at the electrical port and front acoustic port), can be used for the analysis of the performance of either a transmitting or a receiver transducer:

$$\begin{pmatrix} V_1 = V \\ I_1 = I \end{pmatrix} = \begin{pmatrix} A_g & B_g \\ C_g & D_g \end{pmatrix} \begin{pmatrix} V_2 = F_L \\ I_2 = -u_L \end{pmatrix} \tag{3.11}$$

The transfer matrix elements A, B, C, and D of any two port network can be obtained from the short and open circuit conditions (matrix coefficients and terminal variables are functions of frequency):

$$A_g = \left.\frac{V_1}{V_2}\right|_{I_2=0} ; B_g = \left.\frac{V_1}{I_2}\right|_{V_2=0} ; C_g = \left.\frac{I_1}{V_2}\right|_{I_2=0} ; D_g = \left.\frac{I_1}{I_2}\right|_{V_2=0} \tag{3.12a-d}$$

In particular, the transfer matrix coefficients of the piezoelectric transducer element, as represented in Fig. 3.6, can be obtained from circuit analysis, dividing an equivalent circuit into a set of simple cascade networks. Using the KLM circuit model [11], the resulting coefficients can be obtained as the product of the following elementary transfer matrices:

$$\begin{pmatrix} A_p & B_p \\ C_p & D_p \end{pmatrix} = \begin{pmatrix} 1 & j\left(\dfrac{h_{33}^2 \sin\beta t}{\omega^2 Z_0 A} - \dfrac{1}{\omega C_0^S}\right) \\ 0 & 1 \end{pmatrix} \begin{pmatrix} \phi & 0 \\ 0 & \dfrac{1}{\phi} \end{pmatrix} \cdot$$

$$\cdot \begin{pmatrix} 1 & 0 \\ \dfrac{Z_0 + jZ_B \tan(\beta t/2)}{Z_0 A (Z_B + jZ_0 \tan(\beta t/2))} & 1 \end{pmatrix} \cdot \tag{3.13}$$

$$\cdot \begin{pmatrix} \cos(\beta t/2) & jZ_0 A \sin(\beta t/2) \\ \dfrac{j\sin(\beta t/2)}{Z_0 A} & \cos(\beta t/2) \end{pmatrix}$$

where these elementary matrices correspond to the series impedance in front of the transformer, the transformer itself, the transmission line of length *t*/2 terminated by the backing impedance, and the transmission line of length *t*/2 terminated by the load. The transformer ratio of the KLM equivalent circuit is $\Phi = 2h_{33} \sin(\beta t/2)/\omega Z_0 A$.

Generally, the piezoelectric active element has a characteristic impedance much greater than those of the usual loads (i.e. water, tissue, metal). In addition, the electrical impedance of a piezoelectric transducer usually presents a notable miss-match with driver-receiver electronics. Consequently, different acoustic and electrical impedance matching procedures have to be used, looking for a compromise between efficiency and bandwidth. The transfer matrix formalism can be used for the treatment of all components of a transducer system as a set of cascade networks, with the final overall matrix being the product of the elementary ones.

Figure 3.7 presents a global scheme for an ultrasonic pulse-echo process, including an acoustic matching section in the front face and an electrical matching network at the electrical terminals of the transducer. Different configurations of impedance matching schemes can be evaluated using the transmission matrix formula [12, 13]. As an example, the matrix elements for simple series Z_S and parallel Z_P impedances, and mechanical transmission lines, are shown in Table 3.1.

Table 3.1. ABCD-matrix coefficients for simple circuit elements

Series Impedance Z_S	Parallel impedance Z_P	Mechanical transmission line
$\begin{pmatrix} 1 & Z_s \\ 0 & 1 \end{pmatrix}$	$\begin{pmatrix} 1 & 0 \\ 1/Z_p & 1 \end{pmatrix}$	$\begin{pmatrix} \cos\beta l & jZ_l \sin\beta l \\ j\sin\beta l / Z_l & \cos\beta l \end{pmatrix}$

l is the transmission line length, Z_l is the mechanical impedance and β is the propagation constant

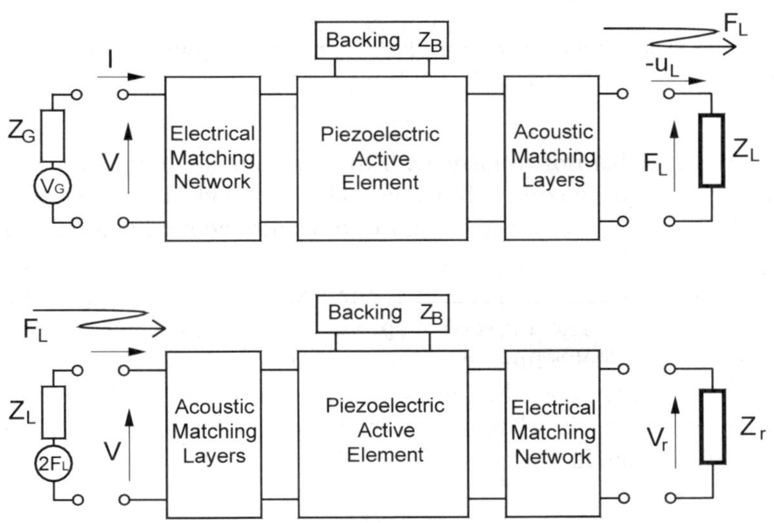

Fig. 3.7. Global scheme for the evaluation of broadband transducers in pulse-echo applications, including electrical and acoustical matching sections

The overall A_g, B_g, C_g and D_g matrix coefficients are obtained by the product of the individual matrices corresponding to the piezoelectric element, the matching layers and the electrical matching. Three of the possible ratios between the terminal variables, the electrical impedance, and the emission and reception transfer functions, are particularly useful in describing the performance of a transducer system, and can be derived from equation (3.11), using the overall matrix coefficients.

The electrical input impedance at the transducer terminals, $Z_{in}(\omega) = V(\omega)/I(\omega)$, accounting for electrical matching networks and acoustic matching layers, is:

$$Z_{in}(\omega) = \frac{A_g Z_L + B_g}{C_g Z_L + D_g} \tag{3.14}$$

The transducer performance can be analyzed separately for emission and reception operations by means of the emission and reception transfer functions. Conventional approaches to piezoelectric emission stage analysis, usually assume a waveform generator with resistive output impedance. If the characteristics of the source V_G, Z_G are known, the emission frequency response F_L/V_G, can be computed:

$$\frac{F_L}{V_G} = \frac{Z_L}{(A_g Z_L + B_g) + Z_G(C_g Z_L + D_g)} \tag{3.15}$$

The reception transfer function, V_r/F_L can be computed from the following expression:

$$\frac{V_r}{F_L} = \frac{2Z_r}{(A_g Z_L + B_g) + Z_r(C_g Z_L + D_g)} \tag{3.16}$$

When the same transducer is used to transmit and receive, the two-way voltage transfer function V_r/V_G (assuming a perfect reflection of the ultrasonic emitted pulse) is the product of the previous emission and reception transfer functions. The impulse response of the transducer system, (response to a Dirac delta function excitation) at the electrical terminals, can be obtained by means of the inverse Fourier transform of the overall frequency response V_r/V_G. The time response to an arbitrary "spike" excitation can be obtained by convolution of the impulse response with the "spike" waveform.

References

1. IEEE Standard on Piezoelectricity, ANSI/IEEE Std 176, (1987)
2. A. Ballato, "Modeling piezoelectric and piezomagnetic devices and structures via equivalent networks", IEEE Trans. Ultrason. Ferroelect., Freq. Contr., Vol. 48, Nº 5, pp. 1189-1240, (2001)
3. C.S. Desilets, J.D. Fraser and G.S. Kino, "The design of efficient broadband piezoelectric transducers", *IEEE Trans. Sonics and Ultrason.*, SU-25, Nº 3, pp. 115-125 (1978).
4. A. Ramos, J.L. San Emeterio and P.T. Sanz, "Improvement in transient piezoelectric responses of NDE transceivers using selective damping and tun-

ing networks", *IEEE Trans. Ultrason. Ferroelec. Freq. Contr.*, 47, pp. 826-835 (2000).

5. G.S. Kino, "Acoustic waves: devices, imaging, and analog signal processing", Prentice Hall, Englewood Cliffs, NJ, (1987).
6. V.M. Ristic, "Principles of acoustic devices", John Wiley and Sons, (1983).
7. B.A. Auld, "Acoustic fields and waves in solids", Krieger Publishing Company, (1990)
8. M.G. Silk, "Ultrasonic transducers for Nondestructive Testing", Adam Hilger Ltd, (1984).
9. J.W. Hunt, M. Arditi, F.S. Foster, "Ultrasound transducers for pulse-echo medical imaging", IEEE Trans. Biomed. Eng., Vol. 30, Nº 8, pp. 453-481, (1983)
10. J.L. San Emeterio, A. Ramos, P.T. Sanz, E. Riera, "Modelling of multilayer piezoelectric transducers for echographic applications. I.-Analysis in the frequency domain; and II.- Equivalent circuits" Mundo Electrónico Nº 186, pp. 85-90; and Nº 187, pp. 159-187, (1988).
11. J.L. San Emeterio, P.T. Sanz, E. Riera and A. Ramos, "Una implementación del modelo KLM para transductores piezoelectricos en modo espesor" *Anales Fisica B*, 84, pp. 48-55 (1988).
12. G.R. Lockwood, F.S. Foster, "Modeling and optimization of high-frequency ultrasound transducers", *IEEE Trans. Ultrason. Ferroelec. Freq. Contr.*, Vol. 41, pp. 225-230 (1994).
13. J.L. San Emeterio, A. Ramos, P.T. Sanz, A. Ruiz, "Evaluation of impedance matching schemes for pulse-echo ultrasonic piezoelectric transducers", *Ferroelectrics* , Vol. 273, pp. 297-302 (2002).
14. W.P. Mason, "Electromechanical transducers and wave filters", Van Nostrand, New York, 1948.
15. M. Redwood, "Transient performance of a piezoelectric transducer", Journal of the Acoustical Society of America, Vol. 33, Nº 4, pp. 327-336, (1961)
16. R. Krimholtz, D.A. Leedom, G.L. Mathaei, "New equivalent circuits for elementary piezoelectric transducers", Electronic Letters, Vol. 6, Nº 13, pp. 398-399, (1970)
17. S.A. Morris, C.G. Hutchens, "Implementations of Mason's model on circuit analysis programs", *IEEE Trans. Ultrason. Ferroelec. Freq. Contr.*, Vol. 33, pp. 295-298 (1986).
18. W.M. Leach, "Controlled-Source analogous circuits and SPICE models for piezoelectric transducers", *IEEE Trans. Ultrason. Ferroelec. Freq. Contr.*, Vol. 44, pp. 60-66 (1997).
19. A. Ramos, J.L. San Emeterio, P.T. Sanz, "Dependence of pulser driving responses on electrical and motional characteristics of NDE ultrasonic probes", *Ultrasonics*, Vol. 38, pp. 553-558 (2000).

$$Q = \frac{L\omega_s}{R} \tag{1.A.38}$$

Considering the relations given by Eq. (1.A.37), the former equation can be rewritten as follows:

$$Q = \frac{\rho l_e^2 \omega_s}{(n\pi)^2 \eta_G} \tag{1.A.39}$$

The motional series resonant frequencies coincide with the natural vibration frequencies without looses given by Eq. (1.26). Consequently, the motional series resonant angular frequency ω_s is given by:

$$\omega_s = n\pi\frac{v_{sh}}{2h_q} \tag{1.A.40}$$

where v_{sh} is given by Eq. (1.23) which is rewritten as follows:

$$v_{sh} = \sqrt{\frac{\bar{c}_{66}}{\rho}} \tag{1.A.41}$$

By substituting Eqs. (1.A.41) and (1.A.40) in Eq. (1.A.39), the following expression for the quality factor of the resonator is obtained:

$$Q = \frac{\bar{c}_{66}}{\eta_G \omega_s} \tag{1.A.42}$$

It can be notice that the former expression coincides with Eq. (1.46) derived from a physical point of view and is consistent with it. However, it is necessary to make clear that the previous expression for the quality factor is related to the motional branch and that the electrical contribution of the parallel capacitor is not included. Therefore the expression obtained for the quality factor must be considered an approximate equation for the quality factor of the resonator at frequencies very close to the motional series resonant frequencies.

References

1. W.G. Hankel, Abh. Sächs. 12, 457 (1881).
2. P. & J. Curie, *Comptes Rendus*, 91, 294 (1880).
3. G. Lippmann, *Annales De Physique Et De Chimie, 5th Series*, 24, 145 (1881).

4 Models for Resonant Sensors

Ralf Lucklum[1], David Soares[2] and Kay Kanazawa[3]

[1]Institute for Micro and Sensor Systems, Otto-von-Güericke-University Magdeburg
[2]Institute de Fisica, Universidade de Campinas
[3]Department of Chemical Engineering, Stanford University

4.1 Introduction

The quartz crystal resonator (QCR), as its acronym implies, is a resonant physical device. Many of its behaviors and properties can be understood physically by examining its resonant behavior. The basic principle of operation for a generic acoustic-wave sensor is a traveling wave combined with a confinement structure to produce a standing wave whose frequency is determined jointly by the velocity of the traveling wave and the dimensions of the confinement structure. The most basic way of resonator modeling consequently requires applying the theory of wave propagation whereby considering material properties and geometric dimensions of the resonator. As another successful way, there is an electrical equivalent circuit often used to characterize the resonance. For these reasons, a closer inspection of the phenomenon of resonance is useful.

4.2 The Resonance Phenomenon

On certain physical systems, the phenomenon of resonance can be used to multiply the effects of a force applied to the system. There are examples in mechanical, electrical and optical systems. When energy in a system is exchanged periodically between two forms, then resonance occurs. For example, in the case of a weight hanging on a rubber band, when the band is stretched, there is potential energy stored in the extended band. Subsequently, as the weight moves, the stored potential energy is exchanged into the kinetic energy of the weight. Following this, the kinetic energy is then transferred back into the potential energy in the band itself. If there are no losses in the system, then this back and forth energy transfer would continue. This is the resonance phenomenon. This energy exchange occurs periodically and is characterized by a resonance frequency. Similarly in an electrical circuit with an inductor and a capacitor, electrical energy can be stored as voltage across the capacitor. This stored energy then produces a current that flows in the inductor, exchanging the energy stored in the electrical field into a stored magnetic energy associated with the magnetic fields produced by the current in the inductor. The equivalent analogy between mechanical systems and electrical systems has been used also to describe the phenomenon of resonance of quartz crystal resonators and other acoustic-wave based sensors.

In the case of the thickness shear mode quartz crystal resonator, the application of alternating voltage across the crystal results in the generation of a shear acoustic wave, causing a distortion of the crystal. When the frequency of the alternating voltage is far from the resonant frequency, the distortion, as measured by the shear displacement of the surface of the resonator, is very small. The importance of the displacement is understood when one considers the force imparted to the surface by a particle rigidly attached. The force is proportional to the acceleration of the particle. This acceleration is given by the product of the displacement and the square of the angular frequency. At resonance, the displacement can exceed the far-from-resonance displacement by 10^5 or more. The force exerted by this particle on the resonator surface is multiplied by 10^5. This accounts for the extremely high sensitivity of the QCM to loaded mass (see Chap. 1). The multiplication factor is called the "quality factor" or Q of the resonator. It measures the ratio of the peak stored energy in the resonant cycle to the mean energy dissipated per cycle. A high Q would then imply a low loss resonance.

4.3 Concepts of Piezoelectric Resonator Modeling

The quartz crystal resonator is the most common device used as acoustic-wave based sensor. The simple geometry of the device and the predominant thickness-shear mode of the propagating wave are propitious conditions for a comprehensive derivation of the acoustic-electrical behavior of quartz crystal devices, including the resonance phenomenon. Other acoustic microsensors introduced in Chap. 2 have more complicated wave propagation pattern; the concepts of piezoelectric resonator modeling are the same. Quartz crystal resonator sensors are therefore indicative of acoustic-wave based sensors to demonstrate the concepts of modeling those sensors.

Quartz resonators are commonly used as frequency reference due to their very high *Q*-factor. A well known sensor application is the measurement of mass deposition (rates) in vacuum deposition technology (gravimetric principle). In most chemical sensor applications a chemically sensitive interface is realized with a coating; the coated quartz crystal hence can be considered as composite resonator. Analyte sorption in this sensitive layer results in a measurable change of properties of this layer, whereas the quartz crystal remains unchanged (see Chap. 7). The *Q*-factor of a quartz crystal with a foreign layer is still high, thus the oscillating frequency is very stable and can be measured with high resolution. Exposure of the resonator to a liquid results in energy loss caused by viscous damping. The decay length of shear waves at frequencies typically for quartz crystal resonators is so small that acoustic energy is dissipated only in a very thin liquid layer adjacent to the driving surface. However, the *Q*-factor is still remarkable high to ensure a significant resonance.

A multilayer structure like that in Fig. 4.1 is the generalization of the single coating case. In the physical understanding acoustic waves travel back and forth in this structure and superimpose. Amplitude and phase of the traveling waves are defined by geometric and material properties of each layer.

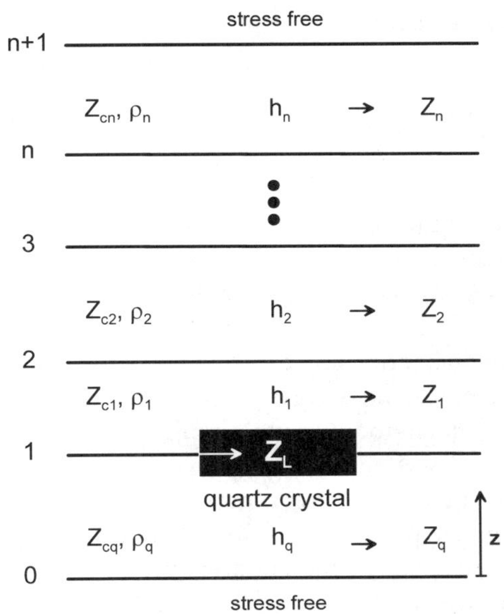

Fig. 4.1. General schema of a quartz crystal resonator with a multilayer coating

The electrical response of such a composite quartz crystal resonator sensor is however governed by the resulting wave in the quartz crystal. This wave is the superposition of the wave reflected off the boundary (1) to the coating and the wave transmitted into the quartz crystal through this interface (assuming the other main surface of the quartz crystal (0) is uncoated).

In the context of Fig. 4.1 the acoustic (thickness-shear) wave travels in normal direction through the layers. Wave propagation is defined by the (complex) wave propagation constant, k, which is a function of the angular frequency, ω, of the propagating wave, the density, ρ, of the layer and its shear modulus, G (Note that the piezoelectrically stiffened modulus of the quartz crystal is a tensor and usually denoted as c (see Appendix B)). As explained above for the boundary (1) between the quartz crystal and the first layer; the acoustic wave is partly reflected off and transmitted through at any boundary of the multilayer structure (Fig. 4.1). The effect of reflection and transmission at the boundary between two materials can be described by (complex) reflection and transmission coefficients, R and T, respectively:

$$R_{i,i+1} = f_R(Z_i, Z_{i+1}) = \frac{Z_{i+1} - Z_i}{Z_i + Z_{i+1}} \tag{4.1}$$

$$T_{i,i+1} = f_T(Z_i, Z_{i+1}) = \frac{2Z_{i+1}}{Z_i + Z_{i+1}} \tag{4.2}$$

where

$$Z = f(\rho, G) = \sqrt{\rho \cdot G} \tag{4.3}$$

is the characteristic impedance of the respective layer.

The complex nature of k, Z, R, T arises from the complex nature of the shear modulus, G, of viscoelastic materials as analyzed in detail in Chap. 12.

The real part of the reflection coefficient, is the known amplitude relation of incident and reflected wave. The imaginary part of the reflection coefficient can be understood as the amplitude relation between the incident and a 90°-phase shifted reflected wave. The real part of the reflection coefficient is governed by the force proportional to particle displacement, whereas the imaginary part of the reflection coefficient is governed by the force proportional to speed of the vibrating particle.

The physical model of a composite resonator considers a piezoelectric plate covered with one or a number of non-piezoelectric layers, each characterized with a set of acoustically relevant parameters. A non-piezoelectric layer may be a thin film of a rigid, pure elastic material, a pure viscous liquid (film) or a film of a viscoelastic material (described in greater detail in Chap. 12). The set of characteristic parameters contains a geometric value, the film thickness, and material properties like film density and the (complex) shear modulus or like film density and complex viscosity. Characteristic impedance of the film material and the wave propagation constant or the wave velocity are other versions to describe film properties.

The vibration behavior of a quartz crystal and the wave propagation in a multilayer arrangement can be derived in a one-dimensional model. This is a commonly accepted approximation. The high aspect ratio between the diameter of a quartz disc and the thickness of the crystal makes this assumption reasonable. As an additional requirement all layers must be uniform and homogeneous. Furthermore, continuity of particle displacement and shear stress at any interface is assumed. However, certain deviations from these assumptions, e.g., the shear amplitude distribution across the surface of a quartz crystal, a non-uniform film or specific interfacial phenomena are therefore not considered in this treatment. Those effects may significantly contribute to the vibration behavior of the quartz crystal and must be considered in more involved resonator models.

The analytical approach to describe acoustic wave propagation (Appendix 4.A.2) uses two waves with unknown amplitudes traveling in opposite direction in each layer. The linear piezoelectric equations together with Newton's equations of motion and Maxwell's equations must be applied for the piezoelectric plate (Eqs. (4.A.1)-(4.A.2)). The appropriate boundary conditions (Eqs. (4.A.10a-f)) must be exploited to calculate the unknown parameters.

The equivalent circuit approach (Appendix 4.A.3) describes the acoustic wave propagation in analogy to electrical waves. The matrix concept uses a three-port element and a transducer impedance matrix to represent the piezoelectric plate (Eqs. (4.A.14)-(4.A.15)). Non-piezoelectric layers are represented by a two-port element and an impedance matrix for each layer (Eqs. (4.A.22)-(4.A.23)). Each matrix is the result of fundamental physical equations and appropriate boundary conditions. The transmission line model and the Mason model are two versions of this approach (see Chap. 3). In contrast to the analytical approach, multilayer arrangements can be treated much easier.

The major results of the physical model of quartz crystal resonator sensors can be summarized as follows:

The electrical impedance or admittance is a function of the electrical capacitance of the quartz crystal formed by the electrodes and the quartz as dielectric material and the so-called motional impedance. The motional impedance is the electrical equivalent of the so-called acoustic load impedance, Z_L, acting at the surface of the quartz plate (Fig. 4.1). This acoustic load impedance must not be mixed up with the characteristic impedance of a material, (Eq. (4.3)). The acoustic load impedance is an effective acoustic impedance, i.e., the characteristic impedance *modified* due to wave propagation.

The quartz crystal resonator sensor response is therefore sensitive to any change in the acoustic load impedance. The acoustic load impedance (change) can be generated from a pure mass (change) of a single rigid film (Eq. (4.10)), a semi-infinite Newtonian liquid (Eq. (4.12)), a single viscoelastic film or a multilayer arrangement. A thin rigid, purely elastic film and a semi-infinite purely viscous liquid are the two special cases, which result not only in a special form of the acoustic load but also in a distinct dependence of the resonant frequency on surface mass (density thickness product) or density viscosity product, respectively (Appendix 4.A.4).

Near resonance the physical model can be developed into a special notation, where the physical parameters can be summarized in lumped equivalent electrical values: motional inductance, motional capacitance and motional resistance. This finally gives rise to the modified Butterworth-Van Dyke equivalent circuit model (see Chap. 1 and Appendix 4.A.4). This model allows the analysis of the electrical behavior of a quartz crystal resonator from electrical measurements without the need of determining the physical properties of the resonator. Some relations are analyzed in the following section.

Another theoretical approach different from the acoustic-wave propagation concept is the energy transfer model. In this model the quartz crystal generates and stores acoustic energy. Acoustic energy trapped in a confined structure explains also the resonance phenomenon described in Sect. 4.2 corresponding to a harmonic oscillator of mass $m = m_q/2$. Alternatively one can also consider an analogue electrical model consisting of a capacitance, C, an inductance, L and a resistance, R in a series circuit (see Fig. 4.2b). If the resonator is coated with another material, e.g. the chemically sensitive film, a small part of this acoustic energy is transferred into this material. This energy is stored in a purely elastic film and partly stored and partly dissipated in a viscoelastic film. Considering the high Q-factor of the quartz crystal, the electrical power applied to the crystal is equal to the developed mechanical power at $\omega = \omega_0$. By taking this into account, the mechanical impedance is analogue to the electrical one. The material medium affects the crystal through its mechanical load impedance, Z_L, a result similar to that derived with the wave-propagation concept.

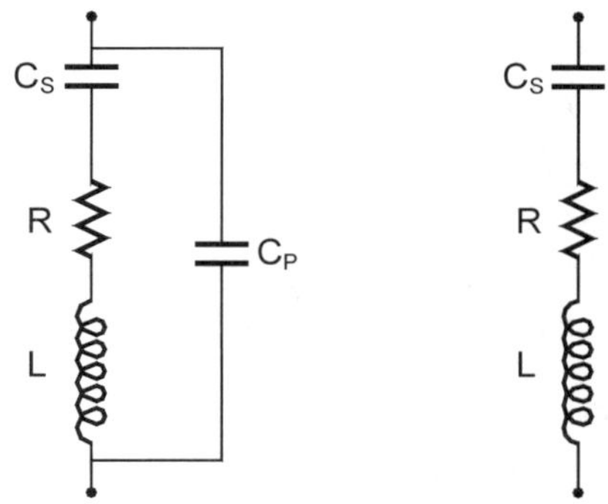

Fig. 4.2. **a** The complete Butterworth-Van Dyke circuit is shown on the left, and **b** the motional branch isolated is shown on the right

4.4 The Equivalent Circuit of a Quartz Crystal Resonator

The Butterworth-Van Dyke circuit (BVD) consists of two parallel branches as shown in Fig. 4.2:

The left hand branch consisting of only the capacitance C_p represents the fixed dielectric capacitance of the resonator. All of the motional information is contained in the right hand branch. The left hand branch does influence the phase of the current relative to the voltage driving the circuit; therefore, to obtain an accurate representation of the motional behavior as a function of frequency, the parallel capacitance must be compensated (Chap. 6). Here the major interest is in the relation between the elements L, C_S and R and the resonance characteristics, such as the resonant frequency, f_s, and the quality factor, Q, so the left hand branch will be neglected and we will study only the behavior of the right hand branch.

The admittance, Y, of the network shown in Fig. 4.2b is defined as the current to voltage ratio and is a function of the applied frequency, f. In terms of the angular frequency ω, defined as $2\pi f$, Y can be expressed as:

$$Y(\omega) = \frac{1}{R + j\left(\omega L - \dfrac{1}{\omega C_s}\right)} \tag{4.4}$$

and its magnitude by

$$|Y(\omega)| = \frac{1}{\left(R^2 + \left(\omega L - \frac{1}{\omega C_s}\right)^2\right)^{1/2}} \tag{4.5}$$

It is seen from Eq. (4.5) that this magnitude is a maximum at the frequency ω_s where $\omega L = 1/(\omega C_s)$. In the most commonly encountered form, this is written:

$$\omega_s^2 LC = 1 \tag{4.6}$$

At this frequency, Eq. (4.4) reveals that the value of the magnitude of Y at resonance has the value

$$Y(\omega_0) = \frac{1}{R} \tag{4.7}$$

and has only a real component, with the imaginary components cancelled. This leads to the conclusion that the phase difference between the voltage and current at this frequency is zero. This is illustrated in Fig. 4.3, where both the magnitude of the admittance and the phase as a function of frequency are shown. These calculations were done taking a circuit with a resonant frequency of 5,000,000 Hz, a resistance of 100 Ω and an inductance of 0.04 H. The phase of the circuit can be seen to pass through the value of zero at the resonant frequency.

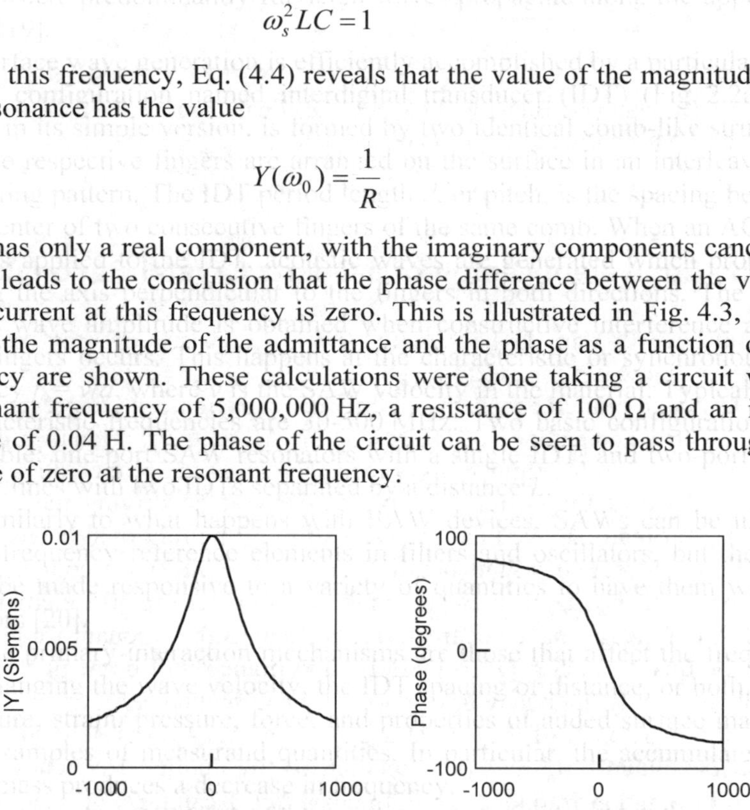

Fig. 4.3. Magnitude of the admittance (left) and its phase behavior (right)

Next we analyze the quality factor. It can be shown that this is related to the half power spectrum of the resonance. At two frequencies, one above ($\omega_{1/2}^+$) and one below ($\omega_{1/2}^-$) the resonant frequency, the power dissipated in the resonance will drop by ½. In terms of the admittance, this occurs when the magnitude of the admittance is decreased by $1/\sqrt{2}$. This is also termed the "3 db points", since the response is down from the maximum by three decibels. Some investigators use dissipation (D) to describe the losses in the resonance and it is very simply related to the quality factor by $D = 1/Q$.

An interesting expression for Q is given by the relation:

$$Q = \frac{1}{R}\sqrt{\frac{L}{C_s}} = \frac{\omega_s L}{R} \tag{4.8}$$

The expression $\sqrt{L/C_s}$ has the units of resistance, and its ratio to the resistance of the circuit yields the quality factor. Typical values for a 5 MHz AT-cut quartz crystal resonator would have $L = 0.04$ H, $C_s = 25$ fF. Thus $\sqrt{L/C_s} = 1.3\text{x}10^6$ Ohms! For a resistance of 10 Ω (typical for a resonator in air), the Q would be $1.3\text{x}10^5$, a large multiplicative factor indeed! If inductance L changes only very little over a set of changing loading conditions (usually less than 1%) the Q is inversely proportional to the resistance R.

From Fig. 4.2b, we see that there are three variables L, C_s and R required to specify the motional impedance. The interpretation in terms of the resonant frequency, ω_s and the quality factor Q has been discussed. It can be instructive to write the starting equations in terms of ω_s, Q and R. While we shall not go through that exercise here, we illustrate its utility by writing the half power frequencies in the following manner:

$$\omega_{1/2}^{+} = \omega_s\left(\sqrt{1+\left(\frac{1}{2Q}\right)^2} + \frac{1}{2Q}\right) \tag{4.9a}$$

$$\omega_{1/2}^{-} = \omega_s\left(\sqrt{1+\left(\frac{1}{2Q}\right)^2} - \frac{1}{2Q}\right) \tag{4.9b}$$

This exact relationship shows that the two half power frequencies are spaced equidistantly from a central frequency which is only negligibly different from the resonant frequency. For example, for a Q of 1000, the central frequency is different from the resonant frequency by only a few parts in ten million.

The central basis for the extreme sensitivity of the QCM is based on its resonant behavior. It is very useful to discuss aspects of the storage and dissipation of energy in the QCM (and its overlays). The BVD-model is one convenient way, if electrical properties of the sensor are of major interest. The physical models should be applied, if the relations of electrical values to material properties of the coating(s) are needed.

4.5 Four Important Conclusions

4.5.1 The Sauerbrey Equation

As shown already in Chap. 1 quartz crystal resonators are very sensitive to mass changes at its surface. The resonator modeling based on acoustic wave propagation recovers Sauerbrey's fundamental equation for a small phase shift of the acoustic wave while propagating through the foreign film. A small phase shift requires exactly Sauerbrey's limitations: a thin film of a rigid material. Under those circumstances, a mass, m_s, uniformly distributed at the surface of a quartz crystal ($m_s = \rho_s h_s$, ρ_s is the surface density of the film $\rho_s = m_s/A$, where A is the effective surface) generates a shift in the resonant frequency

$$\Delta f_s = -2 f_0^2 \frac{\rho_s h_s}{\sqrt{\rho_q c_q}} \tag{4.10}$$

which can be easily rewritten into Eq. (1.66) in Chap. 1 or in Eq. (4.A.37) in the Appendix. Furthermore, replacing f_0 by v, the acoustic wave velocity, and λ, the acoustic wave length $v = \lambda f$, and considering the thickness of the quartz crystal, h_q, of being $\lambda/2$ at mechanical resonance, f_0, Eq. (4.10) can be rewritten as

$$\frac{\Delta f_s}{f_0} = -\frac{\rho_s h_s}{\rho_q h_q} \tag{4.11}$$

4.5.2 Kanazawa's Equation

A second special case is a quartz crystal in contact with a purely viscous liquid (so-called Newtonian liquid) at one surface. Due to the extremely small penetration depth of a shear wave in viscous materials, a liquid film can be considered as semi-infinite. Under those circumstances resonator modeling based on acoustic wave propagation recovers also Kanazawa's fundamental equation:

$$\Delta f_s = -f_0^{3/2} \sqrt{\frac{\rho_l \eta_l}{\pi \rho_q c_q}} \tag{4.12}$$

where ρ_l and η_l are the liquid density and viscosity, respectively. This equation is equivalent to Eq. (4.A.41) in the Appendix.

Equations (4.10) or (4.11) and (4.12) are very important because they show the sensor capability of quartz crystal resonators. Both equations are often applied to calculate absorbed mass in chemical sensor applications or determining density/viscosity of liquids. The modeling presented in the Appendix draws a more complete picture of how acoustic-wave based devices can be applied as sensors, when Sauerbrey's and Kanazawa's equations can be applied and the extended capabilities of those devices in more involved systems.

4.5.3 Resonant Frequencies

The quartz crystal as shown as equivalent circuit in Fig. 4.2a has several resonant frequencies. Oscillators can work at or near one of these resonant frequencies (see Chap. 6). They all depend on the acoustic load in a certain way, however, only the resonance of the motional arm as shown in Fig. 4.2b reflects the change of the acoustic load as predicted from models developed in the Appendix. All other resonance frequencies have a distinct dependence on the motional resistance. Only in the case of a very small acoustic energy dissipation in the sensing film, i.e. a very small motional resistance, R, (e.g. valid for a rigid film), any oscillator should respond with one and the same frequency shift to a certain mass change. Under any other conditions, the motional resonant frequency, f_s, must be selected. Several electronic concepts are available. They are described in Chap. 6.

4.5.4 Motional Resistance

The motional resistance carries information about acoustic energy dissipation in the coating and the surrounding medium. This information is different from the frequency shift, which is related to acoustic energy storage. Therefore the measurement of the equivalent resistance in addition to the frequency shift is optional for all measurements in a gaseous environment and strongly recommended for all applications of acoustic-wave devices in a liquid environment or when viscoelastic materials are used as sensitive film. Both values together allow for a much more assured data interpretation. Under certain circumstances the motional resistance can be even more sensitive to property changes of the analyte than the frequency shift.

Appendix 4.A

4.A.1 Introduction

Acoustic waves can be employed to measure physical or chemical values, like force, film thickness or the concentration of a certain compound in a mixture. Several kinds of devices have been used for generation and detection of acoustic waves and to pick up the relevant information (Chap. 2). The underlying transduction mechanism from the input signal to the output signal contains common and specific features. The common aspect for all devices is their sensitivity to any change of the acoustic properties of themselves or at the device surface. The acoustic properties include intrinsic material parameters (density, elastic moduli) and geometric values (thickness, length of the acoustic path). The acoustic wave traveling in a coated device especially penetrates into the adjacent film, translates and deforms the film, thereby probing its mechanical properties, its thickness and the acoustic properties at the upper film surface. Most physical sensors are based on changes of the acoustic properties of the acoustic device whereas most chemical sensors relay on changes of the acoustic properties of a coating. In any case, the acoustic waves carry the information of interest. The specific aspects are related to the kind of acoustic wave used, the wave propagation in the device and sometimes the electro-mechanical transformation.

In the following a model is described, which is especially useful for acoustic-wave-based chemical sensors. Those sensors obtain their chemical sensitivity and selectivity from a chemically active coating on top of the acoustic device, which interacts with the surrounding environment. This interaction leads to a change in the acoustic-wave propagation, which in turn yields a change of the electrical response of the sensor.

The general concept in modeling acoustic wave sensors is based on the solution of a set of wave equations with regard to suitable boundary conditions between the sensor and the adjacent media. In consequence, the principle behavior of the different types of acoustic-wave devices is similar. Bulk acoustic wave (BAW) devices are typically realized with AT-cut quartz crystals. They vibrate in an almost pure thickness shear mode; therefore they are also called thickness-shear-mode (TSM) devices. Acoustic wave generation and propagation is most concise, therefore a coated quartz crystal resonator is used here as example to demonstrate the physical background of acoustic-wave-based sensors. Surface acoustic waves (SAWs) are coupled compressional and shear waves. Similar to BAWs, the propagation of SAWs generates a periodic displacement field into the adjacent layer. In contrast to BAWs, the surface displacement field has two distinct means of inducing strain in the coating: from the SAW-specific in-plane gradients arising due to the sinusoidal variation in displacement components along the direction of SAW propagation, and analog to BAWs, from surface normal gradients arising from a phase difference between the motion of the upper surface of the layer with respect to the "driven" lower surface.

4.A.2 The Coated Piezoelectric Quartz Crystal. Analytical Solution

The linear piezoelectric equations together with the resulting system of differential equations for the unknown mechanical displacement and electrical potential have been given in Chap.3. They describe in general the behavior of a piezoelectric resonator. The full set of differential equations is, however, difficult to solve for the complete three-dimensional problem, because the unknown displacements and the electrical potential as well as their derivatives with respect to time and location are coupled with each other in these equations. For special geometries there are assumptions which enable an approximate two- or three-dimensional solution [1, 2]. These models intend to characterize the *uncoated* quartz crystal but they cannot solve the problem of a *coated* resonator in general.

Here a one-dimensional solution for AT-cut quartz crystal resonators is presented. Due to the high ratio between the lateral dimensions and the thickness for a typical quartz resonator vibrating in the thickness shear mode it is reasonable to treat the crystal as an infinite plate with a finite thickness [3].

The thickness of the resonator is orientated in the x_2-direction. By using an infinite plate it is assumed that physical properties do not change along the x_1 and x_3-directions. The derivatives along these directions vanish, and only the derivatives along x_2 remain. For the quartz crystal, starting from the general piezoelectric equations, result (see Appendix 1.A):

$$\tau_{12} = c_{66}u_{1,2} + \eta_q \dot{u}_{1,2} + e_{26}\phi_{,2} \tag{4.A.1a}$$

$$D_2 = e_{26}u_{1,2} - \varepsilon_{22}\phi_{,2} \tag{4.A.1b}$$

where τ_{ij} is the component of the mechanical stress tensor, D_i is the electrical displacement (vector), c_{ij}, e_{ij}, ε_{ij} are the components of the material property tensors for mechanical stiffness, piezoelectric constant, and permittivity, respectively, u_i is the mechanical displacement component, and ϕ is the electrical potential. A colon with an index represents the partial derivative of the expression with respect to the specified index. The time derivative is marked as usual as a dot above the variable. A viscous term in the stress-strain relation has been included with the phenomenological quartz viscosity, η_q. This term accounts for losses inside the quartz crystal. The equation of motion and the Maxwell equation for the electrical displacement become

$$\tau_{12,2} = \rho_q \ddot{u}_1 \tag{4.A.2a}$$

$$D_{2,2} = 0 \tag{4.A.2b}$$

With harmonic time dependence, that is the mechanical displacement and the electrical potential vary with exp(jωt), the following differential equations can be given:

$$\left(c_{66} + \frac{e_{26}^2}{\varepsilon_{22}} + \mathrm{j}\omega\eta_q \right) u_{1,22} + \omega^2 \rho_q u_1 = 0 \tag{4.A.3a}$$

$$e_{26} u_{1,22} - \varepsilon_{22} \phi_{,22} = 0 \tag{4.A.3b}$$

The first equation is a wave equation for the unknown mechanical displacement u_1, whereas the second equation couples the mechanical displacement with the electrical potential. Some abbreviations will be used in the following:

$$e_q \equiv e_{26} \quad ; \quad \varepsilon_q \equiv \varepsilon_{22}$$

$$c_q \equiv c_{66} + \frac{e_{26}^2}{\varepsilon_{22}} + \mathrm{j}\omega\eta_q = c_{q0} + \mathrm{j}\omega\eta_q \tag{4.A.4}$$

where c_q is the effective complex shear modulus. With these definitions the differential equations for the quartz crystal are written as

$$c_q u_{1,22} + \omega^2 \rho_q u_1 = 0 \tag{4.A.5a}$$

$$e_q u_{1,22} - \varepsilon_q \phi_{,22} = 0 \tag{4.A.5b}$$

The solution of this wave equation can be written with two components, namely two waves traveling in positive and negative x_2-direction inside the quartz crystal with unknown parameters B_1 and B_2:

$$u_1 = \left(B_1 e^{\mathrm{j}k_q x_2} + B_2 e^{-\mathrm{j}k_q x_2}\right) e^{\mathrm{j}\omega t} \tag{4.A.6}$$

where k_q is the complex wave propagation vector $k_q = \omega\,(\rho_q/c_q)^{1/2}$. For a lossy quartz crystal the imaginary part of k_q represents the decay of the traveling waves.

With the solution from Eq. (4.A.6) the electrical potential (Eq. (4.A.5)), the stress (Eq. (4.A.1a)), and the electrical displacement (Eq. (4.A.1b)) are calculated (with unknown parameters B_3 and B_4) to

$$\phi = \left(\frac{e_q}{\varepsilon_q} \left(B_1 e^{\mathrm{j}k_q x_2} + B_2 e^{-\mathrm{j}k_q x_2}\right) + B_3 x_2 + B_4 \right) e^{\mathrm{j}\omega t} \tag{4.A.7a}$$

$$\tau_{12} = \left(\mathrm{j}k_q c_q \left(B_1 e^{\mathrm{j}k_q x_2} - B_2 e^{-\mathrm{j}k_q x_2}\right) + e_q B_3\right) e^{\mathrm{j}\omega t} \tag{4.A.7b}$$

$$D_2 = -\varepsilon_q B_3 e^{\mathrm{j}\omega t} \tag{4.A.7c}$$

The relations for the coating (acoustic load) are similar to those of the quartz crystal, the index i is used for the layer properties. The coating is assumed to be non-piezoelectric; therefore there is no piezoelectric component. The differential equation for the coating becomes

$$c_i u_{1,22} + \omega^2 \rho_i u_1 = 0 \tag{4.A.8}$$

with c_i and ρ_i being the complex shear modulus and the density of the coating, respectively. As for the quartz crystal, the solution has the form of a wave propagation and is written with parameters C_1 and C_2 as

$$u_1 = \left(C_1 e^{\mathrm{j}k_i x_2} + C_2 e^{-\mathrm{j}k_i x_2}\right) e^{\mathrm{j}\omega t} \tag{4.A.9}$$

where k_i is the complex wave propagation vector inside the coating.

The following boundary conditions have to be applied for the system quartz crystal (thickness h_q) - coating (thickness h_i):

1. continuous displacement $u_1(x_2 = h_q)$ at interface crystal-coating (4.A.10a)
2. continuous shear stress $\tau_{12}(x_2 = h_q)$ (at interface crystal – coating (4.A.10b)
3. vanishing shear stress $\tau_{12}(x_2 = 0) = 0$ at free crystal surface (4.A.10c)

4. vanishing shear stress τ_{12} ($x_2 = h_q + h_i$) = 0 at free coating surface (4.A.10d)
5. driving electrical potential $\phi(x_2 = h_q) = -\phi_0 \exp(\mathrm{j}\omega t)$ at upper electrode (4.A.10e)
6. driving electrical potential $\phi(x_2 = 0) = \phi_0 \exp(\mathrm{j}\omega t)$ at lower electrode (4.A.10f)

These boundary conditions provide six equations for the unknown parameters, B_1, B_2, B_3, B_4, C_1, and C_2. The solution of this system of equations can be obtained with standard matrix methods. It is not given here. The final solution of the one-dimensional problem can be calculated with these six parameters.

Finally some abbreviations are introduced:

$$K^2 = \frac{e_q^2}{\varepsilon_q c_q} \quad ; \quad v_q = \sqrt{c_q / \rho_q} \tag{4.A.11a,b}$$

$$k_q = \omega\sqrt{\rho_q / c_q} = \omega / v_q \quad ; \quad \gamma_q = \mathrm{j}k_q \tag{4.A.11c,d}$$

$$\alpha = k_q h_q = \omega h_q \sqrt{\rho_q / c_q} \quad ; \quad Z_{cq} = \sqrt{\rho_q c_q} = \rho_q v_q \tag{4.A.11e,f}$$

$$\varphi = k_i h_i \quad ; \quad k_i = \omega\sqrt{\rho_i / c_i} \quad ; \quad Z_{ci} = \sqrt{\rho_i c_i} \tag{4.A.11g,h,i}$$

α and φ are the acoustic phase shift inside the quartz crystal and the coating, respectively. Z_{cq} and Z_{ci} are the characteristic acoustic impedance of the quartz crystal and the coating; K^2 is the electromechanical coupling coefficient of quartz. With the static quartz capacitance $C_0 = \varepsilon_q (A/h_q)$ the following relation can be found for the electrical impedance of the coated quartz resonator after some algebraic transformations:

$$Z = \frac{1}{\mathrm{j}\omega C_0}\left(1 - \frac{K^2}{\alpha}\,\frac{2\tan\frac{\alpha}{2} - \mathrm{j}\frac{Z_L}{Z_{cq}}}{1 - \mathrm{j}\frac{Z_L}{Z_{cq}}\cot\alpha}\right) \tag{4.A.12}$$

The electrical impedance of a coated quartz crystal resonator can be calculated from quartz crystal parameters, the frequency, and the acoustic load impedance, Z_L. Eq. (4.A.12) simplifies for the uncoated quartz crystal:

$$Z = \frac{1}{j\omega C_0}\left(1 - \frac{K^2}{\alpha} 2\tan\frac{\alpha}{2}\right) \tag{4.A.13}$$

4.A.3 The Transmission Line Model

The transmission line model can be used to describe both the (piezoelectric) transformation between electrical and mechanical vibration and the propagation of acoustic waves in the system acoustic device-coating-medium in analogy to electrical waves [4, 5]. This model assumes a uniform piezoelectric device and isotropic, homogeneous, uniform layers and a sensor configuration, in which lateral dimensions have no effect on the propagation of waves. The model does not have any restrictions on the number of layers, their thickness and their mechanical properties. The characteristic acoustic parameters and the geometric values of non-piezoelectric layers are summarized in the effective acoustic impedance, which transforms the acoustic properties at one port to the other one. It reflects, in which manner the layer is translated and deformed by the acoustic wave. The complete transmission line model relates the overall system characteristics to the electrical impedance (or admittance) at the electrical port, starting from the front acoustic port with known acoustic properties (usually stress-free corresponding to a short-circuited acoustic port, or a semi-infinite liquid).

The transmission line model allows a formal separation of the acoustic wave propagation inside the acoustic device, including the transformation of mechanical displacement into the electrical signal and vice versa, and outside the acoustic device. In this context the acoustic load (impedance), Z_L, which is a complex number, represents the overall acoustic load at the interface between the acoustic device and the coating. It should not be mixed up with the characteristic impedance, Z_{ci}, which is a material constant. The acoustic load summarizes all acoustically relevant information. It does not play any role, if this load is generated by a simple mass, a single viscoelastic coating, a multilayer arrangement, or a semi-infinite material. Consequently, the acoustic load, Z_L, carries all information, which is related to changes in the chemically sensitive coating, no matter if it is pure mass accumulation, mass accumulation accompanied by material property changes, or only material property changes induced by chemical (e.g. cross-linking) or physical (e.g. phase transition) effects. A change in the acoustic load impedance results in a change of the electrical impedance of the BAW device or a change of sound velocity and attenuation of the propagating SAW. Finally these changes are responsible for the frequency shift and attenuation change of the acoustic device.

The piezoelectric quartz crystal

The piezoelectric transducer can be regarded as a three-port black box, where the two main surfaces form two acoustic ports whereas the electrodes form the electrical port. The independent variables are treated as "currents" and the dependent variables are treated as "voltages" with the following analogy:

mechanical tension $\tau \Leftrightarrow U$ electrical voltage
particle velocity $v \Leftrightarrow I$ electrical current
acoustic impedance $Z = \tau/v \Leftrightarrow Z = U/I$ electrical impedance

The general behavior of a three-port element is described by a transducer impedance matrix, Z_T:

$$\begin{pmatrix} \tau_1 \\ \tau_2 \\ U_3 \end{pmatrix} = Z_T \begin{pmatrix} v_1 \\ v_2 \\ I_3 \end{pmatrix} \tag{4.A.14}$$

The transmission line model implements the fundamental physical equations as well as the boundary conditions already in the transfer matrix giving

$$Z_T = \begin{pmatrix} Z_{cq} \coth(\mathrm{j}k_q h_q) & \dfrac{Z_{cq}}{\sinh(\mathrm{j}k_q h_q)} & \dfrac{e_q/\varepsilon_q}{\mathrm{j}\omega A} \\ \dfrac{Z_{cq}}{\sinh(\mathrm{j}k_q h_q)} & Z_{cq} \coth(\mathrm{j}k_q h_q) & \dfrac{e_q/\varepsilon_q}{\mathrm{j}\omega A} \\ \dfrac{e_q/\varepsilon_q}{\mathrm{j}\omega} & \dfrac{e_q/\varepsilon_q}{\mathrm{j}\omega} & \dfrac{1}{\mathrm{j}\omega C_0} \end{pmatrix} \tag{4.A.15}$$

One of the representations of the transmission line model is the equivalent circuit from Krimholtz, Leedom, and Matthaei, which is referred to as KLM-model [6]. It is presented in Fig. 4.A.1 (big square). The elements are defined by:

$$\frac{1}{N^2} = \frac{1}{A}\left(\frac{2e_q}{\varepsilon_q \omega Z_{cq}} \sin(k_q \frac{h_q}{2})\right)^2 \tag{4.A.16a}$$

$$X = \frac{e_q^{\ 2}}{A\varepsilon_q^{\ 2}\omega^2 Z_{cq}} \sin(k_q h_q) \tag{4.A.16b}$$

$$C_0 = \varepsilon_q \frac{A}{h_q} \tag{4.A.16c}$$

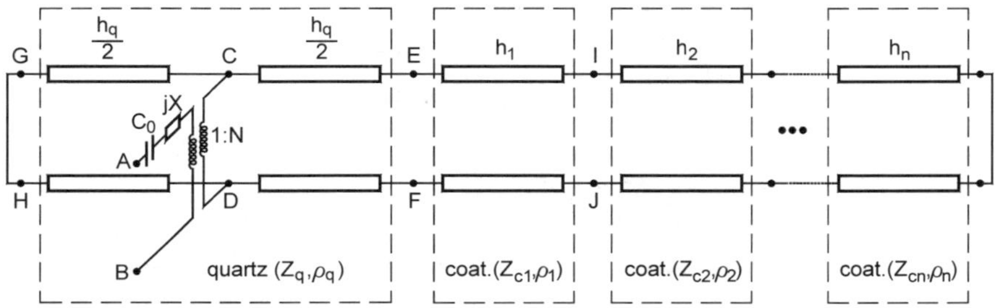

Fig. 4.A.1 Representation of the transmission line model (KLM model)

The equations for the electrical impedance at port AB can be easily derived. The impedance at position CD from the right acoustic port is:

$$Z_r = Z_{cq} \frac{Z_{EF} + Z_{cq} \tanh(\gamma_q\, h_q/2)}{Z_{cq} + Z_{EF} \tanh(\gamma_q\, h_q/2)} \tag{4.A.17a}$$

Similarly, the impedance at position CD from the left acoustic port is:

$$Z_l = Z_{cq} \frac{Z_{GH} + Z_{cq} \tanh(\gamma_q\, h_q/2)}{Z_{cq} + Z_{GH} \tanh(\gamma_q\, h_q/2)} \tag{4.A.17b}$$

The overall impedance at position CD is the parallel arrangement of Z_r and Z_l. The transformer transforms the acoustic into the electrical signal at port AB:

$$Z_{AB} = \frac{1}{\mathrm{j}\omega C_0} + \mathrm{j}X + \frac{1}{N^2} Z_{CD} \tag{4.A.18a}$$

$$\mathrm{j}X = \frac{1}{\mathrm{j}\omega C_0} \frac{K^2}{\alpha} \sin\alpha \tag{4.A.18b}$$

$$\frac{1}{N^2}=\frac{1}{\omega C_0}\frac{4K^2}{\alpha}\frac{1}{Z_{cq}}\sin^2\frac{\alpha}{2} \tag{4.A.18c}$$

In the special case of a single-side coated sensor, i.e. a stress free surface at port GH and an acoustic load acting at port EF ($Z_{EF}=Z_L$), Eqs. (4.A.16) –(4.A.18) yields after some calculations:

$$Z=\frac{1}{\mathrm{j}\omega C_0}\left(1-\frac{K^2}{\alpha}\frac{2\tan\frac{\alpha}{2}-\mathrm{j}\frac{Z_L}{Z_{cq}}}{1-\mathrm{j}\frac{Z_L}{Z_{cq}}\cot\alpha}\right) \tag{4.A.19}$$

which is equivalent to Eq. (4.A.12).

It is possible to separate the impedance Z into a parallel circuit consisting of a static capacitance C_0 and a motional impedance Z_m for the quartz:

$$Z_m=\frac{1}{\mathrm{j}\omega C_0}\left(\frac{1-\mathrm{j}\frac{Z_L}{Z_{cq}}\cot\alpha}{\frac{K^2}{\alpha}\left(2\tan\frac{\alpha}{2}-\mathrm{j}\frac{Z_L}{Z_{cq}}\right)}-1\right) \tag{4.A.20}$$

This motional impedance can be split into two parts:

$$Z_m=\frac{1}{\mathrm{j}\omega C_0}\left(\frac{\frac{\alpha}{K^2}}{2\tan\frac{\alpha}{2}}-1\right)+\frac{1}{\omega C_0}\frac{\alpha}{4K^2}\frac{Z_L}{Z_{cq}}\frac{1}{1-\frac{\mathrm{j}\frac{Z_L}{Z_{cq}}}{2\tan\frac{\alpha}{2}}}=Z_m^0+Z_m^L \tag{4.A.21}$$

The first part represents the unloaded quartz ($Z_L=0$) and should not change during the measurement, the other one is related to the load. This important feature allows for several simplifying approximations presented in Sect.4.A.4.

The Acoustic Load

Non-piezoelectric layers are represented by a two-port circuit. In case of a multilayer loading (Fig. 4.A.1), the surface acoustic load $Z_L=Z_{EF}$ acting at the port EF of the transmission line representing the piezoelectric quartz crystal, would be the resulting impedance from **all** layers placed on the quartz surface. Fig. 4.A.1 shows the respective transmission line. The impedance concept in propagation problems uses a chain matrix technique. The elements of the propagation matrix, $\mathbf{P}_i$, and the transfer matrix, $\mathbf{T}_i$, for each layer of thickness h_i, complex wave propagation constant γ_i, and characteristic acoustic impedance, Z_{ci}, are calculated as follows:

$$\mathbf{P}_i = \begin{pmatrix} e^{-\gamma_i h_i} & 0 \\ 0 & e^{\gamma_i h_i} \end{pmatrix} \qquad \mathbf{T}_i = \begin{pmatrix} \frac{1}{Z_{ci}} & 1 \\ \frac{1}{Z_{ci}} & -1 \end{pmatrix} \tag{4.A.22}$$

The layers are usually acoustically impedance mismatched with respect to the adjacent layers. Considering the layer i as a quadrupole with an input mechanical tension, u_{i+1}, and an input particle velocity, i_{i+1}, and an output mechanical tension, σ_i, and particle velocity, v_i, the transformation can be calculated with the transformation matrix, $\mathbf{M}_i$:

$$\begin{pmatrix} \sigma_i(z) \\ v_i(z) \end{pmatrix} = \mathbf{T}_i^{-1}\mathbf{P}_i^{-1}\mathbf{T}_i \begin{pmatrix} \sigma_{i+1}(z+h_i) \\ v_{i+1}(z+h_i) \end{pmatrix} = \mathbf{M}_i \begin{pmatrix} \sigma_{i+1}(z+h_i) \\ v_{i+1}(z+h_i) \end{pmatrix} \tag{4.A.23a}$$

$$\begin{aligned} \mathbf{M}_i &= \begin{pmatrix} \frac{1}{2}\left(e^{\gamma_i h_i} + e^{-\gamma_i h_i}\right) & \frac{Z_{ci}}{2}\left(e^{\gamma_i h_i} - e^{-\gamma_i h_i}\right) \\ \frac{1}{2Z_{ci}}\left(e^{\gamma_i h_i} - e^{-\gamma_i h_i}\right) & \frac{1}{2}\left(e^{\gamma_i h_i} + e^{-\gamma_i h_i}\right) \end{pmatrix} = \\ &= \begin{pmatrix} \cosh(\gamma_i h_i) & Z_{ci}\sinh(\gamma_i h_i) \\ \frac{1}{Z_{ci}}\sinh(\gamma_i h_i) & \cosh(\gamma_i h_i) \end{pmatrix} \end{aligned} \tag{4.A.23b}$$

Note that both electrodes may also be described in terms of two transmission lines, acting at port CD and EF, respectively. Since the upper electrode works toward a shear stressed layer in the case of a coated quartz crystal, one must expect a noticeable own contribution of this electrode. Nevertheless, the electrodes are acoustically thin layers; their contribution is small and does not change significantly during experiment. Thus, for simplicity, their effect is usually taken into account in an effective quartz crystal thickness.

With $Z = \sigma / v$ the impedance transformation performed with layer i can simply be calculated:

$$Z_i = Z_{ci} \frac{Z_{i+1} + Z_{ci} \tanh(\gamma_i\, h_i)}{Z_{ci} + Z_{i+1} \tanh(\gamma_i\, h_i)} \tag{4.A.24}$$

Equation (4.A.24) can be rearranged with (G replaces c_i in equations (4.A.11h,i))

$$\gamma = \mathrm{j}\frac{\omega}{\sqrt{G/\rho}} \tag{4.A.25a}$$

$$Z_c = \sqrt{\rho G} \tag{4.A.25b}$$

$$Z_i = Z_{ci} \frac{Z_{i+1} + \mathrm{j}\sqrt{\rho_i G_i}\, \tan\left(\omega \sqrt{\frac{\rho_i}{G_i}}\, h_i\right)}{Z_{ci} + \mathrm{j} Z_{i+1} \tan\left(\omega \sqrt{\frac{\rho_i}{G_i}}\, h_i\right)} \tag{4.A.26}$$

The equations in this section are exact within the one-dimensional assumptions. They should be used in all cases, where highest accuracy of the calculation is required and in all cases, where no information is available about error propagation [7].

4.A.4 Special Cases

Although the equations in the previous section are easily to compute on a computer, their comprehensibility is limited. Therefore several approximations are applied to transform these equations into a more convenient form. Some of them are summarized in the following.

The Modified Butterworth-Van Dyke Circuit

Near the resonance of the unloaded quartz sensor, the approximation $\tan(\alpha/2) \approx 4\alpha/(\pi^2\text{-}\alpha^2)$ leads to a simple expression for the unperturbed part of Z_m:

$$Z_m^0 = R_q + \mathrm{j}\omega L_q + \frac{1}{\mathrm{j}\omega C_q} + \frac{1}{\mathrm{j}\omega(-C_0)} = R_q + \mathrm{j}\omega L_q + \frac{1}{\mathrm{j}\omega C_q'} \tag{4.A.27}$$

For small loads ($Z_L/Z_{cq} << 2\tan(\alpha/2)$) the part of Eq. (4.A.21) representing the acoustic load simplifies to:

$$Z_m^L = \frac{1}{\omega C_0} \frac{\alpha}{4K^2} \frac{Z_L}{Z_{cq}} \tag{4.A.28}$$

Here the additional impedance is direct proportional to the load Z_L, thus giving for the motional impedance of the quartz:

$$Z_m = R_q + \mathrm{j}\omega L_q + \frac{1}{\mathrm{j}\omega C_q'} + \frac{1}{\omega C_0} \frac{\alpha}{4K^2} \frac{Z_L}{Z_{cq}} \tag{4.A.29a}$$

$$Z_m = R_q + \mathrm{j}\omega L_q + \frac{1}{\mathrm{j}\omega C_q'} + Z_c \tag{4.A.29b}$$

This expression can be transformed into a modified motional arm of the commonly used Butterworth-Van Dyke equivalent circuit (BVD) as shown in Fig. 4.A.2 lower part. The equivalent circuit elements are defined in a straightforward way; however, it needs the (practically impossible) negative capacitance [8].

$$C_0 = \varepsilon_q \frac{A}{h_q} \tag{4.A.30a}$$

$$C_q = C_0 \frac{8K_0^2}{\pi^2} = \frac{8Ae_q^2}{\pi^2 h_q c_{q0}} \tag{4.A.30b}$$

$$C_q' = \frac{C_q}{1 - \dfrac{8K_0^2}{\pi^2}} \tag{4.A.30c}$$

$$L_q = \frac{\pi^2}{8C_0 K_0^2 \omega_0^2} = \frac{\rho_q h_q^3}{8Ae_q^2} \tag{4.A.30d}$$

$$R_q = \frac{\eta_q / c_{q0}}{C_q} = \frac{\eta_q \pi^2}{8C_0 K_0^2 c_{q0}} \tag{4.A.30d}$$

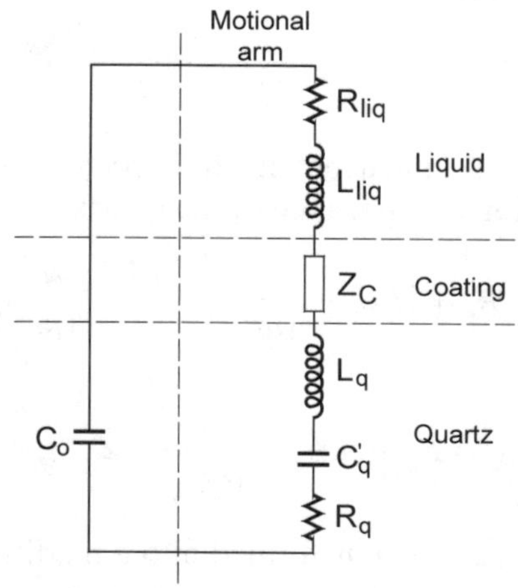

Fig. 4.A.2 Extended Butterworth-van-Dyke equivalent circuit

The Acoustic Load Concept

Unchanged quartz parameters allow for a direct relation between the (complex) acoustic load impedance, Z_L, and two directly measurable values, the frequency shift, Δf_s, and the change in the motional resistance, ΔR_m in the BVD equivalent circuit, the latter equivalent to a change in the quality factor $Q = \omega L/R$ or any other measure for acoustic energy dissipation.

The imaginary part of Z_L is related to the frequency shift taken at the in-phase admittance magnitude, $\mathrm{Re}(Y_{el})$. It is equivalent to the resonance of the motional arm of the BVD equivalent circuit, not to the zero phase crossing of the quartz resonator.

$$\frac{\Delta f_s}{f_0} = -\frac{\mathrm{Im}(Z_L)}{\pi Z_{cq}} \tag{4.A.31}$$

The real part of the acoustic load is related to acoustic energy dissipation:

$$\frac{\Delta R}{2\omega_0 L_q} = \frac{\mathrm{Re}(Z_L)}{\pi Z_{cq}} \tag{4.A.32}$$

The motional resistance change, ΔR_m, is taken as the difference in the reciprocal admittance magnitude at f_s between the coated and uncoated device. The motional inductance of the quartz crystal, L_q, as well as the resonance frequency, f_0 or ω_0, are device-specific and can be calibrated with a separate measurement of the bare quartz. These approximations can be applied for almost all coatings typically used for chemical sensing [9, 10].

Single Film

For a single film with a stress free upper surface ($Z_2 = 0$), usually a quartz crystal resonator with a single coating working in air, Eq. (4.A.26) simplifies to:

$$Z_L = \mathrm{j}\sqrt{\rho G}\tan\left(\omega\sqrt{\frac{\rho}{G}}h\right) \tag{4.A.33}$$

where the shear modulus, G, is a complex value in case of a coating with viscoelastic properties: $G = G' + \mathrm{j}G''$ (Chap. 12). This equation is visualized as a 3-D plot in Fig. 4.A.3 for a 10 MHz quartz crystal, split into imaginary and real part. The film thickness varies between 0 and 1 μm. Typical values of a polymer in the glassy state ($G' \approx 1$ GPa, $G' >> G''$), in the rubbery state ($G' \approx 1$ MPa, $G' > G''$) and in the transition range ($G' \approx G''$) have been taken for the two components of the shear modulus. This treatment allows the combination of the two variables on one axis. A film density of 1000 kg m^{-3} has been taken as a reasonable value. At first glance, one can distinguish between two major regions in both diagrams: an (inclined) rather plane range and a range with significant extreme. The two regions are related to acoustically thin films and acoustically thick films [11]. On acoustically-thin-film conditions the entire coating moves almost synchronously with the acoustic device surface, i.e., the phase shift, φ_f, between the lower and upper film surface is very small. At acoustically-thick-film conditions the upper surface significantly lags behind the device-film interface; the tan-function dominates Z_L. It leads to a rapid decrease followed by a sharp increase of Im(Z_L) (deep groove and high crest in Fig. 4.A.3a), and a dramatic increase of Re(Z_L) (maximum in Fig. 4.A.3b). Equation (4.A.33) can be rearranged to separate the mass contribution to the acoustic load from the acoustic phase shift contribution.

$$Z_L = \mathrm{j}\omega\rho h \cdot \frac{\tan\varphi}{\varphi} = \mathrm{j}M \cdot V \tag{4.A.34}$$

Here, $\varphi = \omega h(\rho/G)^{1/2}$ (Eq. 4.A.11g) is the phase shift the acoustic wave undergoes while traveling through the film, $M = \omega\rho h$ is the mass factor representing the gravimetric response, and $V = \tan\varphi/\varphi$ is the acoustic factor. The latter contains the shear modulus of the film and is responsible for the so-called viscoelastic contributions to the sensor signal [9].

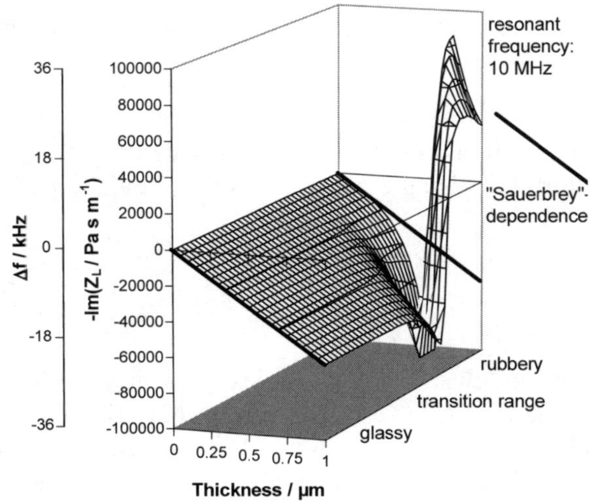

a

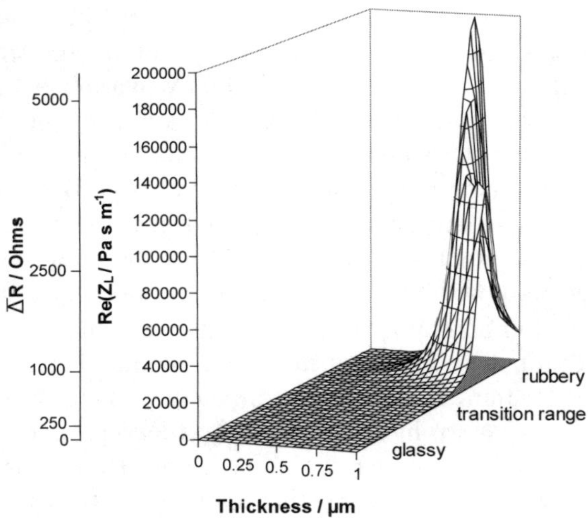

b

Fig. 5.A.3. a Real part, and **b** imaginary part of Eq. (4.A.33) for a 10 MHz quartz crystal.

The Sauerbrey Equation

The Sauerbrey equation assumes a thin rigid film. In terms of Eq. (A.4.34) it reads a small h and a large and real G. It means, that the phase shift, φ, is very small and the tan-function can be approximated by its argument, or, the acoustic factor, V, approaches 1.

The acoustic load of a thin rigid layer is hence

$$Z_L = \mathrm{j}\omega\rho h = \mathrm{j}M \tag{4.A.35}$$

Whereas Eq. (4.A.33) cannot be transformed into simple lumped elements of the BVD-model, the acoustic load of a rigid coating, Eq. (4.A.35), would simplify the complex impedance, Z_c, (Fig. 4.A.2), to an inductance, L_{rig} in series with the inductance L_q:

$$L_{rig} = \frac{1}{4C_0}\frac{\pi}{K^2 Z_{cq}}\frac{1}{\omega_s}\rho h \tag{4.A.36}$$

Applying Eq. (4.A.31) results in an equation similar to Sauerbrey's fundamental equation [12]:

$$\frac{\Delta f_{rig}}{f_0} = -\omega\frac{\rho h}{\pi Z_{cq}} \tag{4.A.37}$$

Furthermore, Eq. (4.A.32) also provides information about the change of the equivalent resistance:

$$\Delta R_{rig} = 0 \tag{4.A.38}$$

A rigid coating does not increase acoustic losses.

The Kanazawa Equation

A semi-infinite pure viscous (Newtonian) liquid, index *liq*, generates a surface acoustic impedance:

$$Z_{liq} = \sqrt{\mathrm{j}\rho_{liq}\omega\eta_{liq}} = (1+\mathrm{j})\sqrt{\frac{\omega\rho_{liq}\eta_{liq}}{2}} \tag{4.A.39}$$

where G in Eq. (4.A.31) is replaced by $G' = 0$, $G'' = \mathrm{j}\omega\eta_{liq}$, whereas $\mathrm{j}\tan\left(\omega\sqrt{(\rho_{liq}/\mathrm{j}\omega\eta_{liq})}h_{liq}\right) = 1$ for $h_{liq} >> \sqrt{2\eta_{liq}/\omega\rho_{liq}}$, the penetration depth of the shear acoustic wave in a viscous liquid. ρ_{liq} and η_{liq} are the liquid density and viscosity. Similar to the Sauerbrey case, Z_{liq} also can be represented as serial elements in the motional arm of the modified BVD-model (Fig. 4.A.2, upper part):

$$L_{liq} = \frac{1}{4C_0}\frac{\pi}{K^2 Z_{cq}}\frac{1}{\omega_s}\sqrt{\frac{\rho_{liq}\eta_{liq}}{2\omega_s}} \tag{4.A.40a}$$

$$R_{liq} = \frac{1}{4C_0}\frac{\pi}{K^2 Z_{cq}}\sqrt{\frac{\rho_{liq}\eta_{liq}}{2\omega_s}} \tag{4.A.40b}$$

Applying Eq. (4.A.31) results in a relation similar to Kanazawa's well known equation [13]:

$$\frac{\Delta f_{liq}}{f_0} = -\frac{1}{\pi Z_{cq}}\sqrt{\frac{\omega\rho_{liq}\eta_{liq}}{2}} \tag{4.A.41}$$

Again, Eq. (4.A.32) provides additional information about the resistance change:

$$\frac{\Delta R_{liq}}{2\omega_0 L_q} = \frac{1}{\pi Z_{cq}}\sqrt{\frac{\omega\rho_{liq}\eta_{liq}}{2}} \tag{4.A.42}$$

Due to the factor (1 + j) in Eq. (4.A.39) the contributions of a viscous liquid to frequency shift and resistance increase are similar.

Martin's Equation

If a quartz crystal with a single, thin and rigid layer works in a Newtonian liquid, the second term of the denominator of Eq. (4.A.24) is small compared to Z_{c1}. The acoustic load of this arrangement simplifies to:

$$Z_{rigliq} = \mathrm{j}\omega\rho h + (1+\mathrm{j})\sqrt{\frac{\omega\rho_{liq}\eta_{liq}}{2}} \tag{4.A.43}$$

The load from the rigid layer and the liquid load are, approximately, additive. Applying again Eq. (4.A.31) results in an equation similar to that published by Martin [14]:

$$\frac{\Delta f_{rig\,liq}}{f_0} = -\frac{2f}{Z_{cq}}\left(\rho h + \sqrt{\frac{\rho_{liq}\,\eta_{liq}}{4\pi f}}\right) \tag{4.A.44}$$

The contribution to the resistance increase arises only from the viscous damping of the liquid, Eq. (4.A.42).

Small phase shift approximation

In case of a small, but not very small phase shift, φ, tan φ can be approximated, giving a simple expression for the acoustic factor, V, and its change, V_ε:

$$V = \frac{\tan\varphi}{\varphi} \approx \frac{\varphi + \frac{\varphi^3}{3}}{\varphi} = 1 + \frac{\varphi^2}{3} = 1 + \frac{M^2}{3Z_{c1}^2} \tag{4.A.45a}$$

$$V_\varepsilon = (V + V_\varepsilon) - V = 1 + \frac{(M + M_\varepsilon)^2}{3Z_{c1}^2} - \left(1 + \frac{M^2}{3Z_{c1}^2}\right) \approx 2(V-1)\frac{M_\varepsilon}{M} \tag{4.A.45b}$$

where M is the mass factor calculated from density and thickness of the coating, M_ε is the mass factor from the mass accumulated in the coating during experiment. With Eq. (4.A.45) the frequency shift due to a single coating, Δf, and due to mass accumulation in the coating, Δf^ε, can be approximated to:

$$\Delta f \propto M\left(1 + \frac{M^2}{3Z_{c1}^2}\right) \tag{4.A.46a}$$

$$\Delta f^\varepsilon \propto M_\varepsilon\left(1 + \frac{M^2}{Z_{c1}^2}\right) \tag{4.A.46b}$$

Note, that the viscoelastic contribution to the frequency shift is three times higher for mass accumulation compared to the viscoelastic contributions from a viscoelastic coating although the material properties are assumed to be unchanged [15].

Whereas viscoelastic properties of a single coating in a gaseous environment contribute to an "extra mass", a soft thin film when adjoined to a liquid gives rise to a "missing mass" [16]. The effective acoustic factor becomes negative even if the acoustic factor of the film alone approaches 1. In such a case the overall acoustic load becomes (L replaces $\sqrt{\omega\rho_{liq}\eta_{liq}/2}$):

$$Z_{soft\,liq} = \mathrm{j}\,M\left(1 - \frac{2L^2(G'' - \mathrm{j}G')}{\rho|G|^2}\right) + (1+\mathrm{j})L = \mathrm{j}\,MV_{eff} + (1+\mathrm{j})L) \tag{4.A.47}$$

The contribution of the liquid is unchanged, whereas the contribution of the film is reduced. In contrast to V in Eq. 4.A.46 the effective acoustic factor, V_{eff}, does NOT depend on the mass factor, i.e., on the film thickness. Suppressing the contribution of the liquid, frequency shift and resistance increase due to a thin soft film can be approximated as follows [17]:

$$\Delta f_{soft\,in\,liq} \propto M\left(1 - 2\frac{L^2}{\rho}\frac{G''}{|G|^2}\right) \tag{4.A.48a}$$

$$\Delta R_{soft\,in\,liq} \propto M\left(2\frac{L^2}{\rho}\frac{G'}{|G|^2}\right) + M^2\left(2\frac{L}{\rho}\frac{G'+G''}{|G|^2}\right) \tag{4.A.48b}$$

References

1. R.D. Mindlin, "High frequency vibrations of plated crystal plates", *Progr. Appl. Mech.*, (1963), 73-84.
2. H. Tiersten, "Linear piezoelectric plate vibrations", Plenum Press, New York, 1969.
3. C.E. Reed, K.K. Kanazawa, and J.H. Kaufmann, "Physical description of a viscoelastically loaded AT-cut quartz resonator," *J. Appl. Phys.*, 68 (1990), 1993-2001.
4. G.S. Kino, "Acoustic Waves: Devices, Imaging and Analog Signal Processing", Englewood Cliffs, N.J.: Prentice-Hall, 1987.
5. H. Nowotny and E. Benes, "General one-dimensional treatment of the layered piezoelectric resonator with two electrodes," *J. Acoust. Soc. Am.*, 82 (1987), 513-521. C. Behling, "The non-gravimetric response of thickness shear mode resonators for sensor applications" *ShakerVerlag*, Aachen (1999)
6. R. Krimholtz, D.A. Leedom, and G.L. Matthaei, "New equivalent circuits for elementary piezoelectric transducers," *Electron. Lett.*, 6 (1970), 398-99.
7. R. Lucklum, C. Behling, R.W. Cernosek, and S.J. Martin, "Determination of complex shear modulus with thickness shear mode resonators," *J. Phys. D: Appl. Phys.*, 30 (1997), 346-356.
8. C. Behling, R. Lucklum, and P. Hauptmann, "Possibilities and limitations in quantitative determination of polymer shear parameters by TSM resonators," *Sensors and Actuators*, A 61 (1997), 260-266.
9. C. Behling, R. Lucklum, and P. Hauptmann, "The non-gravimetric quartz crystal resonator response and its application for polymer shear moduli determination," *Meas. Sci. Technol.*, 9 (1998), 1886-1893.

10. R. Lucklum, C. Behling, and P. Hauptmann, "Role of mass accumulation and viscoelastic film properties for the response of acoustic-wave-based chemical sensors," *Anal. Chem.*, 71 (1999) ,2488-2496.
11. S.J. Martin, G.C. Frye, S.D. Senturia (1994), "Dynamics and Response of Polymer-Coated Surface Acoustic Wave Devices: Effect of Viscoelastic Properties and Film Resonance", Anal. Chem 66, pp. 2201-2219.
12. G. Sauerbrey, "Verwendung von Schwingquarzen zur Wägung dünner Schichten und zur Mikrowägung," *Zeitschrift Physik*, 155 (1955), 206-212.
13. K.K. Kanazawa and J.G. Gordon, "Frequency of a quartz microbalance in contact with liquid," *Anal. Chem.*, 57 (1985), 1770-1771.
14. S.J. Martin, V. E. Granstaff, G.C. Frye, Characterization of a quartz crystal microbalancewith simultaneous mass and liquid loading, *Anal. Chem.* 63 (1991), 2272-2281.
15. R. Lucklum, P. Hauptmann, "The quartz crystal microbalance: Mass sensitivity, viscoelasticity and acoustic amplification", *Sensors Actuators B* 70/1-3 (2000), 30-36.
16. M.V. Voinova, M. Jonson, B. Kasemo, "'Missing mass' effect in biosensor's QCM application", *Biosensors Bioelectronics* 17 (2002), 835-841.
17. R. Lucklum, P. Hauptmann, "Transduction meachanism of acoustic-wave based chemical and biochemical sensors", *Meas. Sci. Techn.* (2003), 1854-1864

5 Interface Electronic Systems for Broadband Ultrasonic Applications

Antonio Ramos and José Luis San Emeterio

Departamento de Señales, Sistemas y Tecnologías Ultrasónicas, Instituto de Acústica (CSIC)

5.1 Introduction

There is a wide variety of applications where broadband piezoelectric systems are used, mainly in order to obtain ultrasonic information for detection or visualization of the internal parts of diverse structures. These applications require external inspections with ultrasonic waves and the use of an echo-graphic procedure. The main application areas are in industry and medicine.

Most broadband piezoelectric applications require the design of very specific interface electronic systems, since the conventional continuous wave (CW) electronic schemes and the conventional analysis methods are not applicable.

In order to obtain a good discrimination of internal structures it is convenient to improve the signal to noise ratio. Therefore it is necessary to guarantee a high efficiency in the ultrasonic process. In addition short ultrasonic pulses must be used in order to obtain good axial resolution.

The above mentioned practical questions impose some technological requirements over the electronic systems used for broadband ultrasonic applications:

1. The use of a pulsed regime for the ultrasonic inspection process, which involves a transient electrical excitation of the piezoelectric transmitter.
2. Sensitivity considerations determine that the transducer should be excited with very short electrical pulses (spikes) of several hundreds of volts at peak amplitude.
3. A high efficiency in emitter and receiver piezoelectric sub-processes, under broadband conditions, is required in order to optimise the cou-

pling between spike generators and piezoelectric emitters, as well as between piezoelectric and electronic receivers.

It is not easy to attain simultaneously all these requirements because parameters influencing the response amplitudes act in a crossed way. Specifically, the optimum setting of one of the parameters depends on the settings of the others.

As a typical example, to obtain a good dynamic range on the received echo-signals, by using broadband piezoelectric arrays in medical imaging [1], it is necessary to drive the array elements with electrical spikes of 300-400 volts and rise-times of less than 30 nanoseconds. In general, the time duration and frequency spectrum of the inspection signals are very relevant parameters for the quality of the result obtained in visualization processes [7].

In this chapter the characteristics required for the high voltage stages driving broadband piezoelectric transducers are explained. The dependence of the driving pulses with matching networks and the external loads is also analysed. A simple diagram to interface transducers with the reception stage is also presented.

5.2 General Interface Schemes for an Efficient Coupling of Broadband Piezoelectric Transducers

It is quite frequent in broadband ultrasonic applications that the electrical excitation and the receiver electronics are not adequately coupled with the piezoelectric transducers. As a consequence, the best ultrasonic performance is not attained in many cases [2-3]. It is difficult to guarantee a high efficiency in the electromechanical conversion processes simultaneously with good characteristics in the whole frequency band, in particular a large bandwidth with a low ripple. These requirements in frequency are related to the generation of short ultrasonic pulses. There are some matching schemes for particular applications [4-5], but there is not a general solution for the matching problem in practical broadband conditions, because of the transient character of the pulsed processes used for transducer driving [6]. Moreover, some internal impedances of the pulser circuits used for this driving are variable in time [7]. Both aspects complicate a theoretical analysis of these coupling problems in the emitter phase.

Driving stages in broadband applications are usually composed of high-voltage ramp generators (with low-value series impedance) in cascade with interface networks to electrically match with transducers. Assuming linear working conditions, the output function of such schemes is the temporal convolution of an exponential function with the impulse response of the coupling circuits IR_{CC} :

$$Driving\ output\ (t) \equiv IR_{CC}(t) * V_o e^{-\tau/t} \tag{5.1}$$

Where, V_o and τ are the amplitude and fall-time of the real ramp function, which follows in practice a time evolution near an exponential curve.

The piezoelectric receiver should also be matched to reception electronics in order to obtain either maximum amplitude in received signals or very short ultrasonic pulses, depending on the kind of application. In the time domain, the final ultrasonic pulse received depends on the convolution of the driving waveform with the impulse responses of emitting and receiving piezoelectric stages.

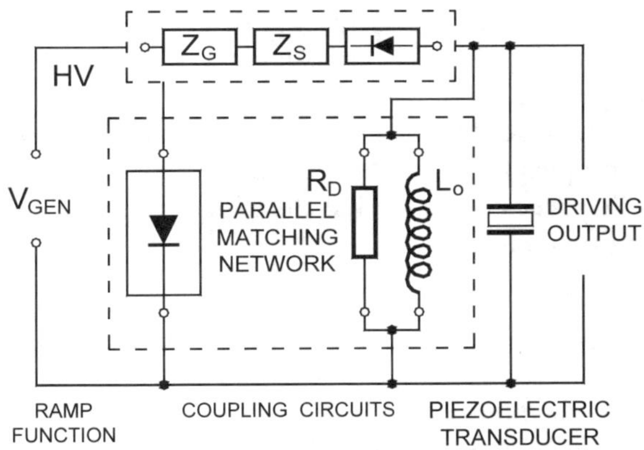

Fig. 5.1. General scheme of interface electronics in pulsed piezoelectric transmitters

Figure 5.1 shows a general diagram of the transmitter section. It includes coupling networks in series and in parallel to make the electronic interface, [8]. In some applications one of these two coupling networks may be absent. The impedances of these networks may vary with time during some phases of the transient process.

The series impedance Z_S normally has a low resistive value and the block Z_G usually is a capacitor fairly larger than the clamped capacitance of the piezoelectric element in the driven transducer. The parallel coupling circuit is formed by: a damping resistor R_D, a coil L_0 and a rectifier block.

The reception matching circuits usually have a simpler structure, because they are often formed by a parallel combination of resistive and inductive components. In this case, the effect desired is only of damping and tuning type.

Other more complex circuital solutions for transducer coupling have also been proposed; for instance in order to improve particular aspects of the transducers band, such as the bandwidth [9].

5.3 Electronic Circuits Used for the Generation of High Voltage Driving Pulses and Signal Reception in Broadband Piezoelectric Applications

Most of the applications of broadband piezoelectric transducers have been developed in frequency bands comprised between 0.5 MHz and 10 MHz. As a consequence the shock excitation of the transducers has to be made by means of high voltage pulses with rise-times lower than 25 nanoseconds. In special applications of ultrasonic detection with very high resolution (for instance using 20 MHz transducers) a rise time around 10 ns is needed as an extreme case.

Other condition present in many cases is a high acoustic attenuation during ultrasonic propagation, increasing strongly with frequency. For this reason it may be necessary to drive piezoelectric transducers with spike amplitudes ranging up to 500 V.

5.3.1 Some Classical Circuits to Drive Ultrasonic Transducers

To obtain high voltage pulses, with short rise-time, different circuits can be used. Typical schemes are shown in Figs. 5.2 and 5.3 giving output pulses with peak amplitude higher than 220 V. A common configuration [10] is detailed in Fig. 5.2. It uses a thyristor device *Th* for the high-voltage switching which produces a fast discharge of the capacitor *C* through the resistor R_D and transducer. The repetition frequency for this discharge generally ranges between 0.5 and 5 kHz. The time evolution of the pulse generated across R_D without transducer loading, during the switch-to-on period in *Th*, can be analysed by the expression:

$$V_{on}(t) = -(V_o / t_{on}) R_D C \left(1 - e^{-t/CR_D}\right) \tag{5.2}$$

where, t_{on} represents the time necessary to get the thyristor to the saturation state.

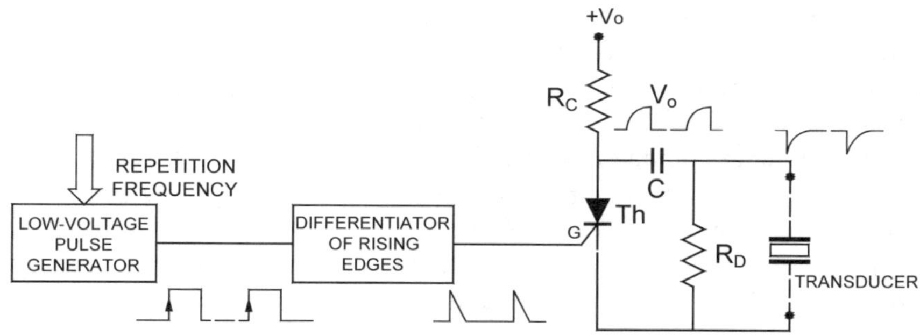

Fig. 5.2. Conventional generator of electrical pulses for driving transducers. Circuit based on thyristor switching

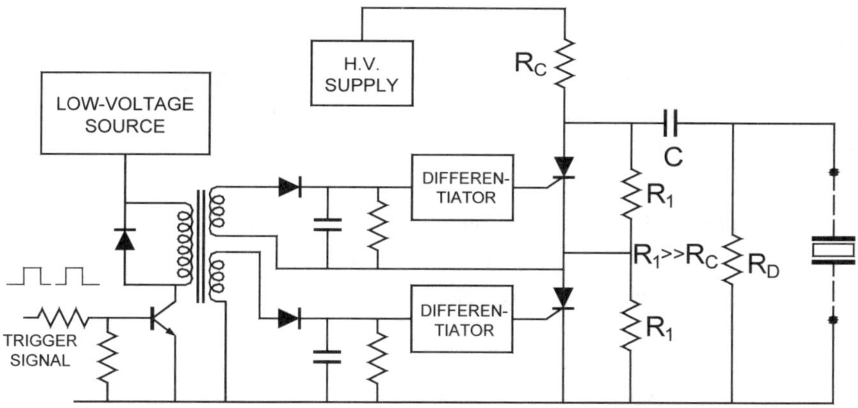

Fig. 5.3. Conventional generator of electrical pulses for driving transducers. Series connection of two SCRs for higher amplitude pulses

The particular circuit in [10] gives a rise-time in V_{on} ranging between 150 and 250 ns. The associated frequency content is clearly insufficient for applications ranging above 2 MHz. This circuit limitation, depending on the effective t_{on}, could be overcome by using a faster thyristor, but such type of devices is only encountered in the low-voltage range. This option is shown in Fig. 5.3 where, with a series connection of two thyristors of reduced voltage, a higher peak value pulse can be generated. By repeating this scheme for several series stages the necessary high voltage pulse could be reached.

Nevertheless, the time tolerances and the -always long- cut-off times in thyristors produce harmful transient impulses, which could distort the received ultrasonic signals. Another disadvantage of these devices is the high current losses in off state.

For these reasons, it is a better option to use other type of device to perform the high-voltage switching. The power mos-fet transistor fulfils all the requirements and does not present the above mentioned limitations. A circuit based on such kind of device was proposed in reference [6]. This circuit can generate pulses of 240 V in peak amplitude over a damping resistance of 300 Ω. The rise-time associated with the output pulse is 100 ns and therefore it can only be used to drive transducers with resonance frequencies below 3 MHz. A faster switching device would be needed for a proper excitation of broadband transducers in most applications.

5.3.2 Electronic System Developed for the Efficient Pulsed Driving of High Frequency Transducers

Figure 5.4 shows the block diagram of a pulser, specially developed for efficient excitation of broadband transducers in a frequency range of 30 MHz [7]. This pulser is a particular implementation of the general scheme in Fig. 5.1.

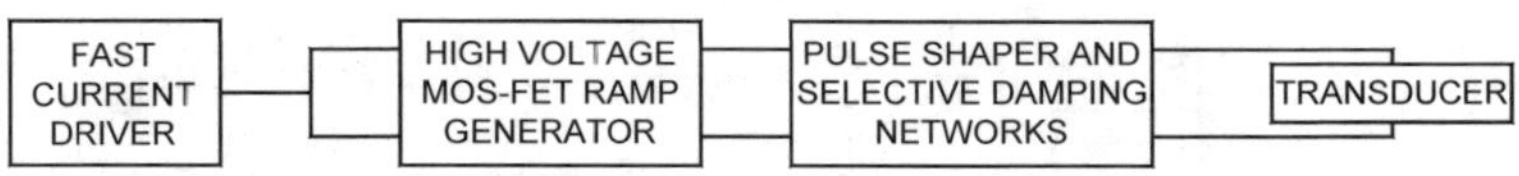

Fig. 5.4. Block diagram of a spike generator developed for an efficient driving of high frequency transducers

The switching of the power mos-fet used as ramp generator was performed by the gate driver detailed in Fig. 5.5 where the complementary transistors T_2 and T_3 supply and carry to ground the bi-directional high currents needed for a fast switching of transistor T_{MF} (see ramp generator circuit in Fig. 5.6). It should be noted that this type of transistors presents a gate-to-source capacitance that can be as high as 1500 pF. As a result, the gate control must be made through very low impedance circuits if an optimisation of transition times is desired.

In the circuit of Fig. 5.6, a zener diode Z appears between the drain of T_{MF} and the capacitor C, producing a light inertia in the electric discharge through T_{MF} and transducer, in such a way that, at the beginning of the discharge, T_{MF} will be closer to total saturation than in the absence of that diode Z. Another important effect of this device is to avoid distortion on the positive half-cycles of the echo-graphic traces collected in ultrasonic reception. Without this zener diode, the positive half-cycles would go to ground through diodes D_1 (see Fig. 5.7), capacitor C and T_{MF}, distorting the signal.

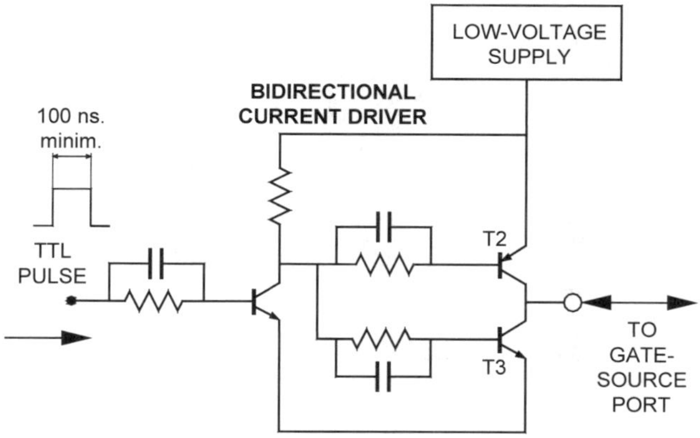

Fig. 5.5. Bi-directional current driver for fast switching of the power mos-fet transistors

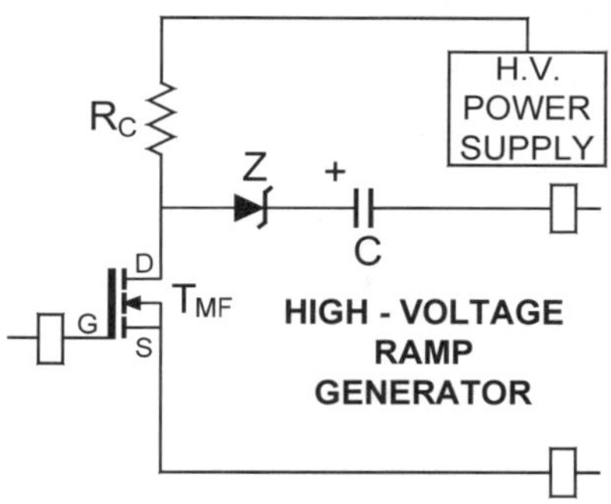

Fig. 5.6. High-voltage ramp generator using a sudden capacitor discharge

The previous circuital blocks are connected in cascade to a block for selective damping and coupling with transducers, whose circuit is detailed in Fig. 5.7. It includes an adjustable coil L_0 in parallel with the damping resistor. The coil setting allows a fitting of the temporal shape in the driving pulse. Diodes D_1 and D_2 prevent oscillatory signal from the C-L_o resonant circuit, because they only permit the flow of the first negative half cycle to the transducer. Therefore, the remaining energy after the launching of this negative pulse will be dissipated abruptly through R_L and D_i. As the diodes D_i have to be rapid, it is necessary to use groups of parallel and series signal diodes, to allow high peak currents (in D_1) and reverse bias close to 500 V (across D_2).

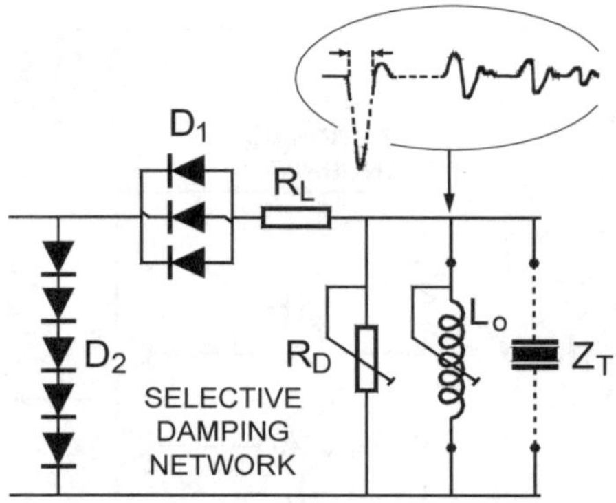

Fig.5.7. Network for selective damping and for the coupling of the ramp generator with the transducers.

The circuit formed by L_o, D_1 and D_2 in the selective damping network, improves the amplitude and shape in the output spike for two reasons:

1. It can efficiently substitute the damping effect of a resistor R_D of low value in order to produce a fast electric damping of the spikes without losses in their amplitude.
2. It can facilitate the formation of driving pulses with trailing slopes being independent of the cut-off instant in transistor T_{MF}. This independence is useful to reduce temporal dispersions related to the usual tolerance in the switching-to-off times of these devices.

Finally the low value resistance ($R_L << R_D$) protects the generator from an accidental short-circuit at the output terminals, and reduces possible signal distortions if positive half-cycles greater than 3.5 V were received. Without R_L, these half-cycles would go directly to ground through D_1 and D_2.

When the impedance of the coil L_0 is relatively high, $\omega L_0 >> R_D$, the behaviour of the output spike $V_{out}(t)$ could be approximately predicted by the following expression:

$$V_{out}(t) = CV_o R_T \left(e^{-t/\tau} - e^{-t/(C+C_T)R_T}\right)\frac{1}{(C_T + C)R_T - \tau} \tag{5.3}$$

where $R_T = \left(A(R_B + R_M)/(2C_T h_{33})^2\right) // R_D$, A and C_T are the area and clamped capacitance in the driven transducer, R_B and R_M are the acoustic impedances loading the faces of the active piezoelectric element and h_{33} is the piezoelectric constant.

5.3.3 Electronic Circuits in Broadband Signal Reception

There are some alternatives to perform the interfacing of the piezoelectric transducers with the signal acquisition systems. Fig. 5.8 shows a typical interface circuit. This circuit corresponds to a simple scheme for the efficient reception of the broadband echo signals produced at the transducer terminals during ultrasonic applications.

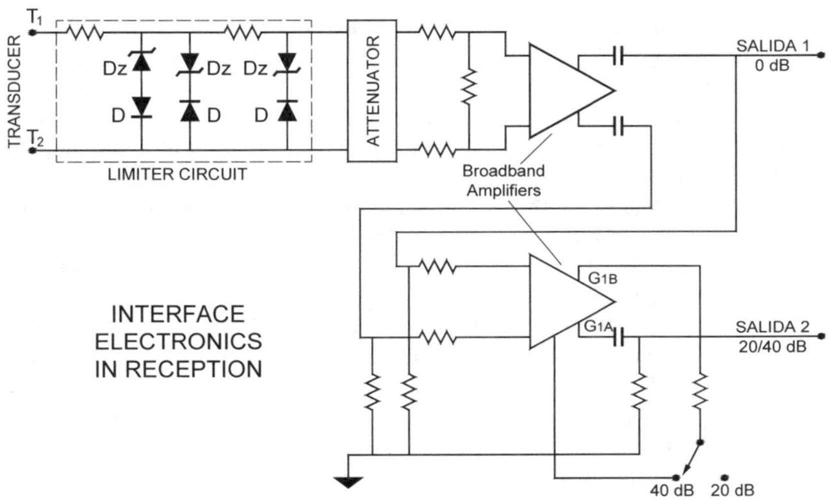

Fig. 5.8. A typical interface circuit between transducers and broadband signal acquisition electronics

Across T_1 and T_2 terminals reactive parallel circuits are often connected in order to perform electrical damping and inductive tuning.

The second stage consists of a bipolar limiter to $\approx \pm 5$ V of three branches to protect the reception electronics from high-voltage spikes launched in transducer terminals during the driving process. This circuit is constituted by three successive parallel mono-polar limiting branches, each one with zener and signal diodes. The two additional branches are needed because of the high value of the current produced in the first branch, originating an elevated conducting threshold in their semiconductors.

The pulsed signals so limited are fed into the input of a broadband signal pre-amplifier, through an optional attenuator step. Figure 5.8 also includes a simple amplifier scheme with two differential steps. The first step performs an impedance matching between the often high-impedance value in transducer terminals and the low input impedance usually selected in low-noise signal amplifier circuits.

References

1. J.W. Hunt, M. Arditi, and F.S. Foster, "Ultrasound transducers for pulse-echo medical imaging", *IEEE Trans. Biomed. Eng.*, Vol. BME-30, No. 8, pp. 453-481, (1983).
2. G. Hayward, "The influence of the pulser parameters on the transmission response of the piezoelectric transducers", *Ultrasonics,* Vol. 23, pp.103-112, (1985).
3. M.E. Schafer, P.A. Lewin, "The influence of front-end hardware on digital ultrasonic imaging", *IEEE trans. Sonics Ultrason.* Vol. SU-31, No. 4, pp. 299-306, July (1984).
4. L.J. Augustine, J. Andersen, "An algorithm for the design of transformerless broadband equalizers of ultrasonic transducers", *J.Acoust. Soc. Am.*, 66, 3, pp. 629-635, (1979).
5. R. Coates and R.F. Mathams, "Design of matching networks for acoustic transducers", *Ultrasonics*, Vol. 26, pp. 59-64, (1988).
6. P. Mattila, and M. Luukkala, "FET pulse generator for ultrasonic pulse echo applications", *Ultrasonics*, 19, pp. 235-236, (1981).
7. A. Ramos, P.T. Sanz, F.R. Montero, "Broad band driving of echographic arrays using 10ns-500ns efficient pulse generators", *Ultrasonics*, Vol 25, pp. 221-228, (1987).
8. A. Ramos, J.L. San Emeterio and P.T. Sanz, "Electrical matching effects on the piezoelectric transduction performance of a through transmission pulsed process", *Ferroelectrics,* Vol. 202, pp. 71-80, (1997).
9. F. Lakestani, J.C. Baboux and P. Fleischmann, "Broadening the bandwidth of piezoelectric transducers by means of transmission lines", *Ultrasonics*, Vol. 13, pp. 176-180, July (1975).
10. J. G. Okyere, and A. J. Cousin, "The design of a high voltage SCR pulse generator for ultrasonic pulse echo applications", *Ultrasonics*, 17, pp. 81 -84, (1979).

6 Interface Electronic Systems for AT Quartz Crystal Microbalance Sensors

Antonio Arnau[1], Vittorio Ferrari[2], David Soares[3] and Hubert Perrot[4]

[1]Departamento de Ingeniería Electrónica, Universidad Politécnica de Valencia
[2]Dipartimento di Elettronica per l'Automazione, Università di Brescia
[3]Institute de Fisica, Universidade de Campinas
[4]Laboratoire Interface et Systèmes Electrochimiques, Université P. et M. Curie

6.1 Introduction

AT quartz crystal microbalance sensors (QCMS) are becoming into a good alternative analytical method in a great deal of applications [1 – 4], with a resolution comparable, in many cases, to chemical techniques used for detecting species and suitable for fluids physical properties characterization [5,6], though simpler and much less expensive.

However, an appropriate evaluation of this analytical method requires recognizing the different steps involved in order to be conscious of their importance and to avoid the possible error propagation if the appropriate care is not taken.

The three steps involved in a QCM system are: 1) Measurement of the appropriate parameters of the resonator. This includes a suitable electronics and cell interfaces for a specific application; 2) Extraction of the effective parameters related to the model selected for the application (2-, 3- or 4-layer model). This is one of the most complicated steps, including mathematical algorithm combined with appropriate measurements in step 1. The most typical situations correspond to 3-layer model (Fig. 6.1), although the analysis of 4-layer model is interesting for understanding the measurements resulting from intermediate processes which occur during the experiment; 3) Interpretation of the physical, chemical or biological phenomena responsible for the change in the effective equivalent parameters of the selected model. This is the last step and the final role of the system.

In this chapter we will focus on step 1, explaining the limitations of different electronic interfaces in relation to the application. In this case, the application will be evaluated as a function of a change in the measuring parameters. Thus, it is first necessary to define the parameters of the QCR system to be measured.

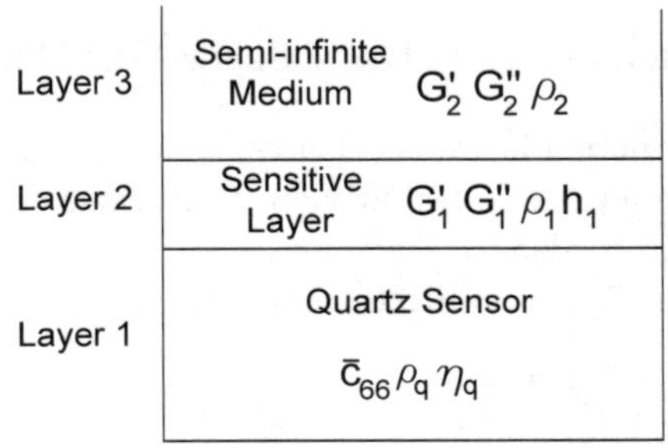

Fig. 6.1. The three layers model

6.2 A Suitable Model for Including a QCMS as Additional Component in an Electronic Circuit

A shear strain is induced in an AT quartz crystal when an alternating-current (AC) voltage is applied across it through opposing electrodes deposited on its surfaces. It generates a transversal acoustic wave propagating through the quartz to the contacting media. The mechanical interaction between the resonator and the contacting media influences the electrical response of the device. This permits the use of the resonator as a sensor device to detect changes in the physical properties of the contacting media.

In order to treat the sensor as a component included in electronic circuits and to be able of analyzing its performance in relation to the external circuitry, an electrical model appropriately representing its impedance would be very useful.

The loaded quartz can be appropriately described by an extended Butterworth Van Dyke (BVD) equivalent circuit (Fig. 6.2a) and, in this way, be included in electronic circuits as an additional component. This model is in fact an approximation of the TLM (Transmission Line Model – Chap. 4, Sect. 4.A.3), but it is enough to describe the problems associated to the different systems in relation to specific applications. For our purposes, it is not necessary to know the expressions relating C_o^*, R_m^q, L_m^q, C_m^q (unperturbed quartz resonator – Chap. 1, Sect. 1.A.1) and R_m^L, L_m^L and C_m^L (loading contribution – Chap. 4, Sect. 4.A.4) to the physical and geometrical properties of the quartz and load and they can be found elsewhere [7,8]. Throughout this chapter we will make use of a BVD equivalent model to study the driver/sensor combination but it will not affect the generality of the results.

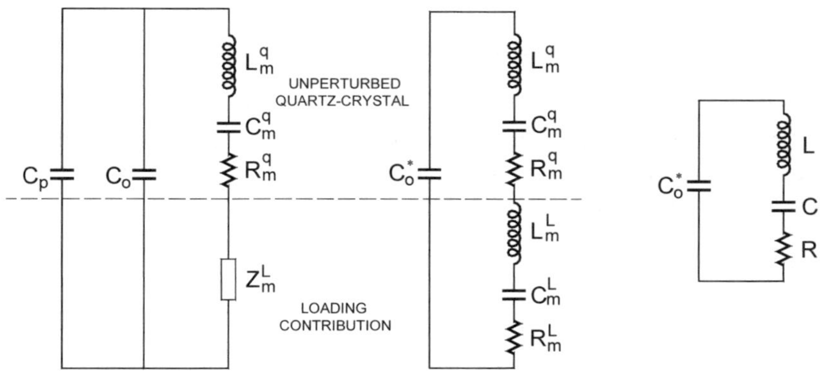

Fig. 6.2. Equivalent circuit models for loaded QCR: **a** general model - Extended-BVD model, and **b** BVD-model

6.3 Critical Parameters for Characterizing the QCMS

To discuss the problem associated to the different electronic systems used to characterize the sensor, it is necessary first to define the parameters to be measured for an appropriate evaluation of the sensor response. These parameters have to be measured in two steps:

STEP 1: Determination of the elements C_o^*, R_m^q, L_m^q and C_m^q for an evaluation of the unperturbed device response.

Generally, values of C_o^*, R_m^q, L_m^q and C_m^q are provided by the manufacturer, but sometimes they are not accurate enough and must be obtained prior to the experiment. Normally, they can be determined with impedance or network analyzers by measuring the electrical response of the unperturbed resonator over a range of frequencies near resonance and fitting the equivalent-circuit model to these data. If an impedance analyzer is not available, the corresponding standard [9], or an alternative method described elsewhere [10], can be used. A more accurate determination of C_o^* can be made at a frequency twice as high as the resonant frequency [11]. From these values, the following parameters of interest can be extracted:

f_s: motional series resonant frequency (MSRF). It is defined as the frequency at which the motional reactance vanishes. It can be determined directly by frequency measurements or by:

$$f_s = \frac{1}{2\pi\sqrt{L_m^q C_m^q}} \tag{6.1}$$

In practice, the frequency corresponding to the maximum conductance typically monitored in impedance analysis is very close in most practical cases to the MSRF (Sect. 6.A.3) [6].

h_q: quartz thickness. It can be determined from (Chap. 1, Eq. (1.A.34):

$$h_q \approx \frac{1}{\omega_s}\sqrt{\frac{\bar{c}_{66}}{\rho_q}}\sqrt{(n\pi)^2 - 8K_0^2} \tag{6.2}$$

where $\bar{c}_{66} = c_{66} + e_{26}^2/\varepsilon_{22}$ is the quartz effective shear modulus, c_{66} is the shear stiffness constant, e_{26} is the piezoelectric constant, ε_{22} is the permittivity, K_o is the lossless effective electromechanical coupling factor, n (n = 1, 3, 5,..) is the harmonic resonance of quartz, ρ_q is the quartz density and $\omega_s = 2\pi f_s$.

C_o: static capacitance (Fig. 6.2a). This capacitance arises from electrodes located on opposite sides of the dielectric quartz resonator. This capacitance does not include parasitic capacitances external to the resonator (C_p) which do not influence the motional parameters [8]. Static capacitance can be determined from the values of C_m^q or L_m^q applying one of the following equations:

$$C_m^q = \frac{8K_0^2 C_0}{(n\pi)^2} \tag{6.3}$$

$$L_m^q = \frac{1}{\omega_s^2 C_m^q} \tag{6.4}$$

A_S: effective electrode surface area. It can be determined from C_o and h_q along with the quartz permittivity from:

$$A_S = \frac{h_q}{\varepsilon_{22}} C_0 \tag{6.5}$$

Its value is necessary in applications involving film thickness.

C_p: parasitic parallel capacitance, external to the resonator. $C_p = C_o^* - C_o$. Its value is useful in applications where the influence of the dielectric properties of the load have to be accounted for.

These parameters have to be measured as reference and must be taken into account for an appropriate load characterization.

STEP 2: Measuring of those parameters of the loaded thickness shear mode (TSM) resonator which can more appropriately characterize the physical and/or geometrical properties of the load on the basis of current models.

From Fig. 6.2a, it is shown that the load contribution is characterized by the elements R_m^L, L_m^L and C_m^L of the motional branch. A change in both L_m^L and C_m^L produces a change in the MSRF. On the other hand, changes in the loading properties are also reflected on changes in the motional resistance R_m^L, which does not produce MSRF changes. Thus, both MSRF and motional resistance are useful and necessary parameters for a complete sensor characterization.

In addition, it is important to state that the majority of the simpler models derived from the most comprehensive TLM, such as the Lumped Element Model (LEM) [12], or the extended BVD model [8] (Chap. 4), assume that the resonator operates around the MSRF. Furthermore, it is important to mention that most of simpler equations used to relate frequency and resistance shifts to the properties of the load have been derived assuming that the resonator is oscillating at its true MSRF. Thus, measurements of loading-induced frequency changes made with the resonator operating at a frequency different from the true MSRF could not agree with the models derived for QCMS. This discrepancy is specially pronounced when the resonator is loaded with heavy damping media.

Another characteristic which makes the MSRF more interesting than other frequencies is that its value is independent of parallel capacitance changes.

For all that mentioned, the MSRF and the motional resistance are the parameters of the loaded resonator to be measured. Then, the problem associated with the measuring system will be discussed in relation to the accuracy in the determination of these parameters of interest.

6.4 Systems for Measuring Sensor Parameters and their Limitations

We will focus this discussion on the interface circuits currently used for sensor characterization which are based on three principles: network or impedance analysis, impulse excitation or decay method, oscillators and parallel capacitance compensation techniques.

6.4.1 Impedance or Network Analysis

Impedance or network analyzers measure the electrical impedance or admittance of the quartz sensor over a range of frequencies near resonance for a complete characterization of the device response. As a test instrument, an impedance analyzer has the following advantages in evaluating the sensor response:

1. The device is measured in isolation and no external circuitry influences the electrical behavior of the sensor.
2. Parasitic influences can be excluded by calibration due to passive operation of the sensor.
3. Differentiated information in relation to diverse contributions of the load can be obtained by measuring both the conductance and the susceptance of the sensor over a range of frequencies around resonance.

However, several inconveniences remain when using this technique for sensor applications [13]:

1. Its high cost and large dimensions of the associated equipment prevent its use for *in situ* or remote measurements.
2. The connection between the sensor and the equipment is sometimes difficult to accomplish such as in electrochemical or biological applications where it is convenient to ground one of the quartz electrodes.
3. It is not suitable for simultaneous multiple sensor characterization. Sometimes a multiplexing interface is used for a sequential connection of different sensors to the same analyzer, but it can perturb the device response.

On the other hand, the impedance analyzer can determine with high accuracy the MSRF and motional resistance of the unperturbed quartz sensors as reference values. The MSRF is obtained by measuring the frequency corresponding to the conductance peak around resonance. The motional resistance is determined as the reciprocal of the conductance peak value. The evaluation of the MSRF and the motional resistance in this way is based on the suitability of the BVD model for characterizing the sensor response. In BVD circuits the relationships between MSRF and maximum conductance frequency and between the motional resistance and the reciprocal of the conductance peak value are exact (Sect. 6.A.3). For an unperturbed resonator, the BVD circuit can very accurately represent the device response. Additionally, the range of frequencies in which the resonance happens is very narrow and therefore the frequency resolution of the instrument is very high. However, for heavy damping loads the quality factor of the device is considerably reduced, which expands the resonance range and reduces the frequency resolution and the suitability of the BVD circuit for representing the sensor response.

The determination of the MSRF and motional resistance by using the mentioned relationships is not as accurate as for the unperturbed situation, but remains accurate enough for these applications in which the sensor can be used [6]. On the contrary, the determination of BVD parameters for high damping loads does not give additional information apart from the parallel capacitance, which can be measured more appropriately at double of the resonant frequency. Furthermore, the MSRF determination from the motional components (Eq. (6.1)) can produce great errors depending on the algorithm used for the motional parameters extraction.

Thus, an impedance analyzer can be used for an accurate determination of the magnitudes considered of interest. The accuracy in the determination of the MSRF is reduced for high damping loads. However, this loss in frequency resolution does not affect the relative accuracy in the motional resistance determination due to the flatness of the conductance peak for high damping loads.

6.4.2 Decay and Impulse Excitation Methods

Impulse excitation and decay methods are based on the same principles. To illustrate the principle let us consider the electrical circuit in Fig. 6.3. At $t = 0$ the switch connects a voltage source (E) with a very low output resistance (r) to the sensor modeled in the figure as a BVD circuit. From this moment the current $i(t)$ changes according to the following expression (assuming $r << R_m$):

$$i(t)=i_{C_o^*}(t)+i_m(t)\approx\frac{E}{r}e^{-\frac{t}{rC_o^*}}+$$

$$\frac{E}{L_m}\frac{1}{\sqrt{\frac{1}{L_mC_m}-\left(\frac{R_m}{2L_m}\right)^2}}e^{-\frac{R_m}{2L_m}t}\sin\sqrt{\frac{1}{L_mC_m}-\left(\frac{R_m}{2L_m}\right)^2}\,t \qquad (6.6)$$

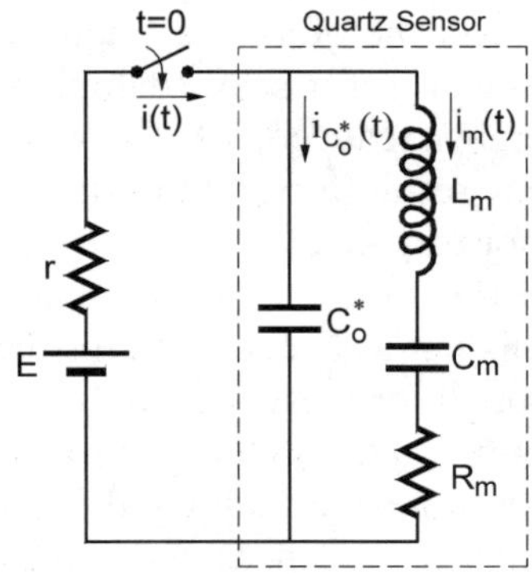

Fig. 6.3. Diagram for illustrating the principle of the impulse excitation method

The first term associated to the current through the parallel capacitance $i_{C_0^*}(t)$, has a time constant $\tau_{C_0^*} = rC_o^*$ while the time constant of the second term associated with the current through the motional branch of the sensor, $i_m(t)$, is $\tau_m = 2L_m / R_m$. Assuming typical values for a 10 MHz loaded AT cut quartz resonator (Table 6.1) and $r = 2\ \Omega$, the decay time constant $\tau_{C_0^*}$ is about 10^6 times smaller than τ_m. Then, the total current $i(t)$, is dominated by the second term in Eq. (6.6) once the first instant has passed. With a proper interface circuit, the damped oscillation can be recorded on a digitizing oscilloscope and subsequently transferred, e.g. via GPIB, to a computer. A numerical fitting of the recorded curve permits to obtain the time constant τ_m and the frequency of the damped oscillation given by:

$$f_m = \frac{1}{2\pi}\sqrt{\frac{1}{L_m C_m} - \left(\frac{R_m}{2L_m}\right)^2} \tag{6.7}$$

The time constant τ_m can be related to the quality factor Q_L by using the approximate expression: $Q_L \approx 2\pi f_s L_m / R_m = \pi \tau_m f_s$ (Chap. 1-Eq. (1.A.38)), where f_s is the MSRF of the loaded QCR. Then, the frequency of the damped oscillations can be written as a function of f_s and Q_L as follows:

$$f_m \approx f_s\sqrt{1 - \frac{1}{4Q_L^2}} \tag{6.8}$$

Table 6.1. Typical values for the BVD parameters of a 10MHz AT-cut QCR

C_0^* [pF]	10
L_m [mH]	7.5
C_m [fF]	33.774
R_m [Ω]	10

Thus, by measuring the time constant and the frequency of the damped oscillations, the quality factor and the MSRF can be determined. It can be considered, without incurring a significant error, that the reduction in the quality factor is proportional to an increase in the motional resistance. In such a way a measurement of this magnitude can be obtained.

In practice, the impulse excitation method is difficult to apply for two reasons:

1. Ideal pulse front slopes are difficult to achieve.
2. Other harmonics different from that desired can be excited. Thus, in order to avoid them, additional circuitry which perturbs the sensor response is necessary.

Instead of the impulse method, the decay method is used in practice [14, 15]. The measuring principle is very similar to the one described. A piezoelectric resonator is excited with a signal generator approximately tuned to the frequency of the desired harmonic. Then, at $t = 0$ the signal excitation is eliminated by opening the appropriate relay (Figs. 6.4 and 6.5). At this moment, the voltage or current (depending on whether the parallel or series resonant frequency is excited according to the electrical setup) [16] decays as an exponentially damped sinusoidal signal, mathematically expressed by:

$$A(t) \approx A_0 e^{-\frac{t}{\tau_m}} \sin(2\pi f t + \varphi), \qquad t \geq 0 \tag{6.9}$$

Where A_o is the amplitude of the magnitude at $t = 0$; φ is the phase and f is the frequency given by:

$$f \approx f_i \sqrt{1 - \frac{1}{4Q_L^2}} \tag{6.10}$$

Where f_i is the MSRF of the loaded QCR, f_s, or the parallel resonant frequency, f_p, given by:

$$f_p = f_s \sqrt{1 + \frac{C_m}{C_0^*}} \tag{6.11}$$

In Eq. (6.10), it has been implicitly assumed that the quality factor is the same for both series and parallel mode, and is given by (Chap. 1, Eq. (1.A.38)):

$$Q_L = \frac{L_m \omega_s}{R_m} \approx \frac{L_m \omega_p}{R_m} \tag{6.12}$$

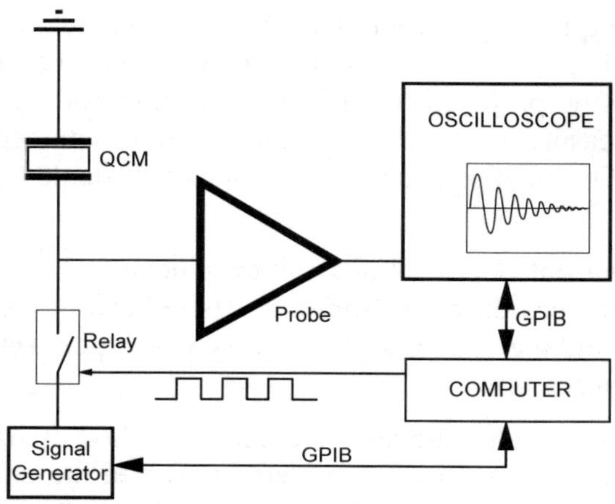

Fig. 6.4. Experimental setup used to measure the parallel resonant frequency and the parallel dissipation factor

In the series excitation mode (Fig. 6.5), the parallel capacitance effect is eliminated by short-circuiting and the frequency of the damping oscillations is very close to the true MSRF. This is one of the principal advantages of the method. The accuracy of the decay method is high, provided that the measurement of the frequency and the envelope are obtained with high accuracy, which becomes complicated for strongly damping loads. This technique reduces the cost of the instrumentation in comparison with network analysis; however, the quality and dimensions of the required equipment still remain high, mainly if an accurate determination of the frequency and the envelope of the exponentially damped sinusoidal is necessary. Therefore, this method is more appropriate for laboratory environment than for sensor applications and becomes more sophisticated when simultaneous multiple sensor characterization at high sampling rates have to be performed.

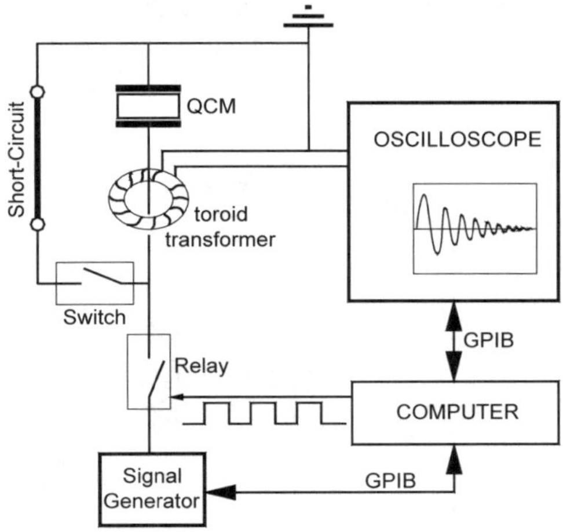

Fig. 6.5. Experimental setup used to measure the series resonant frequency and the series dissipation factor

6.4.3 Oscillators

Basics of LC Oscillators

An LC oscillator consists of an amplifier with positive feedback including an LC resonant circuit as a frequency control element. In the most general case, the resonant circuit can involve transformers and piezoelectric resonators. In the last one, when the oscillator is well-designed, the piezoelectric resonator controls the oscillating frequency and the oscillator is called "*Crystal Oscillator*".

In order to understand the basic operation of an LC oscillator, let us consider the circuit shown in Fig. 6.6. In the initial time the switch S1 is closed and the capacitor is charged at the value of the power supply. Once the capacitor has been charged the switch S1 is opened and the switch S2 is closed. The circuit in Fig. 6.7a shows this situation where $i_g=0$. The current i in the circuit obeys to the following differential equation:

$$i'' + \frac{1}{LC} i = 0 \tag{6.13}$$

This current corresponds to a sinusoidal signal whose angular frequency is $\omega_s = 1/\sqrt{LC}$. This frequency is equal to the electrical resonant frequency of the LC circuit

The phasorial diagram in Fig. 6.8a gives a different point of view. Taking the voltage u as reference, the currents i_L and i_C through the coil and capacitor, respectively, have the same magnitude but opposite sign. Hence:

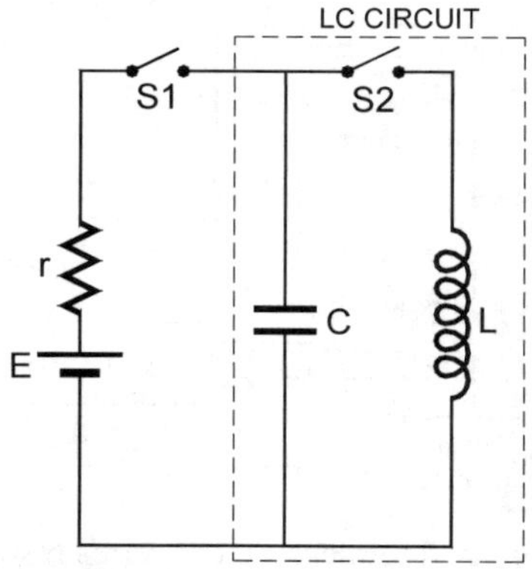

Fig. 6.6. Diagram for illustrating the operation principle of a resonant circuit

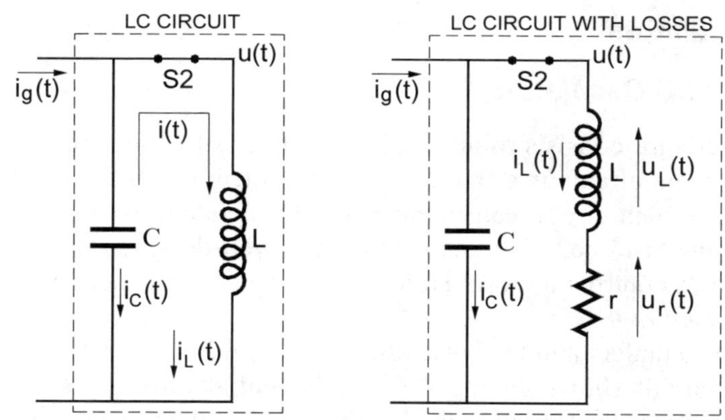

a **b**

Fig. 6.7. Part of the circuit in Fig. 6.6 for explaining the electric resonant phenomenon: **a** without losses, and **b** including losses

$$C\omega\, u = \frac{u}{L\omega} \tag{6.14}$$

The former equation is only valid for the electrical series resonant frequency ω_s. As a consequence, the LC circuit described, which provides a constant amplitude sinusoidal voltage, would constitute the simplest oscillator.

Next, let us consider the unavoidable losses in energy modeled by means of a small resistance in series with the coil, as shown in Fig. 6.7b. The new phasorial diagram shown in Fig. 6.8b makes clear the necessity of an incoming current i_g, provided from an external generator such as an amplifier, in order to maintain the magnitude of the voltage u constant. On the contrary, the amplitude of this voltage would diminish following an exponentially damped sinusoidal signal.

Consequently, it is necessary to supply the current by means of an external circuit for covering the energy losses due to the resistive element. A typical way to connect an amplifier to the resonant LC circuit is shown in Fig. 6.9a, where the capacitor C has been included as two series capacitors C_1 and C_2. In Fig. 6.9a the amplifier could be replaced by a transistor where B=Base, C=Collector and E=Emitter. The resultant circuit and its possible combinations would provide the typical well-known configurations: Colpitts, Hartley, Pierce, Clapp [17, 18]. Despite the simplicity of the circuit shown in Fig. 6.9a, it is extremely useful for making the basic operation of an LC oscillator clear and understanding its limitations.

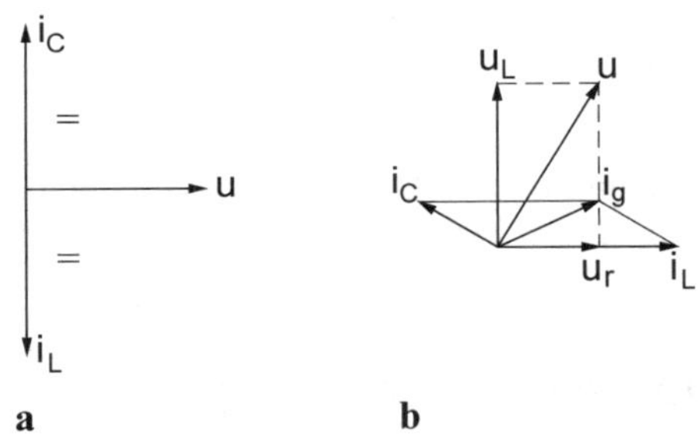

Fig. 6.8. Phasorial diagrams corresponding to the magnitudes involved in the respective circuits in Fig. 6.7

Oscillating Conditions

The circuit shown in Fig. 6.9a can be schematically represented as indicated in Fig. 6.9b, which makes the fulfillment of the following equation for oscillation clear:

$$A\beta = 1 \tag{6.15}$$

The previous equation involves two conditions:

1. The loop gain condition:

$$|A\beta| = 1 \tag{6.16}$$

2. The phase condition:

$$\arg A + \arg \beta = 0 \tag{6.17}$$

This condition is strongly related to the concept of stability, as we will see later. The previous conditions are named "Barkhausen conditions" and their application allows determining the relations among the different oscillator's parameters, including those of the resonator [19].

Circuits in Figs. 6.9a-b can be used for an easy understanding of two typical crystal oscillators configurations: the parallel and series mode crystal oscillators.

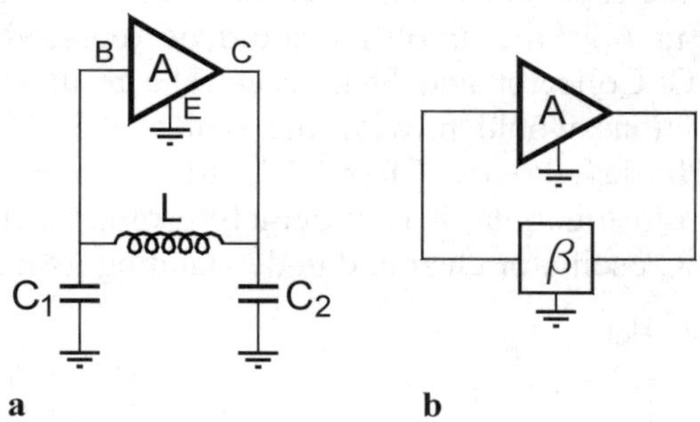

Fig. 6.9. Schematic circuits for illustrating the basic operation of an oscillator: **a** the feedback network is a tank circuit, and **b** the feedback network is considered as a black-box

Parallel Mode Crystal Oscillator

In this configuration the piezoelectric crystal behaves like an inductance and is used in place of the coil in the circuit shown in Fig. 6.9a. This is possible since the impedance of the piezoelectric crystal has an inductive character in the range of frequencies between the two phase-null frequencies corresponding to a specific resonant frequency (see Fig. 6.A.1). Thus, in the parallel mode the phase of the crystal is positive but, in general, it is not known with high accuracy.

In the parallel mode the oscillating frequency corresponds to the electrical resonant frequency of the series circuit formed by the capacitors C_1, C_2 and the equivalent inductance L_{eq} of the piezoelectric crystal at the oscillating frequency ω_o, given by:

$$L_{eq} = \frac{X_{eq}}{\omega_o} \tag{6.18}$$

where X_{eq} is the lossless equivalent reactance of the crystal at the oscillating frequency ω_o, given by (see Eq. (6.A.2)):

$$X_{eq} = \frac{\omega_o^2 LC - 1}{\omega_o^2 (C + C_o)\left(1 - \omega_o^2 \frac{CC_o}{C + C_o} L\right)} \tag{6.19}$$

where L, C and C_o are parameters of the BVD circuit modeling the vibrating crystal in a range of frequencies around a certain resonance.

Consequently, the expression for the oscillating frequency is:

$$\omega_o = \frac{1}{\sqrt{L_{eq} C_t}} \tag{6.20}$$

where $C_t = C_1 C_2 / (C_1 + C_2)$.

Using Eqs. (6.18) and (6.19) in Eq. (6.20) and solving for ω_o, the final expression for the oscillating angular frequency is:

$$\omega_o = \frac{1}{\sqrt{L \dfrac{C(C_o + C_t)}{C + C_o + C_t}}} \tag{6.21}$$

The former expression is equal to the expression for the parallel resonant frequency of a BVD circuit with a parallel capacitance given by: $C_o + C_t$; this is the reason for denominating *parallel crystal oscillator* to that configuration where the crystal behaves like an inductance. However, the oscillating frequency is not the electrical parallel frequency of the crystal at resonance (see ω_p in Eq. (6.A.3)), but a frequency at which the resonator acts like a coil. Such frequency depends on the specific values of capacitors C_1 and C_2, and must be placed between the phase-null frequencies of the crystal at the selected resonance.

Series Mode Crystal Oscillator

In series mode crystal oscillators, the piezoelectric resonator is placed between two points of the feedback path which must be short-circuited for fulfilling the oscillating phase conditions. Thus, the resonator must behave like a short circuit for oscillation. The signal crossing a short circuit does not experience phase shift; therefore, if a series mode crystal oscillator is well designed, the piezoelectric resonator controls the oscillating frequency in such a way that the frequency is maintained near the phase-zero frequency of low impedance.

Let us consider Fig. 6.9b. This scheme will permit to make clear both the operation of a series mode crystal oscillator and the stability concept. To make the explanation easier let us assume that the amplifier has a real and positive voltage gain transfer function[1]. As a consequence, the circuit in Fig. 6.9b will fulfill the phase condition for oscillation at that frequency where the phase-shift through the circuit β is zero. Now, let us suppose that this frequency coincides with the zero-phase frequency corresponding to the lowest impedance of the piezoelectric resonator to be used as frequency control element in the series mode crystal oscillator. The scheme in Fig. 6.10a is a simplified diagram of this oscillator, where the black-box X, in series with the circuit β in the feedback path, represents the resonator. The phase condition for oscillation will be fulfilled when the total phase-shift through the series circuit formed by the resonator and the circuit β is zero. Because the circuit β and the resonator have the same zero-phase frequency, this will be the oscillating frequency; in any case the gain of the amplifier will increase slightly in order to compensate the low losses introduced by the resonator. Consequently, in this ideal situation, the oscillating frequency will be the zero-phase frequency of low impedance of the resonator.

Let us discuss the previous situation when the phase-shift through the circuit β at the zero-phase frequency of the resonator is not zero. This case additionally permits to discuss the concept of stability and to understand the problem associated to the use of oscillators for monitoring the appropriate frequency of the piezoelectric sensor. Fig. 6.10b illustrates this situation where the continuous line represents the impedance phase of the resonator in a range of frequencies around the zero-phase frequency of lowest impedance of the resonator.

Let us suppose now that the phase–shift through the circuit β at the zero-phase frequency of the resonator is $-\alpha$. Because the phase condition for oscillation requires that $\arg X + \arg \beta = 0$, it is necessary that $\arg X = \alpha$. Consequently, the oscillating frequency shifts an amount Δf_1 with regard to the zero-phase frequency of the resonator, as described in Fig. 6.10b.

When the quality factor Q of the resonator is very high, the frequency–shift is very low which is apparent with the sharp slope of the phase transfer function of the resonator. However, the lower the Q the greater the fre-

[1] This ideal condition is not fulfilled in real oscillators where the phase shift through the amplifier depends on the frequency as well. In fact the gain of the amplifier in Fig. 6.9a must be negative since the circuit β in the feedback path provides a 180° phase–shift between the output and input terminals of the amplifier. Then the amplifier must provide an additional phase–shift of 180° for oscillation.

quency-shift, as can be observed in curve 2 in Fig. 6.10b. These fundamental concepts will permit the understanding of the problem associated with the monitoring of sensor parameters, MSRF and motional resistance, with an oscillator.

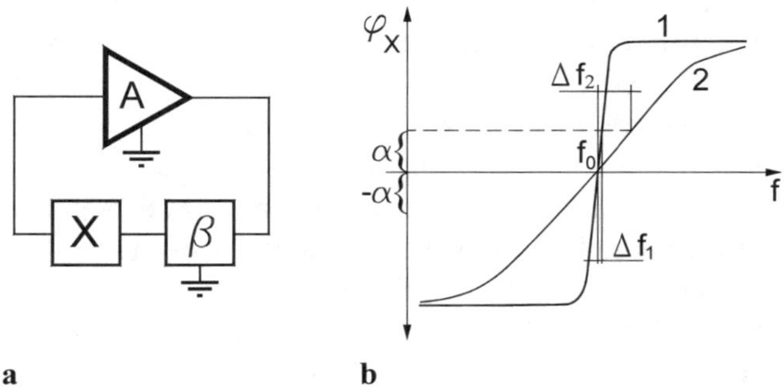

Fig. 6.10. Schematic circuits for illustrating the basic operation of a crystal controlled oscillator: **a** the circuit includes a resonant device X in the feedback circuit, and **b** plot for illustrating the high relationship between the frequency stability and the quality factor of the resonant feedback network

Problem Associated with the MSRF Determination

The systems described before passively interrogate the quartz resonator and, with an appropriate interface circuit, the desired characteristic parameters of the sensor are measured in isolation, i.e., the external circuitry does not interfere with the sensor response. This is the greatest advantage of the methods described. However, in oscillators the sensor takes part of the feedback path and the oscillating frequency is characterized by a certain phase of the quartz impedance, which depends on both the quartz resonator and the external circuitry to the sensor. When real characteristics of the amplifier and the rest of the components in the oscillator must be taken into account, it is difficult to know precisely the phase of the circuitry external to the sensor as a function of the frequency. Therefore, in general, the phase of the sensor for oscillating condition can not be accurately known, which makes the problem associated to oscillators more critical.

When the phase shift provided by the external circuitry in the range of frequencies around resonance can be considered constant, the phase provided by the quartz impedance at which the oscillation takes place can be considered constant as well. Thus, when the resonator properties are altered due to loading effects, the phase of the resonator changes and the oscillating frequency shifts to find the new frequency at which the phase of the sensor fits the oscillating condition. Therefore, due to the operation principle of the oscillator, the true MSRF can not be tracked continuously when an oscillator is used for monitoring load–induced frequency changes, at least in an oscillator in which the phase of the quartz impedance for oscillating condition is maintained constant. This can be easily illustrated using the BVD circuit as sensor–model (Fig. 6.2b). It can be understood that the MSRF only depends on L and C components of the motional branch, but the phase of the complete resonator also depends on the specific values of the motional resistance R and on the parallel capacitance C_o^*. Thus, if the motional resistance changes while remaining constant L and C, the MSRF does not change but the quartz phase changes. Then the oscillating frequency has to change in order to find the quartz phase corresponding to the oscillating condition and this frequency shift supposes a frequency error, which can be significant when the MSRF is taken as a parameter of interest. Fig. 6.11 shows how the oscillating frequency and the frequency shift regarding the frequency of the resonator in the unperturbed state depend on the oscillating phase condition in an oscillator. Sometimes the oscillation ceases because the oscillation phase condition can not be reached.

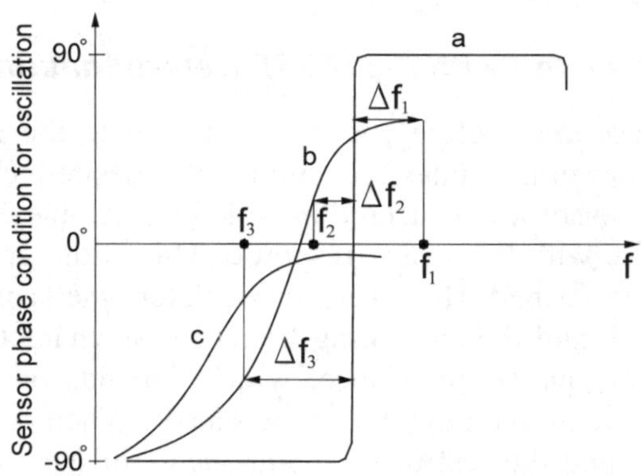

Fig. 6.11. Plot for illustrating the different frequency shifts, with regard to the resonant frequency of the resonator in the unperturbed state, associated with different sensor phase conditions for oscillation in an oscillator

Another consideration which makes clear the selection of the MSRF as a parameter of interest is its independence of the parasitic capacitances in parallel with the sensor. However, due to the operation principle of the oscillator, the stray capacitances in parallel with the sensor always have important effects on the oscillating frequency. This is because an oscillator looks for a certain phase and not for a certain frequency of the sensor. Thus, although a change in the parallel capacitance due to parasitic effects does not change the MSRF, the phase of the sensor changes and, consequently, a shift in the oscillating frequency takes place.

As described, the oscillator frequency is dependent not only on "mass loading" (motional inductance) but on "dissipation" (motional resistance). This must be taken into account and carefully considered when using an oscillator in a specific application, in which the physical properties of the load can change producing changes in both the mass loading and dissipation.

Problem Associated with the Motional Resistance Determination

In a great deal of applications in which the quartz resonator is used as a sensor, an additional magnitude to the MSRF shift is necessary in order to discriminate different physical contributions [8]. In what concerns to oscillators, many designs incorporate an automatic gain control (AGC) system for the measurement of the activity of the quartz sensor at resonance [20]. The AGC system tries to maintain the level of the signal constant in a selected point of the oscillator. With this purpose, it provides a voltage (AGC voltage) that modifies the gain of the amplifier so as to maintain the signal level in the selected point constant in relation to a reference voltage. In many of these designs proportionality between the change in the AGC voltage and the change in the motional resistance is claimed [21-23]. In some of these AGC systems the proportionality is justified only from a physical point of view [22]. In others, mathematical expressions in which some simplifications were made show this proportionality [21], although it is also shown that this proportionality is lost in some cases [24].

It can be shown that even in the most ideal situation it is very difficult to ensure that the voltage shift provided by an AGC system included in an oscillator is proportional to the change in the motional resistance; unless the parallel capacitance of the sensor is suppressed or compensated. This demonstration can be found elsewhere [25].

The most important aspects concerning the problem associated with oscillators as electronic drivers for QCMS have been treated. It has been shown that although the simplicity of an oscillator makes this device very attractive for sensor applications, some limitations remain. These limitations have to be taken into account in the interest of accuracy and they should be kept in mind in any new work that looks for the simplicity and autonomy of oscillators for sensor applications.

6.4.4 Parallel Capacitance Compensation Techniques

These techniques aim at the simplicity of oscillators while overcoming their limitations. From the previous analysis on oscillators, it can be seen that their limitations come from the existence of the parallel capacitance. It can be deduced that, if the parallel capacitance of a QCMS were eliminated, an ideal quartz resonator controlled oscillator with 0° oscillation phase condition would provide continuously the MSRF. Additionally, changes in the motional resistance do not produce changes in the oscillation frequency.

The parallel capacitance can be compensated by the circuit shown in Fig. 6.12 [26, 27] which can be incorporated in an oscillator. The major problem of this technique is the lack of ideality of the transformers and the selection of the capacitance C_v for compensating the parallel capacitance, i.e., the calibration of the parallel capacitance compensation.

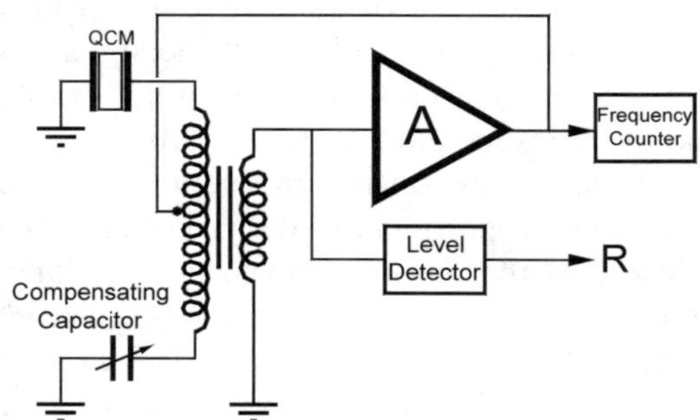

Fig. 6.12. Schematic circuit for parallel capacitance compensation based on a three winding transformer

An additional problem associated with oscillators is the typical loop gain dependence and its relation to the appropriate oscillating phase condition. Phase locked loop techniques permit the system to lock automatically to the appropriate MSRF while maintaining the isolation between the sensor and the electronic interface thanks to a passive interrogation of the resonator [25, 28, 29]. Additionally, these techniques make the calibration procedures of the system and its automation easier, which is also important. A circuit implementing a phase locked loop technique for parallel capacitance compensation is shown in Fig. 6.13 [25].

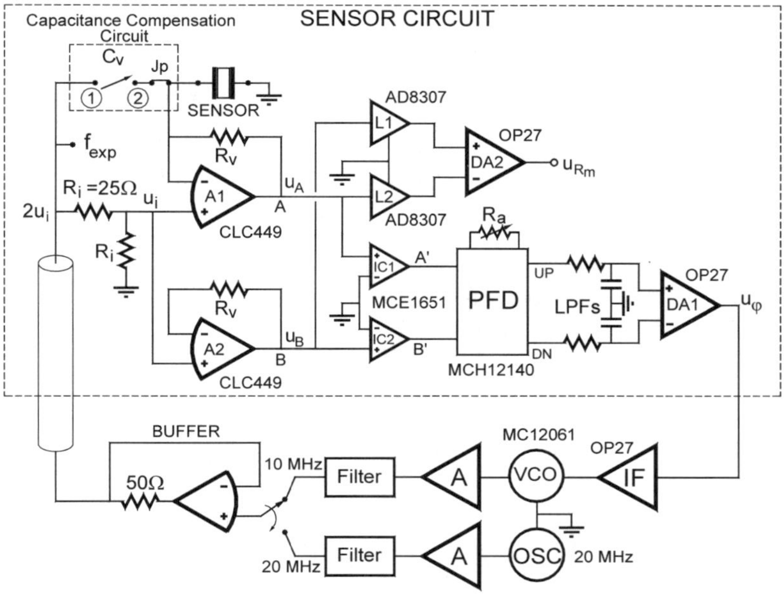

Fig. 6.13. Circuit for parallel capacitance compensation based on a phase-locked loop technique

6.4.5 The Transfer Function Method

In the transfer function method used by Calvo and Etchenique the sensor takes part of an impedance divider where the other impedance in the divider is known [11]. Fig. 6.14 shows the impedance divider where the sensor has been characterized by its equivalent BVD circuit (Fig. 6.2b). A sinusoidal signal applied at the input sweeps a range of frequencies around the resonant frequency of the sensor; the average values of the input and output voltages are measured and acquired with an analog/digital converter and captured with a computer. Thus, an experimental measure of the absolute value of the voltage transfer function is obtained in the range of frequencies considered. The theoretical transfer function is used to find the set of parameters that best fit to the experimental data. It is assumed that the motional capacitance remains constant, the rest of parameters C_o, L and R are obtained by a non-linear fitting. Finally the authors of this technique noticed the advantage of using a capacitor C_m instead of a resistor as the other branch of the voltage divider, as shown in Fig. 6.14. In this context the theoretical transfer function is:

$$\left|\frac{u_o}{u_i}\right| = \frac{\sqrt{\left(\omega L - \frac{1}{\omega C}\right)^2 + R^2}}{\sqrt{\left(\omega L - \frac{1}{\omega C} + \frac{\omega L C_o}{C_m} - \frac{C_o}{\omega C C_m} - \frac{1}{\omega C_m}\right)^2 + \left(R + \frac{R C_o}{C_m}\right)^2}} \tag{6.22}$$

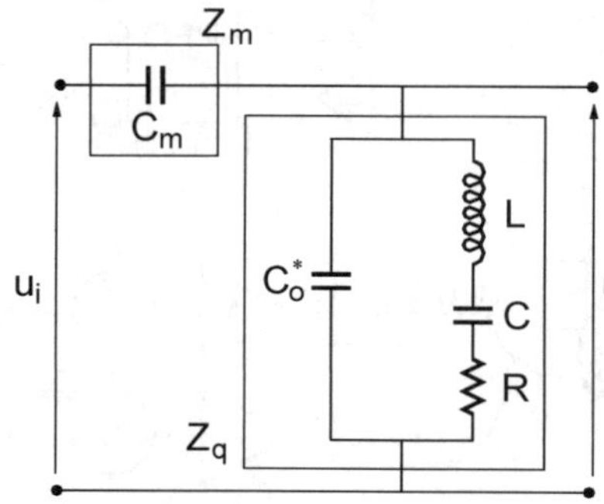

Fig. 6.14. Impedance divider for the transfer function method used in reference [11]

6.5 Conclusions

The problem associated with QCRS interface circuits reveals that an optimal device for sensor characterization does not exist at present. The systems that better characterize the sensor are those which passively interrogate the quartz resonator so that the sensor is measured in isolation. However, these systems based on impedance analyzers or decay/impulse excitation methods include expensive equipment and do not fulfill the autonomy normally required for sensor applications. On the other hand, those systems which fulfill the simplicity, autonomy and low price requirements appropriate for sensor applications have less accuracy in the determination of sensor parameters; mainly, because the external circuitry influences the sensor response which is not measured in isolation. This problem becomes more critical due to the difficulty of calibrating the response of the external circuitry in typical oscillators. Thus, the monitored parameters can not be accurately associated to a certain phase of the sensor.

Consequently, when using an oscillator to measure the frequency shift, it is necessary to consider the change in the motional resistance as well. If the motional resistance changes during experiment, the frequency shift monitored with the oscillator can not be considered equal to the MSRF shift. Thus the associated physical parameters extracted from the monitored frequency shift will include an error which can be significant.

Appendix 6.A
Critical Frequencies of a Resonator Modeled as a BVD Circuit

6.A.1 Equations of Admittance and Impedance

The expressions for the admittance and impedance of the equivalent electric circuit shown in Fig. 6.2b modeling the resonator in a range of frequencies around resonance are:

$$Y = G + jB = \frac{R}{R^2 + X^2} - j\left(\frac{X}{R^2 + X^2} - \omega C_o^*\right) \tag{6.A.1}$$

$$\begin{aligned} Z = R_e + jX_e = & \frac{R}{\left(1 - \omega C_o^* X\right)^2 + \omega^2 C_o^{*2} R^2} + \\ & + j\frac{X\left(1 - \omega C_o^* X\right) - \omega C_o^* R^2}{\left(1 - \omega C_o^* X\right)^2 + \omega^2 C_o^{*2} R^2} \end{aligned} \tag{6.A.2}$$

where

$$\begin{aligned} X = & \left(\omega L - \frac{1}{\omega C}\right) = \frac{1}{\omega C}\left(\frac{\omega^2}{\omega_s^2} - 1\right); \\ \omega_s^2 = & \frac{1}{LC}; \quad \omega_p^2 = \omega_s^2 \frac{C_o^* + C}{C_o^*} \end{aligned} \tag{6.A.3}$$

Typical plots for real and imaginary parts of the impedance and admittance are shown in Figs. 6.A.1 and 6.A.2, respectively.

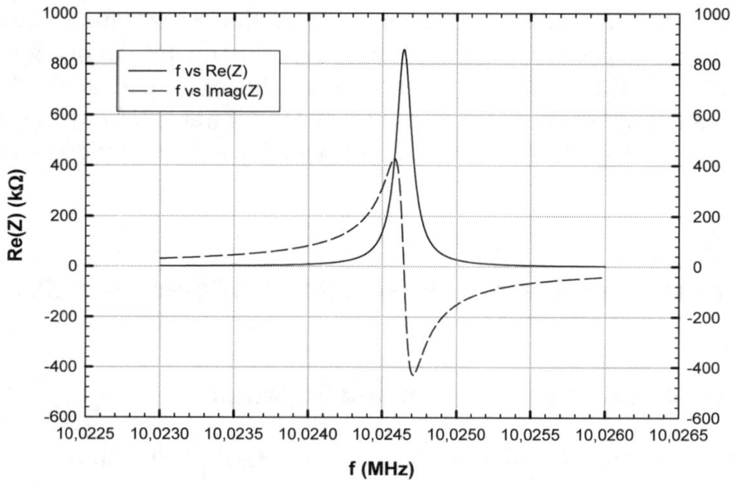

Fig. 6.A.1. Plots for the real and imaginary parts of the impedance of a typical 10 MHz unperturbed quartz resonator around the fundamental resonance

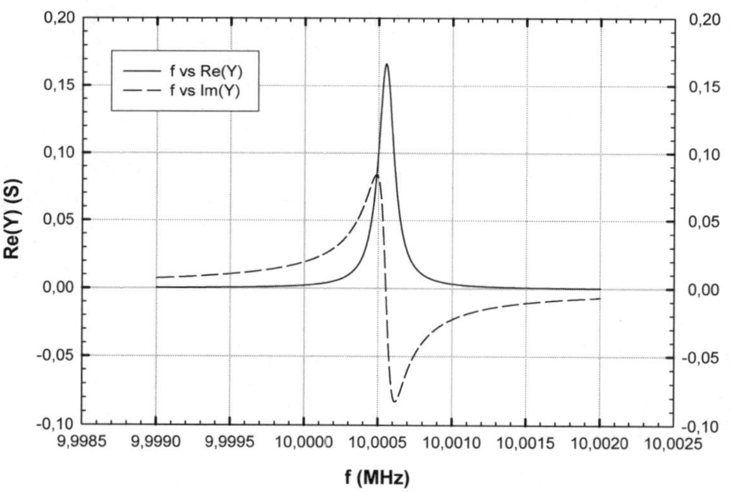

Fig. 6.A.2. Plots for the real and imaginary parts of the admittance of a typical 10 MHz unperturbed quartz resonator around the fundamental resonance

6.A.2 Critical Frequencies

Series and parallel resonant frequencies

The series and parallel electric resonant frequencies are those which cancel the imaginary term in the expressions for admittance and impedance when the losses are neglected; this is, when the parameter R is zero.

The condition for canceling the imaginary part in admittance and impedance is:

$$\frac{\omega^2 R^2 C^2 C_o^*}{C_o^* + C} + \left(\frac{\omega^2}{\omega_s^2} - 1\right)\left(\frac{\omega^2}{\omega_p^2} - 1\right) = 0 \tag{6.A.4}$$

The angular frequencies fulfilling the former equation when the resistance R is neglected are: the series resonant frequency ω_s and the parallel resonant frequency ω_p. When the losses are not neglected, the impedances at resonant frequencies are:

$$Z(\omega_s) = \frac{R}{\omega_s^2 C_o^{*2} R^2} - j\frac{\omega_s C_o^* R^2}{1 + \omega_s{}^2 C_o^{*2} R^2} \tag{6.A.5}$$

$$Z(\omega_p) = \frac{R}{\omega_p^2 C_o^{*2} R^2} - j\frac{1}{\omega_p C_o^*} \tag{6.A.6}$$

Consequently, the impedance has a capacitive character at those frequencies.

Zero-Phase frequencies

The exact resolution of Eq. (6.A.4) gives two frequencies: f_r and f_a at which the imaginary parts of the admittance and impedance cancel. These frequencies at which the electrical behavior of the piezoelectric resonator is purely resistive are:

$$\omega_r = \sqrt{\frac{b - \sqrt{b^2 - 4\omega_s^2 \omega_p^2}}{2}} \tag{6.A.7}$$

$$\omega_a = \sqrt{\frac{b + \sqrt{b^2 - 4\omega_s^2 \omega_p^2}}{2}} \tag{6.A.8}$$

where $b = \omega_s^2 + \omega_p^2 - \dfrac{R^2 C^2 C_o^*}{C_o^* + C}\omega_s^2 \omega_p^2$.

In order to simplify the interpretation of the relative positions among the different critical frequencies with regard to the electrical resonant frequencies f_s and f_p, it is usual to give the following approximate expressions for the zero-phase frequencies as a function of the ratio $r = C_o^* / C$ and the quality factor $Q = \overline{c}_{66} / \omega_s \eta_Q = L\omega_s / R$ (Chap. 1–Eq. (1.A.38)). These expressions are [20]:

$$\omega_r \approx \omega_s \left(1 + \frac{r}{2Q^2}\right) \tag{6.A.9}$$

$$\omega_a \approx \omega_s \left(1 + \frac{1}{2r} - \frac{r}{2Q^2}\right) \tag{6.A.10}$$

Frequencies for Minimum and Maximum Admittance

The exact expressions for the frequencies of maximum admittance f_m and minimum admittance f_n are difficult to obtain. In this case the approximations are very useful and accurate in most cases. However, their validity must be restricted to cases where the quality factor has a high value. These expressions are:

$$\omega_m \approx \omega_s \left(1 - \frac{r}{2Q^2}\right) \tag{6.A.11}$$

$$\omega_n \approx \omega_s \left(1 + \frac{1}{2r} + \frac{r}{2Q^2}\right) \tag{6.A.12}$$

The previous approximate equations for the critical frequencies permit to show graphically, the relative position among these frequencies, as illustrated in Fig. 6.A.3.

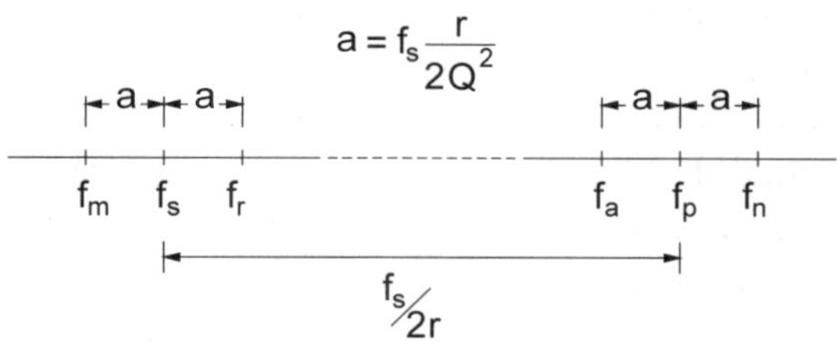

Fig. 6.A.3. Diagram for illustrating the relative positions among the electric resonant frequencies of a quartz crystal resonator

6.A.3 The Admittance Diagram

The admittance Y can be represented by a vector in the complex plane. The real component in the abscissa represents the conductance and the imaginary part in the coordinate axes represents the susceptance. Eq. (6.A.1) permits to write the admittance as the sum of the motional admittance and the admittance corresponding to the parallel capacitance C_o^*, as follows:

$$Y = Y_m + j\omega C_o^* = G + jB \tag{6.A.13}$$

where

$$Y_m = \frac{R}{R^2 + X^2} - j\frac{X}{R^2 + X^2} = G_m + jB_m \tag{6.A.14}$$

As can be deduced from the former equation, the absolute value of the motional admittance describes a circumference in the complex plane. Effectively, the absolute value of the motional admittance is:

$$|Y_m| = \sqrt{\frac{R^2}{(R^2 + X^2)^2} + \frac{X^2}{(R^2 + X^2)^2}} = \sqrt{\frac{1}{(R^2 + X^2)}} = \sqrt{G_m^2 + B_m^2} \tag{6.A.15}$$

Consequently:

$$G_m^2 + B_m^2 = \frac{1}{(R^2 + X^2)} = \frac{G_m}{R} \Rightarrow \left(G_m - \frac{1}{2R}\right)^2 + B_m^2 = \frac{1}{4R^2} \tag{6.A.16}$$

The former equation represents a circumference with the center in $(1/2R, 0)$ and a diameter equal to $1/R$, as illustrated in the continuous plot in Fig. 6.A.4a. Each point in this circumference represents the affix of the vector of admittance. The affix of this vector displaces along the circumference when the frequency is varied from 0 to ∞. The susceptance is positive for frequencies between the zero conductance frequency and the maximum conductance frequency, at which $G_m^{\max} = 1/R$, while it is negative for frequencies greater than the maximum conductance frequency. It can be demonstrated that the maximum conductance frequency is equal to the electrical series resonant frequency for the BVD circuit.

The admittance corresponding to the parallel capacitance, $j\omega C_o^*$, moves up each point corresponding to the affix of the admittance vector at the specific frequency. Because resonance occurs in a very narrow range of frequencies for high Q resonators, the admittance of the parallel capacitance can be considered constant in the range of frequencies around resonance. The result is shown in Fig. 6.A.4a where the locus representing the motional admittance is moved up by the admittance corresponding to the parallel capacitance. The final displaced locus represents the admittance corresponding to the BVD circuit. A piezoelectric resonator can be represented by several BVD circuits, each one corresponding to a specific resonance (see Chap. 1). In this case the locus of a piezoelectric resonator will be represented by a number of loci, each one corresponding to the BVD circuit representing the specific resonance of the resonator. The diameter of each locus is reduced for the resonances corresponding to higher harmonics as illustrated in Fig. 6.A.4b.

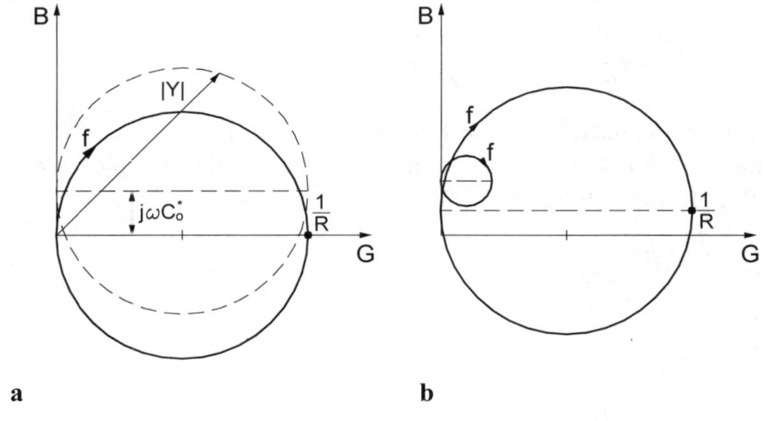

Fig. 6.A.4. Admittance loci of a quartz resonator: **a** admittance locus corresponding to the motional impedance of a BVD circuit (continuous line) and to the BVD circuit including parallel capacitance (dash line); **b** admittance loci corresponding to the fundamental resonance (big locus) and to a higher harmonic (small locus)

References

1. A.W. Czanderna and C. Lu, in *Applications of Piezoelectric Quartz Crystal Microbalances*, edited by C.Lu and A.W. Czanderna (Elsevier, Amsterdam, 1984). Vol. 7, p.1.
2. A. Bund, H. Chmiel, and G. Schwitzgebel, Phys. Chem. Chem. Phys. **1**, 3933 (1999).
3. E.J. Calvo, R. Etchenique, P.N. Barlett, K. Singhal, and C. Santamaría, Faraday Discuss. **107**, 141 (1997).
4. K.Bizet, C. Grabielli and H. Perrot, Analusis EurJAC, **27**, 609 (1999).
5. S.W. Lee, W.D. Hinsberg and K. Kanazawa, Anal. Chem. **74**, 125 (2002).
6. A. Arnau, Y. Jiménez and T. Sogorb, J. Appl. Phys. **88**, 4498 (2000).
7. A. Arnau, Y. Jiménez and T. Sogorb, IEEE Trans. Ultrason., Ferroelect., Freq. Contr. **48**, 1367 (2001).
8. S.J. Martin, V.E. Granstaff and G.C. Frye, Anal. Chem. **63**, 2272 (1991).
9. IEC, Measurement Of Quartz Crystal Unit Parameters By Zero Phase Technique In A Pi-Network (Part 1), International Electrotechnical Commission-IEC Standard, publication 444-1,1986.
10. A. Arnau, T. Sogorb and Y. Jiménez, Rev. Sci. Instrum. **71**, 2563 (2000).
11. W.G. Cady, *Piezoelectricity (An Introduction To The Theory And Applications Of Electromechanical Phenomena In Crystals)*, (Dover Publication, Inc., New York, 1964).
12. R. W. Cernosek, S. J. Martin, A. R. Hillman and H.L. Bandey, IEEE Trans. Ultrason., Ferroelect., Freq. Contr. **45**, 1399 (1998).
13. F. Eichelbaum, R. Borngräber, J. Schröder, R. Lucklum, and P. Hauptmann, Rev. Sci. Instrum. **70**, 2537 (1999).
14. M. Rodahl and B. Kasemo, Rev. Sci. Instrum. **67**, 3238 (1996).
15. M. Rodahl and B. Kasemo, Sensors And Actuators B **37**, 111 (1996).
16. M. Rodahl and B. Kasemo, Anal. Chem. **68**, 2219 (1996).
17. Libro Parzen-Ballato
18. Libro Frerking.
19. H. Ehahoun, C. Gabrielli, M. Keddam, H. Perrot and P. Rousseau, Anal Chem., **74**, 1119 (2002).
20. V.E. Bottom, *Introduction to Quartz Crystal Unit Design,* (Van Nostrand, New York, 1982).
21. K.O. Wessendorf, *Proceedings of the 1993 IEEE International Frequency Control Symposium, 1993* (IEEE, 1993), p. 711.
22. C. Chagnard, P. Gilbert, A. N. Watkins, T. Beeler and D.W. Paul, Sensors And Actuators B **32**, 129 (1996).
23. J. Auge, P. Hauptmann, J. Hartmann, S. Rösler and R. Lucklum, Sensors And Actuators B **24-25**, 43 (1995).
24. S. J. Martin, J. J. Spates, K. O. Wessendorf, T. W. Schneider and R. J. Huber, Anal. Chem. **69**, 2050 (1997).
25. A. Arnau, T.Sogorb, Y. Jiménez, Rev. Sci. Instrum. **73**, Iss 7, 2724 (2002).

26. S. Geelhood, C.W. Frank, and K. Kanazawa, Acoustic Wave Sensor Worshop 3, Taos, New Mexico, 2001.
27. R. Behrends and Kaatze, Meas. Sci. Technol. **12**, 519 (2001).
28. V. Ferrari, D. Marioli, and A. Taroni, IEEE Trans. Instrum. Meas. **50**, 1119 (2001).
29. V. Ferrari, D. Marioli, and A. Taroni, Electron. Lett. **36**, 610 (2000).

7 Modified Piezoelectric Surfaces

Hubert Perrot[1], Ernesto Calvo[2] and Christopher Brett[3]

[1] Laboratoire Interface et Systèmes Electrochimiques, Université P. et M. Curie
[2] Facultad de Ciencias Exactas y Naturales, Universidad de Buenos Aires
[3] Departamento de Química, Universidade de Coimbra

7.1 Introduction

The quartz crystal microbalance (QCM) is turning into an attractive tool for gravimetric measurements and applications and can be found in many research fields such as acoustic sensors (Chap. 2), chemical sensors (Chap. 9) and biosensors (Chap. 10). In general, QCM resonators are covered with noble metals such as gold or silver, usually by evaporation or sputtering, and can be used directly as electrodes or undergo further surface modification. Using such strategies, the study of processes ranging from electroplating to DNA immobilization becomes possible. In this chapter, two separate sequential preparation steps should be distinguished. Metallic deposition (acting as the ultrasonic wave generator) must be first considered as a support for modification. In the second step, organic or biochemical modifications are carried out before testing the modified QCM.

7.2 Metallic Deposition

Metallic deposition can be carried out using two approaches: directly onto the quartz crystal by taking into account the crystalline orientation of the resonator (vacuum techniques) or onto a previously prepared metal layer on the quartz substrate, in general gold. First, we describe the two main vacuum methods: evaporation and sputtering. It should be noted that when high temperatures are necessary to achieve these operations some trouble can occur with respect to the piezoelectric behavior of the quartz crystal. Indeed, when the Curie temperature is overstepped, the piezoelectric properties can be lost. Moreover, the roughness and the porosity of these layers depend on the deposition conditions; it means that the QCM response can be affected by these parameters and therefore, users have to be aware of these effects.

7.2.1 Vacuum Methods

Evaporation (Metals)

This method is old and the first applications date from the 1950s. The system is widely used in microelectronic technology to produce thin metallic films which are necessary for interconnections: all ohmic contacts are based on this technique. Various metals can be used (chromium, copper, silver, gold…). The metal is evaporated at high temperature (500-2000°C) and under vacuum ($10^{-6}/10^{-7}$ torr). Metallic atoms are provided with sufficient energy to be dislodged from the source and to condense onto the substrate (quartz in our case). For quartz, it is necessary to use masks corresponding to the shape of the electrode and, if gold is deposited onto the quartz material, a sub-layer of chromium is necessary in order to improve adhesion. The typical thickness is 20 nm for chromium and 250 nm for gold, with total thickness rarely exceeding 500 nm. The cleaning of the surface is a crucial step as the adhesion of the metallic layer is completely dependent on the quality of the surface.

Sputtering (Metals or Insulating Materials)

Sputtering is a process whereby atoms or molecules of a material are ejected from a target (the material which is to be deposited) by bombardment of high-energy particles [1]. Positive ions derived from an electrical discharge in a gas at low pressure (~ 0.3 Pa) in a vacuum chamber are used and the material of the target is ejected in such a way as to be directed towards the substrate surface. Fig. 7.1 presents a schematic diagram of the device. Many materials can be used, even insulating materials, and it is easy to deposit alternate layers of different materials. The thickness range on the quartz substrate is between a few nm and 200 nm.

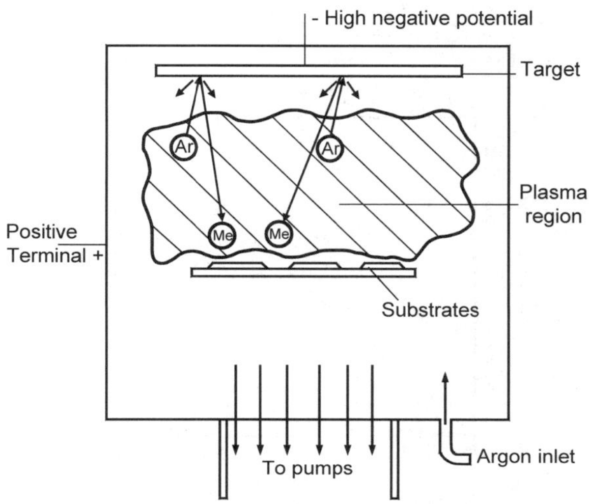

Fig. 7.1. Schematic representation of a sputtering process chamber (according to reference [1])

7.3 Electrochemical Method

A previously prepared electrode (usually gold) on a quartz crystal is necessary and is used as working electrode in an electrochemical cell. The cell contains as electrolyte a salt of the metal to be deposited. The electrochemical process is a reduction reaction of the type:

$$M^{n+} + ne^{-} \rightarrow M \quad (1.1)$$

The reduction is carried out either by control of applied potential (potentiostat) or of applied current (galvanostat). In the latter case, the current is usually maintained constant and the metal thickness can be determined from the quantity of electrons supplied and by using Faraday's law. The film thickness is easily controllable by measuring the current and the time; the thickness range is between 10 nm to 100 μm. Various metals can be deposited in this way: copper, silver, iron, nickel, etc. This approach was used to perform calibration of a QCM by using silver electrodeposition [2]. Figure 7.2 shows a record of the microbalance frequency change over time during silver deposition at various current values: the slope of each curve extracted allows the mass sensitivity of the microbalance, *K*, to be calculated; *K* is defined as:

$$K = \frac{\Delta f_{microbalance}}{\Delta m} \qquad (1.2)$$

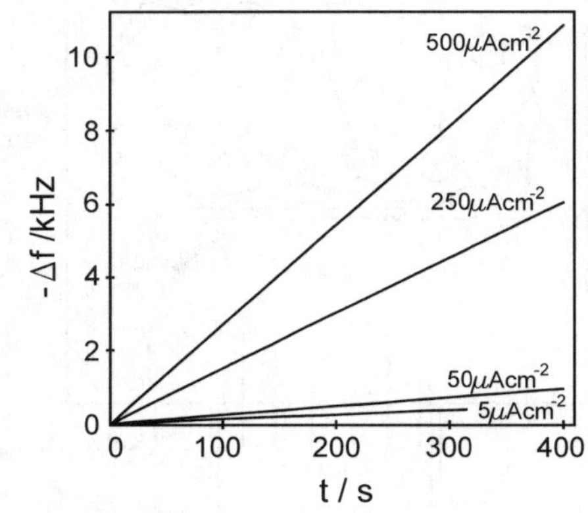

Fig. 7.2. Time variation of the QCM frequency for different current densities S = 0.2 cm^2 (from reference [2])

Figure 7.3 shows the average K values obtained for different experimental conditions of silver deposition onto electrodes of different active surfaces. It can be seen that the theoretical value of K is reached only in the case of the largest active surface area (i.e. 0.64 cm^2 for radius r = 4.5 mm). On the contrary, for smaller electrodes, lower values of K were obtained.

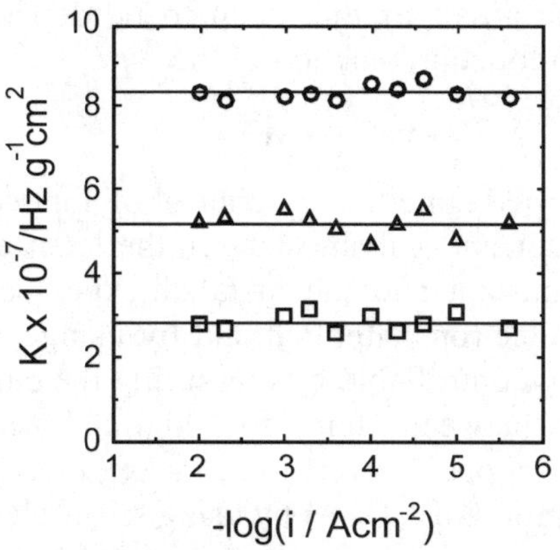

Fig. 7.3. Average sensitivity (K) calculated for different current densities for three different active surface electrodes. r = 4.5 mm, ▲ 2.5 mm and □ 1.5 mm. Theoretical value of K=8.15 10^7 Hz g^{-1} cm^2 for a QCM working at 6 MHz

7.3.1 Technique Based on Glued Solid Foil (Nickel, Iron, Stainless Steel…)

This technique uses a piezoelectric transducer working at low frequencies such as 2 MHz. This frequency is lower than that of the common QCM to allow the crystal to oscillate with much higher loads. The metal sample is a disk cut out from a foil of metal such as stainless steel, iron, nickel, 50 μm thick and 5 mm diameter. The glue is a 2-component epoxy resin with high chemical resistance in concentrated acid solution which, once polymerized, becomes extremely hard and acoustically low absorbing. The resin was insulating, as conductive glues were found not to have suitable acoustic and mechanical properties. Figure 7.4 shows a scheme of the crystal-glue-sample structure. The resonance frequency of the structure is about 1.75 MHz for an initial 2 MHz resonance frequency of the bare quartz. It should be pointed out that in this case the mass changes of the glued solid material can be studied directly, which opens up new possibilities for the investigation of processes on a wider variety of materials. For this technique of metallic preparation, it should be noted that the damping of the quartz crystal is drastically increased; in these conditions, it means that the piezoelectric behavior of the device can be modified (see Chap. 1) and the electronic working conditions have to be adapted (see Chap. 6).

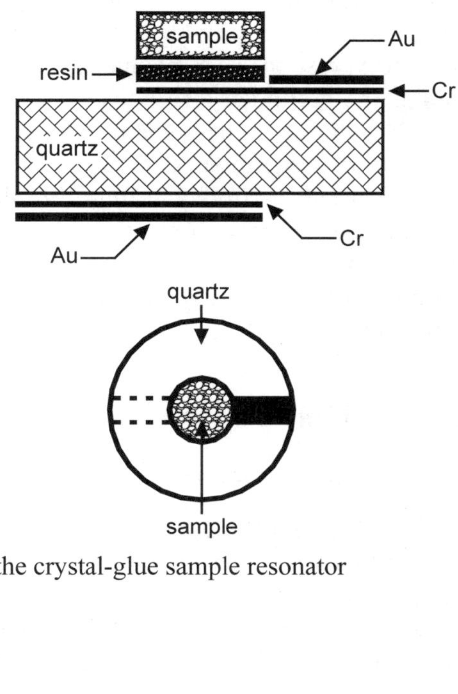

Fig. 7.4. Structure of the crystal-glue sample resonator

7.3.2 Organic Film Preparation

Polymer Electrogeneration (Conducting Polymers: Polypyrrole, Polyaniline…)

The electrogeneration of conducting polymers is widely used in electrochemistry: a polymer film can be generated directly onto the gold electrode of the quartz. Three methods of electropolymerisation are available: potentiostatic (constant applied potential), galvanostatic (constant applied current) or cyclic voltammetry (potential scanning). The thickness and the morphology of the polymer film are easily controllable by changing the value of the applied potential, the value of the applied current, or the value of the scan rate as well as by the composition of the electrodeposition bath. As an example, a schematic representation of the electropolymerisation of pyrrole can be depicted by the two following reactions [3]:

$$\mathrm{Py} \rightarrow \mathrm{Py}^{+} + \mathrm{e}^{-} \tag{2.1}$$

$$\mathrm{Py}^{+} + -\mathrm{Py} \rightarrow -(\mathrm{Py})_2 + 2\mathrm{H}^{+} + \mathrm{e}^{-} \tag{2.2}$$

Following this, the redox behavior of the deposited film, the insertion and removal of counter ions, the effect of doping, etc., can be studied through microbalance measurements: by using fast QCM. The microrheology of these materials must be well defined. Because polymers are not rigid materials, the QCM response can be due to the viscoelastic properties of the film as well as the gravimetric response. Thus, interpretation of the QCM response purely in terms of mass changes may be completely wrong, i.e., the classical Sauerbrey equation is not valid [4] (see Chap. 16). The electroacoustic approach can partly solve these problems [5]. When conducting polymer films are deposited onto the quartz resonators, it should be noted that under some conditions, the viscoelastic contribution of the material can be important (see Chaps. 4, and 16). In this case, the pure mass response can be drastically affected or overestimated.

7.3.3 Monolayer assemblies

SAM Techniques (Thiol Molecule)

This procedure, which began to be used recently, is based on the adsorption of alkanethiols onto a gold electrode [6] as shown in Fig. 7.5. A covalent bond is created between the gold surface and the thiol groups: the alkane chains are maintained close to perpendicular over the surface owing to Van-der-Waals interactions (Fig. 7.6). The result obtained is, in general, close to a monolayer. Originally, this method was used for the QCM in order to investigate the influence of this surface treatment on molecular slipping [7]. It is now also used to investigate the formation and properties of different alkanethiols at applied potential, e.g. [8], and to prepare biosensors [9] (see Chaps. 9 and 10).

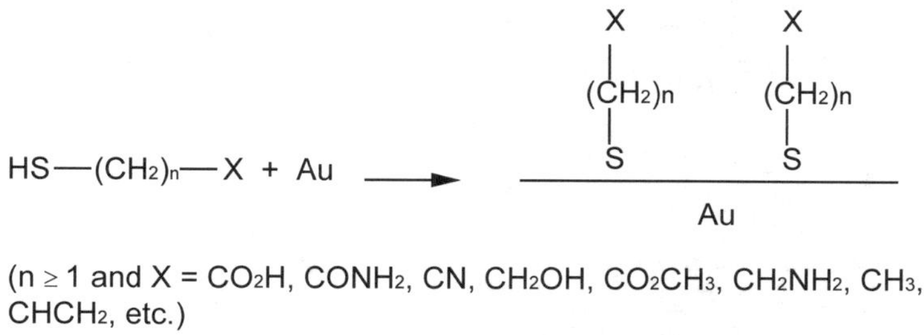

Fig. 7.5. Generation of a SAM over a gold electrode: general reaction (from reference [6])

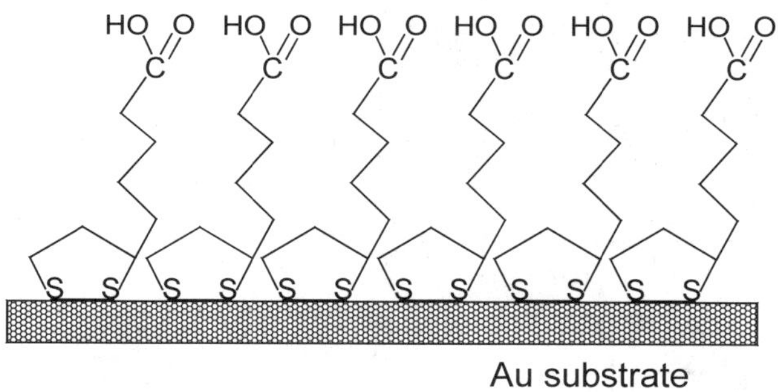

Fig. 7.6. Diagram of the thioctic acid monolayer on the Au substrate (from reference [23])

Langmuir-Blodgett Method

The Langmuir-Blodgett (LB) is a simple method which allows molecular assemblies to be built up [10] through the deposition of one complete monolayer of amphiphilic molecules, i.e. molecules having a hydrophilic head and a hydrophobic chain. The film is first prepared on a water surface and then transferred to a solid substrate as shown in Fig. 7.7. Applications are various and LB films can be deposited directly onto the QCM active surface. These techniques can be coupled and lead, for example, to interesting smell sensors [11].

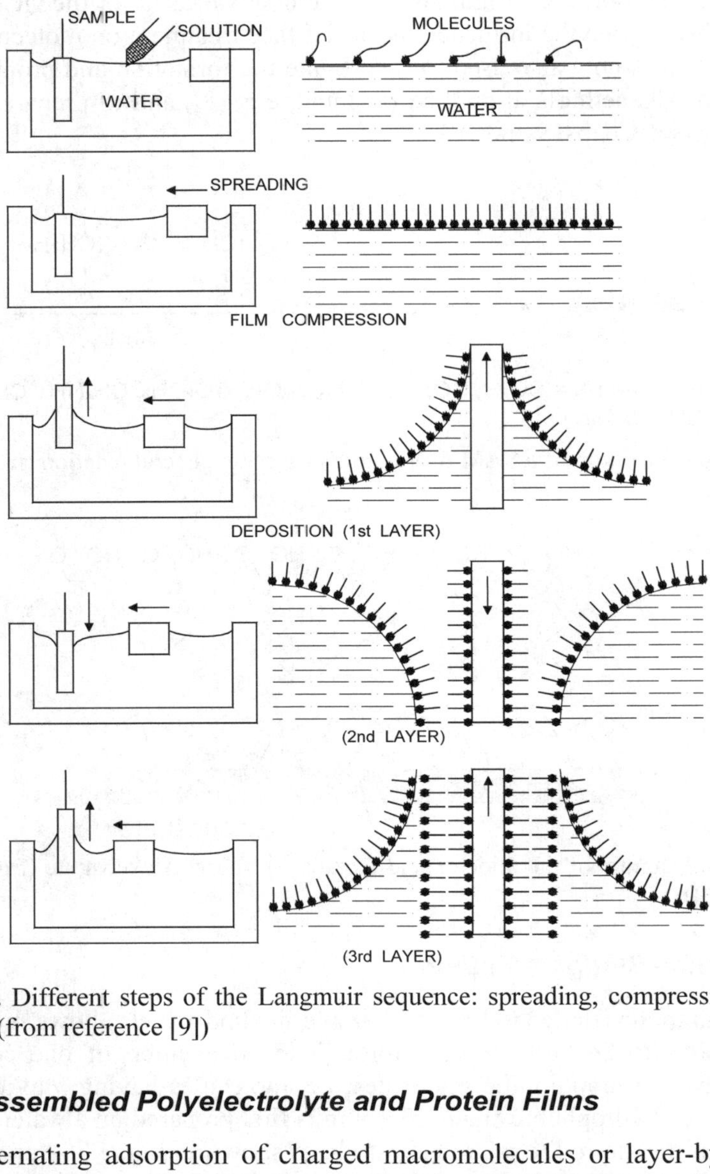

Fig. 7.7. Different steps of the Langmuir sequence: spreading, compression and transfer (from reference [9])

Self-Assembled Polyelectrolyte and Protein Films

The alternating adsorption of charged macromolecules or layer-by-layer adsorption has recently emerged as a promising simple method to fabricate controlled and highly ordered molecular assemblies in a pre-designed architecture. The QCM would, in principle, permit quantification of adsorbed protein, measurement of protein uptake kinetics in real time and determination of the binding constants from adsorption isotherms. The electroacoustic approach allows the Sauerbrey range validity to be determined more or less easily as has been shown [12, 13].

7.4 Biochemical Modifications

Several techniques are available for immobilizing active biomolecules, including direct immobilization and entrapment.

7.4.1 Direct Immobilization of Biomolecules (Adsorption, Covalent Bonding)

Two main types of biomolecules can be immobilized: enzyme and antibodies (or antigens). Biological assemblies can be used, with different degrees of sophistication, and the QCM appears to be a very interesting tool to follow the various interactions. One possibility is based on direct adsorption of the antibodies onto a QCM gold electrode: this method leads to interesting results since the activity of the antibodies is maintained at a high level. The principle is depicted in Fig. 7.8; in this case, a reference sensor is used in order to subtract non-specific interactions [14]. Using a QCM, the limit of detection can reach 10^{-10} M for the corresponding antigen (the detected species) in solution [15].

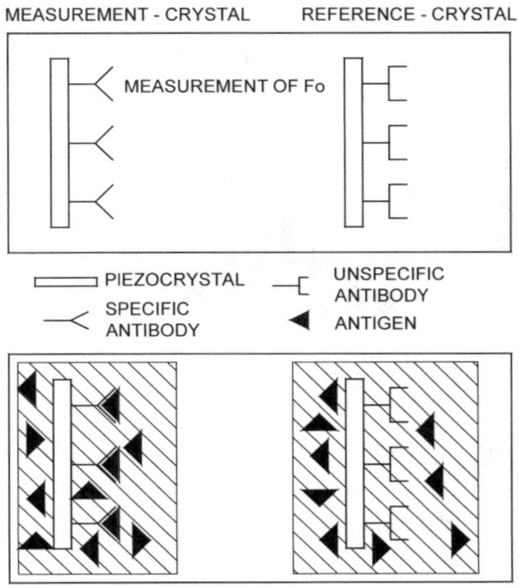

Fig. 7.8. Schematic of measurement procedure for immunodetection with adsorbed antibodies onto a QCM surface (from reference [11])

7.4.2 Entrapping of Biomolecules (Electrogenerated Polymers: Enzyme, Antibodies, Antigens...)

This approach is widely used for making biosensors (se Chaps. 9 and 10). In general, the active biomolecule is entrapped inside a polymer structure. Either the biomolecules are mobile in the polymer film or the biomolecules are covalently bonded to the polymer chains. Various polymers can be used but mainly conducting polymers are employed. An example is a quartz crystal microbalance bioaffinity sensor for avidin in a flow cell, based on an electropolymerized biotinylated polypyrrole film. The functionalization of an Au quartz electrode was achieved by controlled potential oxidation of a biotinyl pyrrole monomer in CH_3CN leading to chemical grafting of the biotin molecule to the electrogenerated polypyrrole. In the presence of avidin solutions, gravimetric measurements indicated the reproducible formation of a compact avidin layer via the specific avidin-polymerized biotin interaction. Avidin concentrations in the range $1.5 \cdot 10^{-9}$ to $3.7 \cdot 10^{-5}$ M can be measured by this method [16].

7.4.3 DNA Immobilization

This application using QCM devices is relatively recent [17] and the idea is similar to the immunosensor. It means that a DNA probe has to be immobilized onto the quartz surface or one of its metallic electrodes. All the chemical methods are available to do this: covalent bonding, adsorption or more sophisticated methods. For example, Biotin-DNA can be immobilized with avidin as shown in Fig. 7.9 [18].

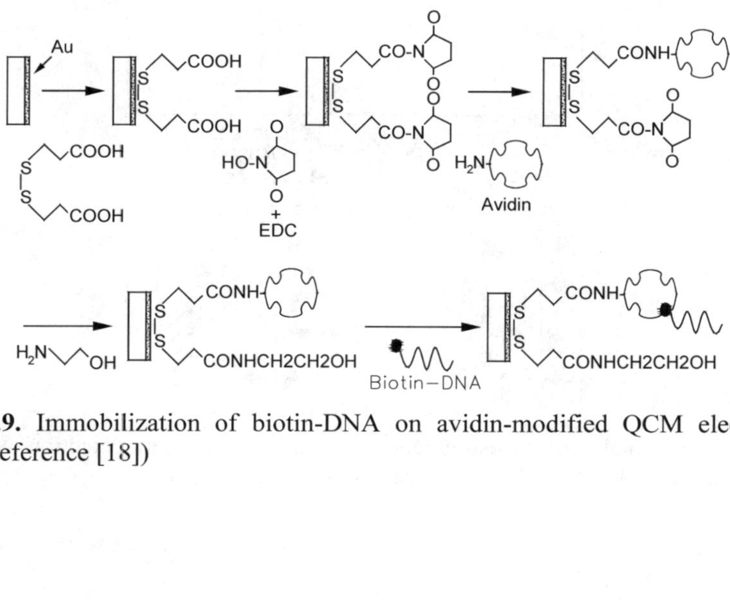

Fig. 7.9. Immobilization of biotin-DNA on avidin-modified QCM electrodes (from reference [18])

The kinetics of hybridization is studied via in situ frequency measurements. Very recent results show interesting potentialities of the QCM to achieve DNA detection [19, 20, 21, 22].

References

1. Swann S (1988) Phys. Techno., 19 67.
2. Gabrielli C, Keddam M and Torresi R (1991) J. Electrochem. Soc., 138 2657.
3. Scharifker BR, Garcia-Pastoriza E and MarinoW (1991) J. Electroanal. Chem., 300 85.
4. Etchenique RA and Calvo EJ (1999) Electrochem. Commun., 1 167.
5. Garcia-Jareno JJ, Gabrielli C and Perrot H (2000) Electrochem. Commun., 2 195.
6. Nuzzo RG, Dubois LH and Allara DL (1990) J. Am. Chem. Soc., 112 558.
7. Ferrante FF, Kippling AL and Thompson M (1994) J. Appl. Phys., 76 3448.
8. Brett CMA, Kresak S, Hianik T and Oliveira Brett AM (2003) Electroanalysis, 15 557.
9. Fung YS and Wong YY (2001) Anal. Chem., 73 5302.
10. Barraud A (1987) British Polymer J., 19 409.
11. Shiratori SS, Kohno K and Yamada M (2000) Sensors and Actuators B, 64 70.
12. Calvo EJ and Wolosiuk A (2002) J. Am. Chem. Soc., 124 8490.
13. Calvo EJ, Forzani ES and Otero M (2002) Anal. Chem., 74 3281.
14. Guilbault GG, Hock B and Schmid R (1992) Biosens. Bioelectron., 7 411.
15. Bizet K, Gabrielli C and Perrot H (2000) Appl. Biochem. Biotechnol., 89 139.
16. Cosnier S, Perrot H and Wessel R (2001) Electroanalysis, 13 971.
17. Okahata Y, Matsunobu Y, Ijiro K, Mukae M, Akira M and Makion M (1992) J. Am. Chem. Soc. 114 8299.
18. Caruso F, Rodda E and Furlong DN (1997) Anal. Chem., 69 2043.
19. Dupont-Filliard A, Roget A, Livache T and Billon M (2001) Anal. Chim. Acta, 449 45.
20. Willner I, Patolsky F, Weizmann Y and Willner B (2002) Talanta, 56 847.
21. Zhou XC, Huang LQ and Yau Li SF (2001) Biosens. Bioelectron., 16 85.
22. Tombelli S, Mascini M and Turner APF (2002) Biosens. Bioelectron., 17 929.
23. Cheng Q and Brajter-Toth A (1992) Anal. Chem., 64 1998.

8 Sonoelectrochemistry

Christopher Brett

Departamento de Química, Universidade de Coimbra.

8.1 Introduction

Sonoelectrochemistry is the study of the effects of the combination of ultrasonic radiation with electrode processes occurring at surfaces of electrodes immersed in a solution in an electrochemical cell [1-3]. The benefits of this coupling have been recognised for many years, since 1934 when insonation was used to increase the rate of water hydrolysis [4].

Ultrasound introduces energy into the system, both in homogeneous solution and on solid surfaces. Resulting phenomena include cavitation, acoustic microstreaming, mixing, emulsification, and enhanced hydrodynamics. There are also effects on suspended solids, particles, nucleation-enhancing species and on dissolved gases. All of these can influence an electrochemical process.

Ultrasound itself has many recognised applications. Examples are:

1. Biological studies: cell disruption and homogenization.
2. Chemistry: increasing rates of chemical reactions.
3. Dentistry: cleaning and drilling of teeth.
4. Medicine – ultrasonic imaging (see Chap. 15).
5. Materials and engineering: non-destructive testing (see Chap. 14), drilling, grinding and cutting, welding of thermoplastics.
6. Pulse-echo techniques: location of mineral and oil deposits, sonar (see Chap. 14).
7. Industrial cleaning and dispersion of pigments in paints.

Different frequency ranges are employed:

1. 20 kHz – 100 kHz: power ultrasound.
2. 1 – 10 MHz: diagnosis and chemical analysis.

In chemical processes, ultrasound can lead to increased transport, interfacial cleaning and thermal effects. In the rest of this chapter we shall be concerned with the influence of power ultrasound on electrode processes.

8.2 Basic Consequences of Ultrasound

The passage of ultrasonic waves through a liquid leads to two main effects:

1. Acoustic streaming – a time-independent flow of liquid.
2. Cavitation – the generation, expansion and collapse of vapour-filled cavities in the liquid.

The way in which this occurs depends on the shape, composition and size of the reactor.

In general, the influence on electrode processes can be summarized as:

1. Enhancing the rate of mass transport to the electrode and the transport-limited current.
2. Reducing adsorption phenomena.
3. Enhancing the rate of many chemical reactions.
4. A cleaning and eroding effect on the electrode surface caused by cavitation.
5. Formation of radical species during cavitation which can influence the electrode reaction mechanism.

Cavitation is a highly energetic process. Sound waves consist of alternating compression and rarefaction cycles. During rarefaction the negative pressure is sufficient to overcome the fluid's intermolecular forces and produce cavitation bubbles. In the power ultrasound region, i.e. 20 - 100 kHz, these bubbles are usually short-lived and collapse, releasing large amounts of energy locally, as shown in Fig. 8.1. This may happen close to a solid surface, causing microjets to hit the surface, leading to surface erosion and cleaning.

8.3 Experimental Arrangements

There are two basic experimental arrangements.

1. Immersion of the electrochemical cell in an ultrasonic bath with various transducers bonded to the walls and base, such that the ultrasonic radiation is relatively widely distributed and of low intensity (1-2 W cm^{-2} at each transducer face).
2. Focussing the ultrasonic field on the electrode surface from a single transducer via an immersed horn, such that radiation intensities up to several hundred W cm^{-2} can be achieved. In these arrangements the horn tip is placed above the electrode, it being possible to vary the distance and usually the power of the horn. An example of a horn

with the tip immersed in a thermostatted electrochemical cell is shown in Fig. 8.2. Thermostatting is necessary to avoid solution heating.

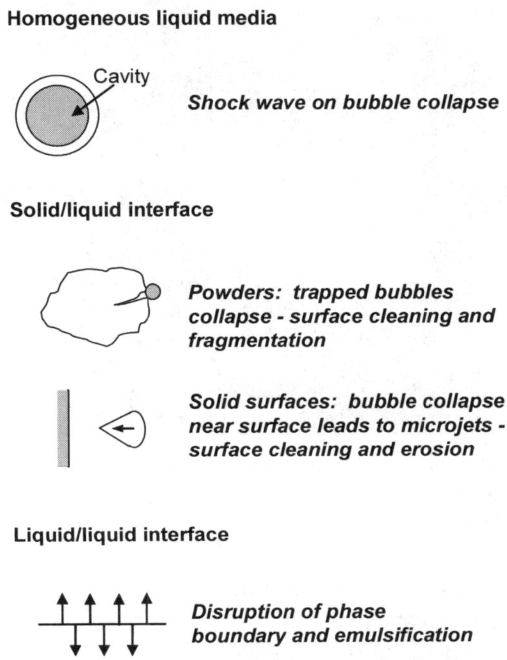

Fig. 8.1. Possible sites of reaction induced by cavitation

Transducers for horns are usually of the prestressed piezoelectric design, between two and four piezo elements (e.g. prepolarised lead titanate zirconate) being bolted between a pair of metal ends. The length is around 70 mm at 20 kHz and maximum peak displacements range between 15 - 20 μm. The function of the horn is to magnify or amplify this motion as a resonant element in the compression mode. It must therefore be exactly half a wavelength long although multiples are possible. Three designs are used in commercial horns – linear taper, exponential taper or stepped. Several materials for horn construction are used to be sufficiently resistant to cavitation and chemical attack, the best being titanium alloy. With a view to scale-up of sonochemical reactors, radially vibrating horns have been recently investigated [5].

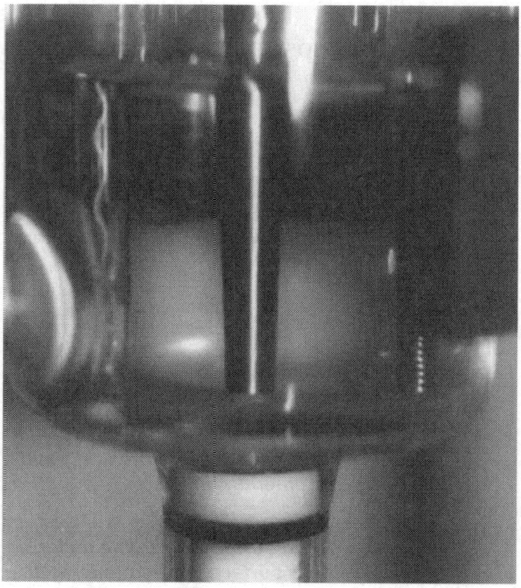

Fig 8.2. A 20kHz ultrasound horn titanium alloy probe tip, immersed in a thermostatted electrochemical cell, directly above an electrode at a distance of ~1 cm. Cavitation bubbles can be seen on the probe tip

The ultrasound field created at the horn tip in solution leads to acoustic streaming which manifests itself as a jet of solution directed towards the electrode surface; thermostatting is necessary to avoid heating of the solution inside the cell. In most sonoelectrochemical experiments cavitation is avoided as much as possible, first, because it destroys the electrode surface and secondly, because it reduces the reproducibility of the experiments carried out since the formation and collapse of bubbles is not sufficiently controllable. Cavitation has been characterised electrochemically, by luminescence and photographically [6].

Important parameters which can be varied are, for a given ultrasonic frequency: ultrasonic power, cell geometry, distance between the horn tip and the electrode and chemical composition of the system.

8.4 Applications

Some examples of the different types of application of sonoelectrochemistry will be briefly indicated. The reader is referred to the literature for a broader account of these, together with more detailed examples [3, 7].

8.4.1 Sonoelectroanalysis

Given the fact that the horn probe increases mass transport and therefore enhances currents, convenient strategies can be developed for the measurement of electroactive species at trace level through preconcentration by deposition [8]. This can be particularly useful for the measurement of species in environmental matrices which usually contain organic components which adsorb and block the electrode surface rapidly. Although many of these problems could be solved with sample digestion, this alters sample speciation and precludes the probing of the sample under as close to natural conditions as possible. One of the first applications was the measurement of trace cadmium and lead ions at Nafion-coated mercury thin film glassy carbon electrode using the pre-concentration technique of anodic stripping voltammetry, the Nafion serving to prevent surface fouling and prevent the mercury thin film from being washed away [9]. Since then a number of different systems have been studied at bare and protected electrodes, mainly for trace metal analysis, see [10] and references therein.

8.4.2 Sonoelectrosynthesis

Some of the earliest applications of ultrasound were in organic and organometallic electrosynthesis [11]. A well-studied example on mechanism change was the Kolbe reaction for oxidation of carboxylate anions to hydrocarbons [12]. The effect of ultrasound is to alter the relative importance of different mechanistic pathways and thus the product distribution, owing to the change in mass transport (in general, there is no evidence for intervention of radical species generated by cavitation), shown recently with oxidation of the pesticide diuron [13]. Interesting benefits can also be obtained by acoustic emulsification, which can increase yields by enhancing the contact between species soluble in two different solvents, one usually aqueous, and the reaction product generally being extracted into the organic medium [14]. This strategy is also more environmentally friendly since it enables a smaller quantity of organic solvent to be employed.

8.4.3 Ultrasound and Bioelectrochemistry

Biological molecules usually adsorb on electrode surfaces, making it more difficult to measure their electrochemical parameters. Insonation permits an increase of the mass transport, enhancing the current and also reduces adsorption. In one approach, application of ultrasound in differential pulse voltammetry until just before the oxidation waves of nucleosides and nucleotides appeared allowed their characterization, e.g. [15]. In a recent paper ultrasound was applied whilst recording an electrochemical impedance spectrum, in this way avoiding adsorption of the molecules and their products, and permitting the calculation of the rate constant of the oxidation of guanine and adenine nucleic acid bases in solution [16].

8.4.4 Corrosion, Electrodeposition and Nanopowders

A number of recent studies have been undertaken concerning the influence of power ultrasound on the erosion and corrosion of different metal surfaces, using bursts or pulses of ultrasound over an extended period, most showing a synergistic effect between the two phenomena [17, 18]. Important aspects investigated concern the influence of ultrasound on the corrosion mechanism, particularly of mass transport, the surface cleaning effect, enhancement of corrosion through cavitation and the ease of re-formation of any protective oxide layer.

Open circuit potential, polarization curves and electrochemical impedance spectroscopy as well as microscopy techniques have all been used to study the processes.

Recently, it was found that ultrasound can have a beneficial effect on the corrosion of steels, in that the reformation of corrosion products after ultrasonic radiation leads to a more robust oxide layer. On other metals such as aluminium, where the rate of corrosion is limited by the rate of oxide formation rather than metal dissolution, the rate of corrosion remains as before, the metal surface having been roughened [19].

Electrodeposition is also influenced by ultrasound, particularly affecting the balance between formation of nucleation centers and growth phases. This has been little exploited so far, but a recent example concerns the deposition of lead dioxide on glassy carbon, which is useful for the use of these electrodes in anodes for batteries, waste water treatment, ozone generation or electrosynthesis [20].

If the horn probe tip is made into the electrode itself – a sonotrode – the deposits formed do not have the opportunity to grow and are detached from the electrode. This is an attractive way to form nanopowders of metals or semiconductor oxides of uniform size (diameter ~ 5 nm) which can be used for a variety of applications such as in chemical synthesis, see [21] and references therein.

8.4.5 Waste Treatment and Digestion

The treatment of waste by degradation of pollutants is a relatively new application of ultrasound and can be combined with electrochemical processes to increase the efficiency, usually by anodic oxidation, e.g. [22]. It is also one of the methods currently used for the digestion of complex samples prior to chemical analysis.

8.4.6 Multi-frequency Insonation

The combination of two or more ultrasonic frequencies (one in the kHz power region and the other in the MHz region) enhances the cavitation yield [23]. The mechanism for this probably involves the production of new bubbles by the low-frequency stimulating field which enhance the higher frequency acoustic cavitation. There can also be a resonance effect which could have useful consequences.

8.5. Final Remarks

Sonoelectrochemistry has been demonstrated in the last two decades to be a powerful tool in enhancing the rate of electrochemical processes, including the degradation of pollutants, in electroanalysis and in diagnosis of corrosion. Its continued use in the future seems certain.

References

1. Mason TJ and Lorimer JP (1988) Sonochemistry: theory, applications and uses of ultrasound in chemistry, Ellis Horwood, Chichester, UK
2. Mason TJ (1990) Sonochemistry: the uses of ultrasound in chemistry, RSC, Cambridge, UK
3. Walton DJ and Phull SS (1996) Advances in Sonochemistry, 4 205
4. Moriguchi N (1934) J. Chem. Soc. Jpn., 5 751
5. O. Dahlem O, Reisse J and Halloin V (1999) Chem. Eng. Sci., 54 2829
6. P.R. Birkin PR, Power JF, Abdelsalam ME and Leighton TG (2003) Ultrasonics Sonochem, 10 203
7. Compton RG, Eklund JC and Marken F (1997) Electroanalysis, 9 509
8. Madigan NA, Murphy TJ, Fortune JM, Hagan CRS and Coury Jr LA(1995) Anal. Chem., 67 2781
9. Matysik FM, Matysik S, Oliveira Brett AM and Brett CMA (1997) Anal. Chem., 69 1651

10. Hardcastle JL and Compton RG (2001) Electroanalysis, 13 89
11. Walton DJ, Iniesta J, Plattes M, Mason TJ, Lorimer JP, Ryley S, Phull SS, Chyla A, Heptinstall J, Thiemann T, Fuji H, Mataka S and Tanaka Y(2003) Ultrasonics Sonochem, 10 209
12. Walton DJ, Chyla A, Lorimer JP, Mason TJ and Smith G (1989) J. Chem Soc. Chem. Commun., 603
13. Macounova K, Klima J, Bernard C and Degrand C (1998) J. Electroanal. Chem., 457 141
14. Wadhawan JD, Marken F and Compton RG (2001) Pure Appl. Chem., 73 1947
15. Oliveira Brett AM and Matysik FM (1997) J. Electroanal. Chem., 429 95
16. Oliveira Brett AM, Silva LA and Brett CMA (2002) Langmuir, 18 2326
17. Kwok CT, Cheng FT and Man HC (2000) Mat. Sci. Eng. A 290 145
18. Neville A and McDougall BAB (2001) Wear 250 726
19. Morais NLPA and Brett CMA (2002) J. Appl. Electrochem., 32 653
20. González-García J, Sáez V, Iniesta J, Montiel V and Aldaz A (2002) Electrochem. Commun., 4 370
21. Durant A, Delplancke JL, Libert V and Reisse J (1999) Eur. J. Org. Chem., 2845.
22. Azprykowicz L, Juzzolino C and Kaul SN(2001) Water Res., 35 2129
23. Feng R, Zhao Y, Zhu C and Mason TJ (2002) Ultrasonics Sonochem, 9 23

9 Chemical Sensors

Ernesto Julio Calvo

Facultad de Ciencias Exactas y Naturales, Universidad de Buenos Aires

9.1 Introduction

We are surrounded by a complex environment and therefore need to monitor aspects of this environment in real time in connection with pollution, health and safety, etc. Like the human senses of olfaction, sight, touch, hearing and taste link us with the external world, chemical sensors allow us to monitor and record the chemical variables of the external environment in the same way we monitor temperature, pressure, etc.

A chemical sensor is a device which responds to a particular analyte molecule in a selective way in the presence of other molecules through a chemical reaction and can be used for the qualitative and quantitative determination of the analyte in the complex sample.

After recognition of a molecular entity, the chemical sensor transduces the signal through a physicochemical process into an electrical signal which can be further amplified and processed.

While Chap. 2 deals with different types of acoustic transducers, this tutorial focuses on the chemical sensing layer that affects the molecular recognition and the different transducers that can be employed. Different types of chemical sensors include different transducing methods to obtain processable electrical signals, so we have a variety of chemical sensors:

1. Electrochemical (potential or current).
2. Optical (optical waves).
3. Mass sensitive (acoustic wave sensing).
4. Heat (calorimetric).
5. Magnetic.

The present trend is to integrate the chemical sensor with microelectronics and sample handling. This has led to impressive achievements in miniaturization using silicon micromachining methods and more recently soft lithography techniques [1, 2]. In addition, the recent explosive growth of microfluidics [3] has allowed liquid samples to be transferred by electrophoretic driving forces into contact with an array of chemical microsensors giving rise to the concept of lab-on-chip. This is a fully integrated chemical laboratory in a microchip with applications as wide as health care and remote medicine, environmental control and food analysis. Applications in the area of explosive early detection, chemical and biological warfare has recently emerged as a fast growing field.

Patients requiring rapid blood analyses in emergency rooms, intensive or critical care units as well as in the operating theatres can undergo rapid blood tests with only a few drops of whole blood with the point-of-care systems. In 1985 Dr. I. Lauks developed the I-Stat point-of-care blood analysis system [4]. It comprises a single-use cartridge containing an array of electrochemical sensors capable of measuring blood oxygen, CO_2 and pH. Also it measures electrolytes such as potasium, sodium, chloride, calcium ions, glucose, urea nitrogen and hematocrit in few minutes with only a few drops of whole blood. The development of I-Stat was made possible by the incorporation of microfabrication techniques from the microelectronic industry in a 5 x 4 mm chip.

There are also examples of commercial systems to monitor glucose in a tiny drop of whole blood of a Diabetes Mellitus patient. Glucose levels in the clinical range 20-600 ng dL^{-1} can now be measured in 20 seconds with a drop from 0.3 to 3.5 μL of whole blood using commercial integrated chemical sensors [5, 6]. Implantable glucose sensors have also been developed.

An alternative novel way of chemical sensors is the electronic nose [7]. This olfaction machine is an electronic device capable of detecting and recognizing complex odors. Unlike molecular recognition, electronic noses work on the principle of multiple signal detection and pattern recognition. The signal can be generated by chemical detection interfaces with changes of conductivity of a solid state or conducting polymer prompted by the interaction of analyte molecules from a complex sample. The signal transduction can be affected by a variety of physicochemical processes, i.e. metal-oxide gas sensors, conducting polymer gas sensors, bulk and surface acoustic wave gas sensors, field effect chemo-transistors, electrochemical gas sensors, etc. They are normally used in a multisensor array to generate a signal pattern.

The recognition is not selective at the molecular identity but rather to a signal pattern recognition and multicomponent analysis. It can be applied to health care or food analysis, i.e., to very complex samples such as coffee, beer, wine, etc., freshness of fish or to smell a cow to determine its health status. The signal generation can be affected by electrochemical, electrical (resistance), acoustic (bulk acoustic and surface waves), etc. pro-

processes, normally with an array of detectors (metal oxide or conducting polymer detectors) each one slightly different from the rest.

The general requirements that a chemical sensor must fulfill are:

1. Sensitivity

The limit of detection of a given substance is related to the minimum concentration capable of producing a signal that can be discriminated from the noise level (i.e. by the signal-to-noise ratio) and the sensitivity is related to the slope of a signal-to-concentration plot at a given concentration. It is desirable that such a sensor response plot be linear.

2. Selectivity

High selectivity is needed to discriminate qualitatively and quantitatively a given component in a complex sample matrix, i.e., detection of glucose in whole blood where other molecules like ascorbate, urate, etc., may interfere. Selectivity can be achieved by molecular recognition, i.e., by molecular shape as key and lock problem or by pattern recognition when a complex pattern of transduced signals is obtained like in electronic noses.

3. Stability

The molecules that comprise the chemical recognition interface may not be stable enough and the chemical sensor response decays with time. If the inventory of chemo-receptor or molecules that recognize the analyte is not in excess with respect to the stoichiometric value – which happens when the sensing response is determined by kinetics – then the stability is a critical factor to guarantee a reproducible response. On the other hand, if the signal is determined by mass transport of the analyte, the decay in sensing molecules should not be determining.

9.2 Electrochemical Sensors

Electrochemical reactions take place at electrode-electrolyte interfaces and provide a switch for electricity to flow between two phases of different conductivity, i.e. the electrode (electrons or holes are the charge carriers) and solid or liquid electrolyte (ions are the main charge carriers). Sensors based on the transduction of chemical information into an electrical signal by an electrochemical reaction are very well suited for chemical sensors [8].

Among electrochemical sensors we can distinguish potentiometric and amperometric sensors depending on the mode of the electrochemical process as follows [9-11].

9.3 Potentiometric Sensors

Potentiometric sensors work at equilibrium or quasi-equilibrium conditions with very little perturbation of the analyte concentration adjacent to the sensor surface. In these sensors the molecular recognition of the analyte results in changes of surface charge at the electrode-electrolyte interface due to changes of concentration which are amplified by the high electrical capacitance of electrochemical interfaces. Usually, one obtains a logarithmic response of the electrode potential to the analyte activity, a_S:

$$E = A \log a_S + B \tag{9.1}$$

where $A = RT/nF$ with R the universal gas constant, T the absolute temperature, n the number of electrons exchanged in the electrochemical recognition reaction and F the Faraday constant (98,485 coul mol^{-1}). The ion activity is related to the ion concentration by the mean ionic activity coefficient, $\gamma_{\pm}$, which itself can be dependent of the concentration c_S, $a_S = \gamma c_S$.

As a simple example let us consider water sparingly soluble salt like AgCl deposited on a silver wire that undergoes a solubility equilibrium:

$$AgCl(s) \rightarrow Ag^{+}(aq) + Cl^{-}(aq) \tag{9.2}$$

with an equilibrium ionic product that relates the ionic activities:

$$K_{sp} = a_{Cl^-} . a_{Ag^+} \tag{9.3}$$

and therefore according to Nernst equation [10-11],

$$E = E^{O}_{Ag/Ag^+} + \frac{RT}{F} \ln a_{Ag^+} = E^{O}_{Ag/AgCl} - \frac{RT}{F} \ln a_{Cl^-} \tag{9.4}$$

with, $E^{O}_{Ag/AgCl} = E^{O}_{Ag/Ag^+} + (RT/F) \ln K_{sp}$ and $RT/F = 59.1$ mV (at 25°C) which results in the slope A of Eq. (9.1) if the ion activity coefficient can be regarded as independent of ion concentration.

From Eq. (9.4) it can be seen that the linear relationship fails at very low ionic concentration when the ion concentration falls in the range of ionic solubility of sensing layer $s = a_{Ag+}$, i.e., when the analyte activity $a_{Cl} \approx K_{sp}/s$.

Other examples of ion specific electrodes are pH electrodes where a difference in proton activity across a special thin glass membrane (between the external analyte and an internal reference) produces a difference in electrode potential measured with respect to a pH independent reference electrode immersed in the same electrolyte solution. The membrane potential is given by [11]:

$$E = \frac{RT}{z_i F} \ln a \frac{a_i^{ext}}{a_i^{\text{int}}} \tag{9.5}$$

where a_i^{ext} and a_i^{int} are the activities of the ion *i* of charge z_i respectively in the external and internal solution to the membrane. The membrane potential is measured with respect to a reference electrode immersed in the internal solution the potential of which is independent of ion *i*.

In 1967, a liquid membrane ion selective electrode for calcium was introduced, and many other sensors followed after this one. In 1979, the liquid membrane component was incorporated as plasticizer of PVC with 1% ionophore and a solid membrane calcium ion sensor was developed. Other examples are sodium, ammonium ion specific electrodes which recognize ions by size, i.e. crown ethers or antibiotic molecules (i.e. validomicyn K complex) which can host ions (guests) of a given size in a cavity when supported in a PVC membrane. Solid state ion selective electrodes like PbS-Ag_2S selective to sulphide ion were also developed.

Alternatively, a pH glass electrode can be covered by a gel containing an enzyme such as urease, which catalyzes the hydrolysis of urea into ammonia which enters into an acid-base equilibrium:

$$NH_3 + H_2O \rightarrow NH_4^+ + HO^- \tag{9.6}$$

Therefore, a biosensor responsive to urea concentration can be built by simply measuring changes in pH with the modified glass electrode.

Integration with microelectronics can be achieved with ion-selective field effect transistors (ISFET). This chemo-electronic device is an adaptation of the IGFET (insulated gate field effect transistor) and can be manufactured in very small size, which allows the fabrication of sensor arrays for multi-species sensing. An ISFET consists of a p-type semiconductor in contact with two regions of n-type semiconductor (the source and the drain). A layer of insulator is deposited on top of this structure and on top of it an ion selective membrane replaces the metal gate material of an IGFET. When this sensing layer is in contact with the sample solution, the ion recognition by the selective membrane results in a specific potential at the membrane surface, with respect to the reference electrode immersed in the solution. The current that flows between source and drain depends on the gate potential generated by the ion selective membrane and, therefore, senses the ion concentration integrated in the hybrid circuit. The insulating layer itself (SiO_2, Si_3N_4) responds to changes of pH. When an enzyme like urease is used in the chemical sensing layer, the enzyme field effect transistor (ENFET) results.

9.4 Amperometric Sensors

In amperometric sensors, dynamic conditions prevail with a net electrochemical reaction (i.e. oxidation or reduction of sample species) taking place at the electrode-electrolyte interface:

$$O + ne \rightarrow R \tag{9.7}$$

where oxidized species O are reduced by electrons at the electrode surface into the reduced species R and a net current flows through the external circuit. The number of moles of O, n, converted into R is related to the electrical charge, q, which flows at the interface by the Faraday law,

$$q = nF \tag{9.8}$$

where F = 96,485 coulombs/mol is the Faraday constant. The electrical current per unit area that flows at the electrode-electrolyte interface due to reaction in Eq. (9.7) is given by:

$$i = \frac{I}{A} = \frac{1}{A}.\frac{dq}{dt} = \frac{F}{A}\frac{dn}{dt} \tag{9.9}$$

and is proportional to the rate of the electrochemical interfacial reaction.

Since removal of O molecules from the electrode surface prompts mass transport of O from the bulk electrolyte, the mass transport limited current is given by:

$$i_L = \frac{nFAD_i c_i}{\delta} \tag{9.10}$$

where A is the electrode area, D_i the diffusion coefficient of the reactant species of concentration c_i and δ is a mass transport layer where the concentration gradient is constant.

As an example, we will describe the Clark electrode or oxygen probe [12] to measure dissolved oxygen that is based on the electrochemical oxygen reduction reaction:

$$O_2 + 4H^+ + 4e \rightarrow 2H_2O \tag{9.11}$$

or

$$O_2 + 2H^+ + 2e \rightarrow H_2O_2 \qquad (9.12)$$

with $n = 4$ and $n = 2$ respectively at appropriate electrode potentials. The electrode is separated from the sample by a Teflon membrane which is permeable to oxygen but not to water or other molecules and encloses a thin electrolyte layer where an oxygen concentration diffusion gradient determines the oxygen flux and therefore the diffusion limiting current given by Eq. (9.10) which results proportional to the dissolved oxygen concentration, and δ in that case is simply the thickness of the membrane and liquid electrolyte film.

An extension of Clark's oxygen probe is the enzyme electrode introduced later by Clark himself [13]. In this device, an enzyme (commonly an FAD oxidoreductase like glucose oxidase that catalyzes the aerobic oxidation of glucose to gluconic acid) immobilized at the inner side of the hydrophilic diffusion membrane converts glucose (or other substrate) and oxygen both diffusing from the external analyte solution through the semi-permeable membrane to produce hydrogen peroxide which can be detected electro-chemically on a metal electrode.

$$H_2O_2 \rightarrow O_2 + 2H^+ + 2e \qquad (9.13)$$

This type of sensors, however, is limited in anaerobic samples by the supply of oxygen and this has been solved by using artificial redox mediators instead of the natural cofactor oxygen to shuttle the electrons into the redox enzyme and being regenerated at the electrode surface. The redox shuttle or redox mediator may be a soluble molecule added to the electrolyte or integrated into a surface layer with the enzyme. Cass et. al. reported a glucose biosensor using ferrocene to mediate the electron transfer between the redox site of glucose oxidase (GOx) and an electrode surface [14]. Since then, many efforts have been put on optimization of the mediated enzyme biosensors.

Several strategies have been followed to integrate the redox mediator to the enzyme in amperometric biosensors as Schuhmann reviewed [15]. Intelligent biochips of the future will make use of the present technology of enzyme, immune antigen-antibody recognition, genomic (DNA hybridization) sensing integrated with microfluidics and microelectronics.

There are already several successful commercial devices to measure glucose with a small drop sample of whole blood; these glucometers work on the principle of mediated enzyme electrodes with glucose oxidase [5, 6]. Biosensors find important applications in biomedical analysis and remote medicine (i.e. rapid glucose measurement in whole blood, detection of cholesterol, ethanol, etc.), food analysis (i.e. freshness of fish), and environmental control (detection of pollutants).

Molecular switches or molecular transistors in amperometric mode were introduced by Mark Wrighton in 1985 [16, 17] and then the concept has been expanded into enzyme electrodes by Bartlett and Birkin in 1994 with the introduction of a microelectrochemical enzyme transistor responsive to glucose [18]. In these devices a conducting polymer like poly(aniline) thin layer bridges the gap between two electrodes (drain and source of the molecular transistor). A sensing layer on top of the conducting polymer film carries out the recognition reaction injecting charge into the conducting polymer and thus changing it from the conductor state to the insulator state or from the insulator state into the conductor. The amount of charge that circulates is a measure of the number of analyte molecules transformed and therefore, since $q = i.t$, the time to reach the midpoint in the signal time course measures the concentration of analyte. Bartlett et al. have described a system comprising a poly(aniline) thin film containing poly(vinylsulfonate) or poly(styren-sulfonate) as the only anion present in the polymerization solution [19].

9.5 Optical Sensors

Optical sensors rely on the optical transducer of the signal and comprise ultraviolet, visible and infrared spectrophotometry in transmission or reflectance modes. The relationship between the incident light intensity and the transmitted radiation is given by the Lanbert-Beer law:

$$A_\lambda = \varepsilon_\lambda . l . c \tag{9.14}$$

where $A_\lambda = \log_{10} I_o/I$ is the absorbance at a given wavelength λ (being I_o and I the incident and transmitted light intensities); ε is the molar absorptivity; l is the optical path through the absorbing sample and c is the molar concentration of absorbing analyte. The transmitted light is then detected by a photodiode or phototube which transduces light into an electrical signal.

More recently "optrodes" were introduced (the term optrode or optode was coined in analogy with "electrode" in electrochemical sensors) with fiber optics covered at one end with reagent immobilized in a gel or membrane which produces a change in color in the presence of the analyte. The light is then transmitted to the solution and partly reflected and detected in a bifurcated fiberoptics. The advantages of "chemosensor optodes" are that they do not require an extra reference like the reference electrode in potentiometric sensors and that are insensitive to electrical noise, but double transduction chemical-optical-electrical is needed at the end signal.

In addition to absorption-transmission optical sensors, luminescent sensors detect light emission that results from the absorption of a photon by a molecule and then the excited state decays with either fluorescent or phosphorecent emission. These techniques are highly sensitive and the emitted light of higher wavelength against zero background enhances the sensitivity and signal-to-noise ratio allows extraction of chemical information from complex samples.

Optical sensors can be used to monitor pH, metal ions, oxygen, glucose and other enzyme substrates, etc., by appropriate choice of the chemical sensing layer. They are limited, however, to transparent and colorless samples.

Biosensors based on evanescent waves have had a high impact in recent years, for instance surface plasmon resonance (SPR) biosensors. Surface plasmon resonance arises from the interaction of light with suitable metal or semiconductor surface which generates a quantum optical-electrical phenomenon. Under certain conditions, the photon energy can be transferred to the surface of the metal as packets of electrons called plasmons. This energy transfer occurs at a particular wavelength of light, when the quantum energy carried by the photon exactly matches the quantum energy level of the plasmons.

The incident light is almost completely absorbed at the wavelength that excites plasmons, that is the resonance wavelength. Using monochromatic light at the resonance wavelength and changing the incidence angle, the maximum absorption occurs at a particular angle of incidence. The interaction between the plasmon's electric field and analyte molecules within this field determines the resonant condition (wavelength at constant angle or conversely plasmon resonant angle at constant wavelength). Therefore, any change in composition adjacent to the surface plasmon alters the plasmon resonance angle and the SPR angle shift is directly and linearly proportional to the change in surface composition. This has been used to detect biomolecules such as antibody-antigen or single stranded DNA hybridization with success. Optical waveguides and fiber optics work in a similar way differing only in geometry.

9.6 Acoustic Chemical Sensors

Bulk acoustic wave sensors (BAW) and surface acoustic waves sensors (SAW), described in Chap. 2, can be used as chemical sensors if the acoustic wave travels through a chemical sensing layer that affects the sound propagation, when the molecules of the analyte react with this sensing layer [20].

Acoustic wave devices with an applied sorbent layer as vapor sensor represent one of the first applications of acoustic wave chemical sensors. King used quartz crystal microbalance (QCM) in 1964 [21]. In 1979 acoustic sensors were extended to surface acoustic wave devices [22].

The sorbent layer, often a polymer layer, acts to collect and concentrate vapor molecules from the gas phase onto the sensor surface [23]. The adsorbed molecules are then detected by their effect on the velocity of the acoustic wave (i.e. added mass and viscoelastic effects on the surface layer). The partition coefficient quantifies the distribution of vapor molecules between the gas phase and the sorbent phase,

$$K = \frac{c_S}{c_V} \tag{9.15}$$

where c_S is the analyte concentration in the surface sensing layer and c_V represents the concentration in the vapor. The sensor responds to the analyte molecules in the sorbent phase and not in the gas phase; therefore the acoustic device detects c_S. It is necessary to calibrate the sensor against the gas phase with known c_V and thus the partition coefficient is the calibration parameter.

A number of analytes have been sensed with acoustic wave devices such as water, sulphur dioxide, ammonia, aliphatic and aromatic hydrocarbons, hydrogen sulphide, mercury, carbon monoxide, etc. An interesting example is the self-assembled monolayers coupled to SAW devices used effectively as chemical sensors for the detection of vapor phase of pesticides and chemical warefare such as organophosphonates, which proves the high sensitivity of acoustic detection [24].

Piezoelectric sensors have also been employed in contact with liquid samples in flow injection analysis, immunosensors and DNA hybridization. However, in this case the attenuation of the acoustic wave by the viscous liquid also enters into the generation of the sensor response.

9.7 Calorimetric Sensors

Calorimetric sensors are based on the measurement of the heat produced by the molecular recognition reaction. The amount of heat produced is correlated to the reactant concentration. Thermistor based sensors are devices that detect changes in electrical resistance as a result of temperature changes with high accuracy.

Calorimetric sensors have been described for enzyme reactions detecting glucoses, urea, gases, etc. The thermal conductivity gas sensor has long been employed as a detector in gas chromatography (GC) and works on the basis of a heated W-Re wire filament to measure relatively high concentrations in gases. Pellistors are based on the same principle but the platinum coil used as heater and temperature sensor (resistance thermometer) contains a catalyst to enhance a combustion process.

References

1. T. Vo-Dinh, B. Cullum, D.L. Stokes, Sensors and actuators B, 3654, 2001, 1-10.
2. Y. Xia, G. M. Whitesides, Angew. Chem. Int. Ed., 1998, 37. 550-575.
3. Micro Total Análisis Systems, A.van den Berg, P. Bergvedl (Eds.), Kluwer, Boston, 1995.
4. Commercial biosensors. Applications to clinical, bioprocesses and environmental samples", G. Ramsay (Ed.), J. Wiley & Sons, N.Y. 1998.
5. www.medisense.com
6. www.therasense.com
7. Electronic Noses. Principles and applications, J.W. Gardner, P.N. Bartlett, Oxford University Press, Oxford 1999.
8. E.J. Calvo in Encyclopedia of Electrochemistry. Vol. 2 Interfacial Kinetics and Mass Transport, Wiley-VCH, Weinheim, 2003, 3.
9. Chemical sensors, R.W. Cattral, Oxford University Press, Oxford, 1997.
10. Electrochemical Methods. Fundamentals and Applications, A.J. Bard, L.R. Faulkner, J. Wiley & Sons Inc., 2nd. Ed. 2001.
11. Instrumentation and Electroanalytical Chemistry, P. Unwin in Encyclopedia of Electro-chemistry. Vol. 3, Wiley-VCH, Weinheim, 2003.
12. L.C. Clark Jr., Trans. Am. Sopc. Artif. Intern. Organs, 2, 1956, 41-48.
13. L.C. Clark Jr., C. Lyons, Ann. N.Y. Acad. Sci., 105, 1962, 20-45.
14. Cass, A.E.G., Davis, G., Francis, G.D., Hill, H.A.O., Aston, W.J., Higgina, I.J., Plotkin, E.V., Scott, L.D.L., Turner, A.P.F. Anal. Chem. 56, 1984,671.
15. W. Schuhmann, Rev. Mol. Biotechnology, 82, 2002, 425-441.
16. J.W. Thackeray, H.S. White, M.S. Wrighton, J. Phys. Chem. 1985, 89, 5133-5140.
17. G.P. Kittlesen, H.S. White, M.S. Wrighton, J. Am. Chem. Soc., 1985.
18. P.N. Bartlett, P.R. Birkin, Anal. Chem., 66, 1994, 1552.
19. P.N. Bartlett, J.H. Wang, E.N.K. Wallace, Chem. Comm.1996, 359.
20. D.S. Ballantine, R.M. White, S.J. Martin, Acoustic Wave Sensors, Academic Press, San Diego, USA, 1997.
21. W.H. King, Anal. Chem, 1964, 36, 1735.

22. H. Wohltjen, R.E. Dessy, Anal. Chem., 1979, 51, 1465.
23. J.W. Grate, M.H. Abraham, Sens. Actuators B, 1991, 3, 85.
24. L.J. Kepley, R.M. Crooks, A.J. Ricco, Anal. Chem., 1992, 64, 3191.

10 Biosensors. Natural Systems and Machines

Stanislav Stipek[1] and Ernesto Calvo[2]

[1]Institute of Medical Biochemistry, Charles University in Prague
[2]Facultad de Ciencias Exactas y Naturales, Universidad de Buenos Aires

10.1 Introduction

All living organisms are connected to the environment by some kind of chemical recognition and signal generation system. For instance, we make use of our senses: olfaction, sight, touch, taste and hearing to recognize either molecules (olfaction and taste) or physical stimuli (photons, sound waves, pressure, etc.). Likewise, each individual cell in unicellular or multicellular organisms senses stimuli from its surrounding environment (extracellular fluids, other cells and so on) through receptors on the cell surface.

Humans have been able to copy these functions in machines, which can be connected to electronic circuits and computers to report on the quality of the outside world. In this chapter, the mechanisms that Nature has developed in living systems will be first addressed and afterwards the "biosensing" machines that mimic those functions will be addressed in order to analyze a variety of different types of sample (environment, clinic, food, industrial, etc.).

10.2 General Principle of Cell Signaling

The living cell releases some kinds of extracellular *signaling molecules* or it presents them on its surface. Each other cell is able to response to a particular set of these molecular signals in a cell-specific way [15].

The term extracellular signaling molecules include proteins, peptides, steroids and eicosanoids, catecholamines, thyroxin and nitric oxide. In physiological classification they are hormones, growth factors and neurotransmitters. These signals can act over either short (neurotransmitters in the synaptic cleft) or long distance (hormones in the blood).

There are two basic properties of any signaling molecule:

1. It is bound as a *ligand* with a high affinity by specific protein (*receptor*) in the target cell.
2. The lifetime of the signaling molecule in the functional compartment is very short. It is rapidly destroyed or removed by the neighboring cells. In this way the concentration of the signals can be adjusted very quickly. The essence of control is *reversibility*.

The signal molecule binding to a receptor results in cellular response. The *lipid-soluble hormones* cross the cell membranes and bind to appropriate receptor inside the cell (in cytosol or in nucleus). Specific binding site appears on the complex of signal molecule with the receptor that recognizes a specific site of the DNA in a cell nucleus and so the hormone influences the activity of the genes modulating the expression of genetic information stored in the DNA. But *hydrophilic signal molecules*, which cannot cross the membrane lipid bilayer of the target cell, perform most of the cell-cell communications. These ligands do not enter the target cell but as the *first messenger* they interact with *membrane receptors*, which are embedded in the cell membrane having extracellular, transmembrane and intracellular domains, acting as signal transducers (Fig. 10.1). When the extracellular domain binds the signal the cytoplasmic domain undergoes a conformational change, which induces a cascade of intracellular signaling events in the intracellular *signaling pathway*. The last acceptor of such signaling pathway is a target molecule controlling appropriate cell response, such as change in metabolism, cell division, programmed suicidal death (apoptosis) etc.

According to the *mechanism of signal transduction* there are three classes of membrane receptors. Some receptors as *ligand-gated ion channels* promptly change the ion permeability of a membrane. Other class of receptors is functionally *linked with special GTP-binding proteins (G-proteins)*, which can interact only with an activated receptor conveying the signal by changing their own conformation. In this way, the GTP-binding proteins induce the production of some intracellular signaling molecules or they change the membrane permeability. Membrane receptors of the third class are the *enzyme-linked receptors*. When activated they can modify other molecules mostly by phosphorylation of some amino acids.

The signaling pathways consist of a set of various *intracellular signaling molecules*. Some of them are *small second messengers*, which are rapidly formed (cyclic AMP, diacylglycerol) or released from cellular stores (Ca^{2+}) in response to the receptor activation and so they multiply and convey the signal to other compartments of the cell. It is possible to cancel the signal by the rapid destruction of the messenger molecules or by removing these molecules from the cell compartment. The other intracellular signals are *intracellular signaling proteins*. Some of them are enzymes. When activated, they produce a large number of messenger molecules (*amplification of the signal*). Others convert the signal in another form or just convey the signal to the next component of the signaling pathway. In most cases, the activation mechanism is *conformation change* of the signaling protein. It can be introduced by *phosphorylation* catalyzed by a *protein kinase* (signaling enzyme transferring the phosphate from ATP to serine, threonine or tyrosine residue of the target protein). Such signal can be quickly canceled by removing the phosphate by another enzyme - *phosphatase*.

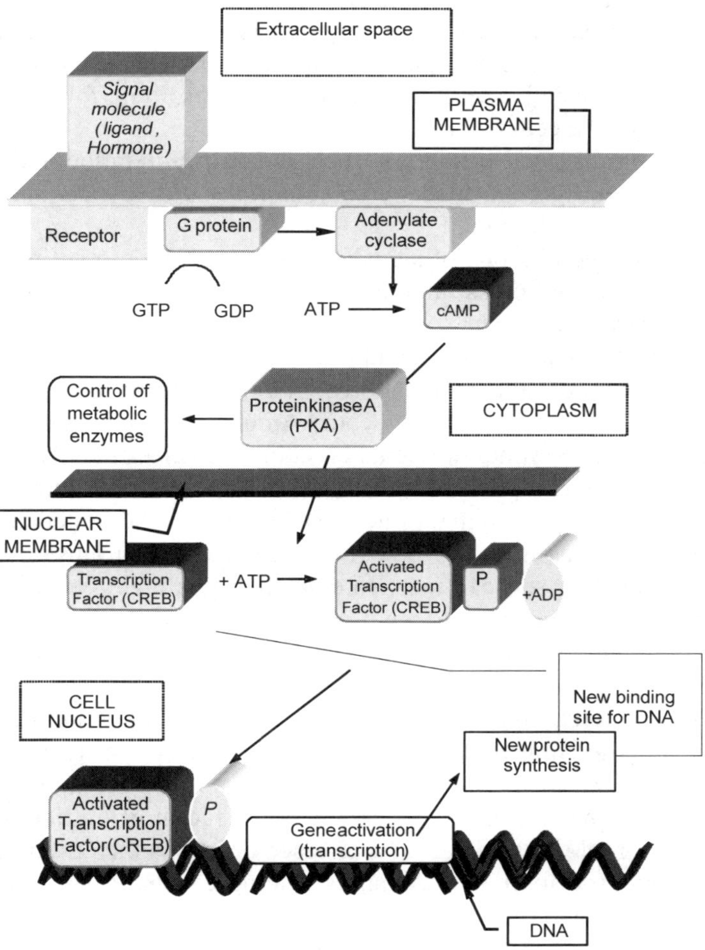

Fig. 10.1. Beta-adrenergic G-protein signaling pathway

One can summarize that molecules in a signaling pathway accept the signal mostly as a change in conformation resulting in modification of its affinity to substrate or to the next molecule in the cascade. In this way, the signal molecule acquires a new ability to activate (or inactivate) the next molecule in the cascade.

Some of the signaling proteins behave like *molecular switches*. When they receive a signal they switch from an inactive to an active state, until another signal or process switches them off.

The *last target molecule modified by a signaling pathway is the factor controlling the cell response to the extracellular signal.* When the final acceptor of information is an enzyme catalyzing the special reaction of a metabolic process, the result of the signaling is, for example, the release of a nutrient from the cellular storage. In the case that the signaling pathway ends at a transcription factor, then the information changes the activity of a gene, and consequently synthesizes a protein controlling, for example, the cell division or the cell differentiation.

But the complex cell behaviors, such as cell proliferation and cell survival, are generally controlled by *combinations of extra cellular signals* rather than by a single signal.

Nowadays several kinds of signaling pathways are known in the living cell. Interactions and cross talks of these pathways result in efficient *molecular information network* controlling the cell behavior and its function according to the surrounding conditions and requirements. The network enables to change cellular functions, morphology, movement, division (proliferation) and programmed death. Among the hundreds of various signaling proteins are the individual signaling proteins (mainly protein kinases) that serve as input-output devices, or "microchips", integrating the signaling process. Such signal protein has to be phosphorylated at multiple special sites to be activated. One set of phosphorylated groups serves to activate it while other sets can inactivate it. The signaling protein can collect the information from several signal pathways and only in certain specific signal combinations the protein conveys the information downstream. It works as a *protein microchip*.

Transfer activity, signaling specificity, binding affinity and other properties *of the signaling molecules* in the cell information network *can be modulated by microenvironment of the cellular compartments*. The metabolism of animal cell is characterized as the recast of nutrient chemical structure to simpler molecules. The released energy is then used for necessary synthetic, transport, motoric and signaling events. Protons and reducing forms of oxygen (water but also reactive oxygen species like peroxide, superoxide and free radicals) are released in addition to energy. *The proteins have sensors* for the concentration of protons (acid-base status) in the environment and sensors for the redox status in the cell. Redox status is defined by the capacity of biological antioxidant system, the accessibility of reducing equivalents and by the intensity of oxidative load to which a cell is exposed.

Changes in proton dissociation of carboxyl-, amino-, amido-, and other groups influence the space conformation of signaling proteins resulting from the modulation of their functions. These reactive groups can be considered as *acid-base sensors*.

The protein activity is modulated also by oxidation and reduction of other sensitive groups, like –SH groups and iron-sulfur complexes. These groups are *redox sensors* on the proteins, as they capture the information about the redox status in the cell compartment. Sulfhydryl groups oxidized to disulfide bridges provide another conformation of protein than reduced -SH groups. Similarly Fe^{3+} in Fe-S complexes influence the protein conformation in other way than complexes with Fe^{2+}. In response to oxidative stress (imbalance in oxidant generation, antioxidant defense and repair of oxidative damage) some bacteria use the protein redox sensors as transcription factors activating several defensive genes to synthesize antioxidant enzymes like hydroperoxidase, glutathione reductase, glutaredoxin.Binding of hormone epinephrine or glucagon to the receptor converts the G-protein to the active form and one of its subunits activates the formation of cyclic adenosine monophosphate (cAMP) by adenylate cyclase. The cAMP produced activates protein kinase A which phosphorylates enzymes metabolizing glycogen (in liver cell) or it is transported into the nucleus (in endocrine cells), where it phosphorylates (activates) cAMP-response element binding protein (CREB), which gets the right conformation to be a transcription factor for specific genes.

10.3 Biosensors

Biosensors can be defined as analytical devices incorporating a biological material (e.g. enzymes, antibodies, nucleic acids, tissues, microorganisms, organelles, cell receptors, etc.), a biologically derived material or a biomimetic one intimately associated with or integrated within a physicochemical transducer or transducing microsystem [1].

Professor L.C. Clark Jr. has been the father of the *biosensor* concept since, in 1956, Clark published a paper on the oxygen electrode [2] and then based on this O_2 sensor he made a landmark address in 1962 at a New York Academy of Sciences symposium in which he described how "to make electrochemical sensors (pH, polarographic, potentio-metric or conductometric) more intelligent" by adding "enzyme transducers as membrane enclosed sandwiches". The use of some 50 oxido-reductases and 250 dehydrogenases would allow expanding the range of analytes that could be measured in the body, or body fluids.

The concept was illustrated by an experimental device in which glucose oxidase (GOx) was entrapped at a Clark O_2 electrode using dialysis membrane; thus a decrease in the measured O_2 concentration was proportional to glucose concentration.

Clark and Lyons [3], coined the term *enzyme electrode*, which in many reviews has been attributed to Updike and Hicks [4], who expanded on the experimental detail necessary to build functional enzyme electrodes for glucose. Guilbault and Montalvo [5] were the first to describe in detail a potentiometric enzyme electrode based on a urea sensor in which urease was immobilised at an ammonium-selective liquid membrane electrode. A commercial product based on Clark's ideas appeared on the market in 1975 commercialized by Yellow Springs Instrument Company as a glucose analyser based on the amperometric detection of hydrogen peroxide, the product of the aerobic enzymatic reaction of glucose and GOx. Many other enzyme biosensor-based laboratory analysers followed this initiative.

Biosensors have been applied to a wide variety of analytical problems such as medicine, the environment, food, the process industries, security and defense. There is at present a growing interest in antiterrorism to be able to rapidly detect dangerous material "in situ" with high sensitivity. The emerging field of bioelectronics seeks to exploit biology in conjunction with electronics in a wider context, encompassing, for example, biomaterials for information processing, information storage and actuators.

An interesting source of commercial biosensors is the book edited by G. Ramsay "Commercial Biosensors" where the reader can find information on the fundamentals and applications of optical, acoustic, and electrochemical biosensors applied to health care, bio-processing and environmental samples. Also, the journals Biosensors (Elsevier) and Biosensors and Bioelectronics (Elsevier) are a frequent source of updated scientific information.

In general we need to achieve highly *selective* molecular recognition in complex analytical matrices like natural water, blood, urine, etc. with a multitude of molecules that may compete for the sensing reaction. Thus *selectivity* is achieved by *molecular recognition* with biological material in a lock-and-key fashion.

Following the molecular recognition of the target analyte, we need to generate a signal with a *transducer* that may be optical, electrochemical, piezoelectric (acoustic), thermometric, or magnetic. Chemical reactions at electrodes are electrochemical *transducers* that turn biochemical information into an electrical signal.

The *sensitivity* of the biosensor depends on the transducer and the *amplification* stage of the signal, while the *limit of detection* depends on the signal-to-noise ratio.

Depending on the nature of the molecular recognition event, biosensors can be based on:

1. Antigen/antibody (immunological)
2. Enzyme/substrate (enzymatic)
3. Receptor/hormone
4. S-DNA hybridization (genomic)
5. Interaction at the sensing layer.

Biosensors usually yield a digital electronic signal which is proportional to the concentration of a specific analyte or group of analytes in a complex matrix sample. The electrical signal that results from the molecular recognition and *transduction* may, in principle, be *continuous*, devices can be configured to yield *single measurements*, like in glucometers to assess the glucose level in diabetes mellitus patients.

In 1975, Divis [8] suggested to employ bacteria as the biological element in *microbial electrodes* for the measurement of alcohol. A major research effort in Japan and elsewhere sprung from Divis paper with applications in biotechnology and environment.

Lubbers and Opitz [9] coined the term *optode* in 1975 to describe an optical fiber-sensor with immobilised indicator to measure carbon dioxide or oxygen. They extended the concept to develop an optical biosensor for alcohol by immobilising alcohol oxidase at the end of an optical fiber oxygen sensor [10]. Commercial optodes are now showing excellent performance for *in vivo* measurement of pH, pCO_2 and pO_2, but enzyme optodes have not succeed yet. Clemens et al. [11] incorporated an electrochemical glucose biosensor in a bedside artificial pancreas in 1976 and this was later marketed by Miles as the Biostator. A new semi-continuous catheter-based blood glucose analyser has been introduced by VIA Medical. In 1982 a major breakthrough *in vivo* application of glucose biosensors was the first needle-type enzyme electrode for subcutaneous implantation reported by Shichiri et al. [12].

Liedberg et al. [13] described the use of optical evanescent waves in surface plasmon resonance to monitor affinity reactions in real time following attempts to fixing antibodies to a piezoelectric or potentiometric transducer. A direct *immunosensor* was marketed in 1990 by BIACORE™ (Pharmacia, Sweden) with a commercial device which was announced to be able to find out how much, how fast, and how strong the immuno-affinity reaction could detect in real time.

La Roche introduced in 1976 the *lactate analyzer,* LA 640, with hexa-cyanoferrate as soluble mediator to shuttle electrons from lactate dehydrogenase to an electrode. This was a landmark in the development of *amperometric* enzyme electrodes. But, the greatest impact was from a much cited paper in 1984 by Tony Cass et. al. who proposed the use of ferrocene derivatives as a soluble integrated mediator for oxido-reductases [14]. This became the basis for the first disposable screen-printed enzyme electrodes launched by MediSense in 1987 with a pen-sized meter for home blood-glucose monitoring. With redesigned electronics into credit-card and computer-mouse style formats MediSense boosted sales reaching US$175 million by 1996 when Abbott bought the company. Boehringer Mannheim and Bayer have competing mediated biosensors and the combined sales of the three companies dominate 85% of the world market for biosensors and are rapidly displacing conventional reflectance photometry technology for home diagnostics. A recent "one-touch" device from Terasense employs 20 microliter blood sample from a mosquito-like bite on the forehand and utilizes an integrated osmium redox polymer mediator that "wires" the enzyme molecules to the electrode.

The next generation of biosensors comprises integrated systems with electronics and fluidic separation integrated in a single device with improved sensitivity, stability and selectivity towards a multitude of analytes. The success of single analyte sensors has been followed by the formulation of arrays of sensors useful, for instance, in critical care where commercially-available hand-held instruments provide clinicians with information on the concentration of several analytes such as glucose, lactate, urea, creatinine and so on in blood samples, and bench-top instruments on the ward that can measure even more analytes.

New techniques for the fabrication of integrated devices with micro fluidics arrays of thousands to a million sensors per cm^2, such as photolithography, microcontact printing and layer-by-layer self-assembled techniques have emerged. Local probe techniques such as SECM and dip-pen lithography are promising in their ability to "write" proteins to surfaces with very high resolution and high density in nanoliter vials and individually addressable electrodes.

The actual trend indicates that analytical chemistry and sensor technology will be increasingly associated with microelectronics. Miniaturization, massive use and simplicity of biosensors will cause, most probably, a proliferation of biosensors integrated with micro-electronics and telecommunications. An interesting example is the remote medicine, where a patient can be bio-chemically monitored remotely.

10.3.1 Molecular Transistor

Future intelligent biochips are supposed to acquire and process biochemical data and transmit them from remote locations to monitor our personal health, the food we eat and our environment. The future of biomolecular electronics is already here with bio-molecular transistors, molecular switches, microfluidic devices, integrating the molecular recognition, transduction, amplification and data processing in miniaturized devices.

10.3.2 Analogy and Difference of Biological System and Piezoelectric Device

Piezoelectric biosensors and transductors accept the information from the surrounding environment, but they convert it to a change of quartz plate oscillation. The following part of the signaling pathways electronically integrating and displaying the piezoelectric information certainly differs from the signaling molecular network of the living cell. But *the initial steps performed by the piezoelectric plate surface can (or should?) use similar principles as the natural cell signaling system.*

Self-assembled monolayer (SAM) of organic molecules containing free anchor groups such as thiols, disulphides, amines, silanes, or acids ordered in high degree on the plate surfaces *mimics the cellular microenvironment of lipid bilayer structures* providing novel substrates for immobilized biomolecules (antibodies, enzymes, nucleic acids) or biological systems (receptors, whole cells) [16, 17]. The term "self-assembly" involves the arrangement of atoms and molecules into an ordered or even aggregate of functional entities without the intervention of mankind towards an energetically stable form. In nature, many supramolecular structures (like cellular organelles, membranes, viral particles) are formed by self-assembly process. Even the cell is able to produce appropriate sets of compounds step-by-step according to a program controlled by cell signaling network.

The piezoelectric biosensors use the same molecular binding mechanism known in molecular biology. The *DNA sensors* estimate the same base pairing in nucleic acids (hydrogen bonds), which is used in the cell to store and/or express genetic information. In artificial *immunosensors, enzyme electrodes and specific devices measuring any ligand* in liquid or gas environment mimic the conditions in the tissue. Very specific and very strong interactions between substrate and active center of enzyme, antigen and antibody, ligand and receptor, protein and nucleic acid and interactions between signal proteins are based on the multiple simultaneous weak binding of the very well fitting surfaces of macromolecules with appropriate space conformation and arranged reacting groups. *Any change from the optimal conformation of the biomolecule cancels part of the weak interaction and weakens the specific intermolecular binding.* This dependence of protein affinity on the conformation of the macromolecules is the basic principle of control, signaling and specificity in supramolecular biological events.

The "interpretation" of the molecular interaction can differ in signal transduction in piezoelectric device and in the cell signaling. Living cells recognize the specific interaction on the base of conformation changes in the activated molecule or complex. In the detection layer of piezoelelectric plate the interaction can be interpreted as an increase of mass and consequent change in the plate oscillation frequency, or as a change in the physical properties of the detection layer (impedance, density, friction, conductivity?).

In any case, the correlation of the biological processes and technical proposals of the biosensors and transducers could be a good tool for the development of efficient devices useful in medicine and biotechnology.

Table 10.1. Comparison of steps of signal pathways in cell and biosensor.

Step	Biological cell	Biosensor
Signal (ligand, 1st messenger, oxidoreductase)	Biomolecule (hormone, growth factor, antigen, light, NO, metabolite)	Very similar Molecular recognition layer
Layer	Membrane (lipid bilayer)	Molecular recognition layer
Receptor	Transmembrane or intracellular protein	Protein, antibody or hapten, nucleic acid, micro-organisms, tissue, etc.
Signal transduction through the layer	Conformation change (+ phosphorylation) resulting in change of binding ability	Electrochemical, optical, acoustic, magnetic, thermal, etc.
Signal detection and effect	Change in metabolic enzyme results in altered metabolism Gene regulation protein binding or release results in altered gene expression Cytoskeletal proteins result in altered cell shape or movement	Display, recording and storage, feedback control based on species concentrations.

References

1. Turner, APF, Karube, I and Wilson, GS (1987) *Biosensors: Fundamentals and Applications*. Oxford University Press, Oxford. 770p
2. Clark, LC Jnr (1956) *Trans. Am. Soc. Artif. Intern. Organs* 2, 41-48
3. Clark, LC Jnr. (1962) Ann. NY Acad. Sci. 102, 29-45
4. Updike, SJ and Hicks, JP Nature 214, 986-988
5. Guilbault, GG and Montalvo,(1969) J JACS 91, 2164-2569
6. (Commercial biosensors. Applications to clinical, bioprocess and environment samples, Ed. G. Ramsay, vol 148, Series of Monographs on Analytical Chemistry, J. Wiley and Sons, Inc., N. York 1998)
7. Cooney, CL, Weaver, JC, Tannebaum, SR, Faller, SR, Shields, DV and Jahnke, M (1974) *In: "Enzyme Engineering"* (Eds. E.K. Pye and L.B. Wingard Jnr.) 2, 411-417. Plenum, New York.
8. Mosbach, K and Danielsson, B (1974) *Biochim. Biophys. Acta*. 364, 140-145
9. Divis, C (1975) *Annals of Microbiology* 126A, 175-186
10. Lubbers, DW and Opitz, N Z(1975) *Naturforsch. C: Biosci*. 30c, 532-533
11. Voelkl, KP, Opitz, N and Lubbers, DW (1980) *Fres. Z. Anal. Chem*. 301, 162-163
12. Clemens, AH, Chang, PH and Myers, RW (1976) *Proc. Journes Ann. de Diabtologie de l'Htel-Dieu*, Paris
13. Shichiri, M, Kawamori, R, Yamaski, R, Hakai, Y and Abe, H (1982) *Lancet* ii, 1129-1131
14. Liedberg, B, Nylander, C and Lundstrm, I (1983) *Sensors and Actuators* 4, 299-304
15. Cass, AEG, Francis, DG, Hill, HAO, Aston, WJ, Higgins, IJ, Plotkin, EV, Scott, LDL and Turner, APF (1984) *Anal. Chem*. 56, 667-671
16. Alberts B, Johnson A, Lewis J, Raff M, Roberts K, Walter P (2002) *Molecular Biology of the Cell*. 4th edition. Garland Science, New York
17. Calvo EJ, Danilowicz C, Wolosiuk A (2002) *Molecular "Wiring" Enzymes in Organized Nanostructures*. J.Am.Chem.Soc. 124:2452-2453
18. Chaki NK, Vijayamohanan K(2002) *Self-assembled monolayers as a tunable platform for biosensor applications*. Biosensors & Bioelectronics 17: 1-12

References

11 Fundamentals of Electrochemistry

Christopher Brett

Departamento de Química, Universidade de Coimbra

11.1 Introduction

Electrochemistry involves chemical phenomena associated with charge separation, usually in liquid media. The charge separation often leads to charge transfer, which can occur homogeneously in solution, or heterogeneously on electrode surfaces. In order to ensure electroneutrality, two or more charge transfer half-reactions take place simultaneously, in opposing directions: oxidations (loss of electrons or increase in oxidation state) and reductions (gain of electrons or decrease in oxidation state). In the case of heterogeneous redox reactions, these are separated in space, usually occurring at different electrodes immersed in solution in a cell. The electrodes are linked by conducting paths both in solution (via ionic transport) and externally (via electric wires etc.) so that charge can be transported and the electrical circuit completed. If the cell configuration permits, the products of the two electrode reactions can be separated. When the sum of the free energy changes at both electrodes is negative the electrical energy released can be harnessed (batteries, fuel cells). If it is positive, external electrical energy can be supplied to overcome the positive free energy and oblige electrode reactions to take place and convert chemical substances (electrolysis).

It is clearly useful to be able to investigate the fundamentals of these electrode processes in the laboratory using careful control of the electrode reactions.

11.2 What is an Electrode Reaction?

Electrode reactions involve the transfer of charge to or from an electrode; the electrode process may include charge transfer steps and chemical reactions of the products or intermediates formed. The reactions take place in the interfacial region between electrode and solution, the region where charge distribution differs from that of the interior of the phases. Different types of electrode reaction are: oxidation or reduction of a solution-soluble species, oxidation or dissolution of the electrode material, and electrodeposition of a species from solution by reduction.

The simplest case is a one-step reaction with the electrode material itself inert, i.e. the electrode acts only as a source (for reduction) or a sink (for oxidation) of electrons transferred to or from species in solution. This reaction can be written

$$\mathrm{O}+n\mathrm{e}^{-} \rightleftharpoons \mathrm{R} \tag{11.1}$$

where *O* and *R* are the oxidized and reduced species, respectively. At each electrode, charge separation can be represented by a *capacitance* and the difficulty of charge transfer by a *resistance*.

For electron transfer to occur, there must be a correspondence between the energies of the electron orbital where transfer takes place in the donor and acceptor. In the electrode, this level is the highest filled orbital, which in a metal is the Fermi energy level, E_F. In soluble species it is simply the orbital of the valence electron to be given or received.

- For a reduction, there is a minimum energy that the transferable electrons from the electrode must have before transfer can occur, which corresponds to a sufficiently negative potential (in volts).
- For an oxidation, there is a maximum energy that the lowest unoccupied level in the electrode can have in order to receive electrons from species in solution, corresponding to a sufficiently positive potential (in volts).

The change in charge distribution from the bulk in this region means that the relevant energy levels in reacting species and in the electrode are not the same as in the interior of the phases, and soluble species need to adjust their conformation for electron transfer to occur. These effects should be corrected in a treatment of kinetics of electrode processes.

It must be remembered that there must always be an electrical circuit, i.e. with two electrodes. If a reduction process occurs at one of the electrodes (the cathode) then oxidation occurs at the other electrode (the anode). Note also that by external control of the potential (see below) the process occurring at the electrode of interest (the working electrode) can be changed from reduction to oxidation and at certain values of controlled potential it may be a mixture of both processes.

11.3 Electrode Potentials

It is useful to have a scale which measures the potentials, i.e., energies, necessary for the transfer of electrons associated with a particular electrode reaction. By convention, electrode reactions are written as reductions. Each has an associated standard electrode potential, $E^{\ominus}$, measured relative to the standard hydrogen electrode (SHE) with all species at unit activity ($a_i = 1$), at 25 °C and at 1 bar pressure.

For half-reactions at equilibrium, the potential, E_{eq}, is related to the standard electrode potential through the *Nernst equation*

$$E_{eq} = E^{\ominus} + \frac{RT}{nF} \ln \frac{\Pi a_{O_i}^{\nu_i}}{\Pi a_{R_i}^{\nu_i}} \tag{11.2}$$

where ν_i are the stoichiometric numbers of each species in the electrode reaction. For example, in the reaction

$$MnO_4^- + 8H^+(aq) + 5e^- \rightleftharpoons Mn^{2+} + 4H_2O \tag{11.3}$$

the stoichiometric number of MnO_4^- is 1, of H^+(aq.) 8 etc. (and n is 5).

The tendency for the reduction to occur, relative to the SHE reference, is given by

$$\triangle G^{\ominus} = -nF\,E^{\ominus} \tag{11.4}$$

under standard conditions, where $\triangle G^{\ominus}$ is the change in free energy. The more negative the value of $\triangle G^{\ominus}$ the more easily the reduction occurs. Thus, for example, Group IA metals, which have very negative values of $E^{\ominus}$, tend to oxidise.

It is often useful to be able to employ concentrations, c_i, instead of activities, particularly since electrochemical experiments are carried out with known concentrations of species in solution and the number of electrons transferred is directly proportional to the amount (i.e. the local concentration) and not the activity. The Nernst equation is rewritten as

$$E_{eq} = E^{\ominus'} + \frac{RT}{nF} \ln \frac{\Pi [O_i]^{\nu_i}}{\Pi [R_i]^{\nu_i}} \tag{11.5}$$

in which $E^{\ominus'}$ is the formal potential, dependent on the medium since it includes logarithmic activity coefficient terms as well as $E^{\ominus}$. The normal relation between activities and concentrations of species is $a_i = \gamma_i c_i$, where the proportionality coefficient, γ_i, is the *activity coefficient*. It should be recognised that activities can also be influenced by "third party" species which do not form part of the reaction scheme.

If the oxidised and reduced species involved in an electrode reaction are in equilibrium at the electrode surface, the Nernst equation can be applied. The electrode reaction is then known as a *reversible* reaction since it obeys the condition of thermodynamic reversibility. Clearly, the applicability of the Nernst equation, and therefore reversibility, depends on the time allowed for the electrode reaction to reach equilibrium.

11.4 The Rates of Electrode Reactions

An electrode reaction can be divided into several steps. The first step is that the species that will react must diffuse from bulk solution, the rate of which is often described by the mass transfer coefficient, k_d. When this mass transport is slower than the kinetics of the electrode reaction, the reaction is described as *reversible*. It can be seen immediately that increasing the mass transport, usually by externally imposed convection, can lead to a reversible reaction being influenced by kinetics and moving into the so-called *quasi-reversible* regime.

The kinetics is usually expressed by the following equations, for an oxidation

$$k_a = k_0 \exp[\alpha_a nF(E - E^{\ominus\prime}) / RT] \tag{11.6}$$

where k_0 is the standard rate constant, the value of k_a when $E = E_{eq}$. α_a is the anodic charge transfer coefficient whose value lies between 0 and 1 and is often around 0.5 for metallic electrodes, being a function of the position of the activation energy barrier. For a reduction, in a similar way

$$k_c = k_0 \exp[-\alpha_c nF(E - E^{\ominus\prime}) / RT] \tag{11.7}$$

with α_c the cathodic charge transfer coefficient.

Thus, the criterion for a *reversible* reaction is

$$k_0 >> k_d \tag{11.8}$$

and for an *irreversible* reaction

$$k_0 << k_d \tag{11.9}$$

the situation where the electrode reaction cannot be reversed. A high kinetic barrier has to be overcome, which is achieved by application of an extra potential (extra energy) called the *overpotential*.

Quasi-reversible reactions exhibit a behaviour intermediate between reversible and irreversible reactions, the overpotential having a relatively small value, so that with this extra potential, reactions can be reversed.

Rates of electrode reactions, measured experimentally as the current passed, are directly proportional to the product of rate constant and reagent concentration. The dependence of current, I, on potential is exponential when kinetically controlled, i.e., there is a linear relation between lg I and the potential - this is the *Tafel relation.* However, the consumption of reagents leads to concentration gradients and the rate of reaction begins to be limited by mass transport of reagent, as the potential applied becomes more negative (reduction) or more positive (oxidation). Eventually a transport-limited current is reached.

For reversible reactions only thermodynamic and mass-transport parameters can be determined, for quasi-reversible and irreversible reactions both kinetic and thermodynamic parameters can be measured. It should also be noted that the electrode material can influence the kinetics of electrode processes; it should not be assumed that electrode materials are inert. For example, platinum and gold form oxides at potentials of ~ 0.8 V vs. the saturated calomel electrode.

More complex electrode processes involve consecutive electron transfer or coupled homogeneous reactions. The theory of these reactions is also more complicated, but they correspond to a class of real, important reactions, particularly involving organic and biological compounds. The electrode material can play an even more important role in these cases.

11.5 How to Investigate Electrode Reactions Experimentally

In order to study electrode reactions, reproducible experimental conditions must be created which enable minimization of all unwanted factors that can contribute to the measurements and diminish their accuracy. Normally it is desirable to suppress migration effects, confine the interfacial region as close as possible to the electrode, and minimise solution resistance. These objectives are usually achieved by addition of a large quantity of inert electrolyte (around 1 mol dm^{-3}), the electroactive species being at a concentration of 5 mM or less.

The electrochemical cell contains three electrodes: the working electrode, the auxiliary electrode and a reference electrode. The auxiliary electrode completes the electrical circuit and should be inert; it is often put in a separate compartment connected to the working electrode compartment by a salt bridge or a porous frit. The reference electrode serves to control the potential of the working electrode and should not pass current. Commonly used reference electrodes are shown in Table 11.1, given the impracticalities in normal use of the standard hydrogen electrode. The calomel electrode (Hg_2Cl_2) in the third line is usually used with saturated KCl and is then known as the saturated calomel electrode (SCE).

Control of the experiment is achieved using a *potentiostat* instrument. The potential of the working electrode is applied externally and controlled with respect to the reference electrode.

The current which passes through the electrical circuit linking the working electrode and the auxiliary electrode as a result of the applied potential is measured by means of a current follower in the instrument and appropriate signal treatment. Currents due to oxidation (anodic currents) have a positive sign and currents due to reduction processes (cathodic currents) have a negative sign. The study of the current vs. potential profile is called *voltammetry* and the data usually presented as a plot of current, I, on the y-axis vs. applied potential, E, on the x-axis. The example of the cyclic voltammetry of ferrocyanide ion is shown in Fig. 11.1 and demonstrates an oxidation process on the forward scan ($Fe(CN)_6^{4-} \rightarrow Fe(CN)_6^{3-}$) and the reverse reduction process on the inverse scan of potential. Information can be gained from the position of the peaks on the potential scale and their height and width.

If the working electrode is a coating on a piezoelectric crystal, then mass changes will usually occur over the same range of potentials as oxidation and reduction.

Occasionally, the current at the working electrode is controlled using a *galvanostat*, and the reference electrode is employed to measure the potential of the working electrode, usually as its variation with time (chronopotentiometry).

Table 11.1. Reference electrode electrochemical reactions in water solvent

Electrochemical reaction	$E^{\ominus}$ / V
$AgBr + e^- \rightleftharpoons Ag + Br^-$	0.071
$AgCl + e^- \rightleftharpoons Ag + Cl^-$	0.222
$Hg_2Cl_2 + 2e^- \rightleftharpoons 2Hg + 2Cl^-$	0.268
$HgO + H_2O + 2e^- \rightleftharpoons Hg + 2OH^-$	0.098
$Hg_2SO_4 + 2e^- \rightleftharpoons 2Hg + SO_4^{2-}$	0.613

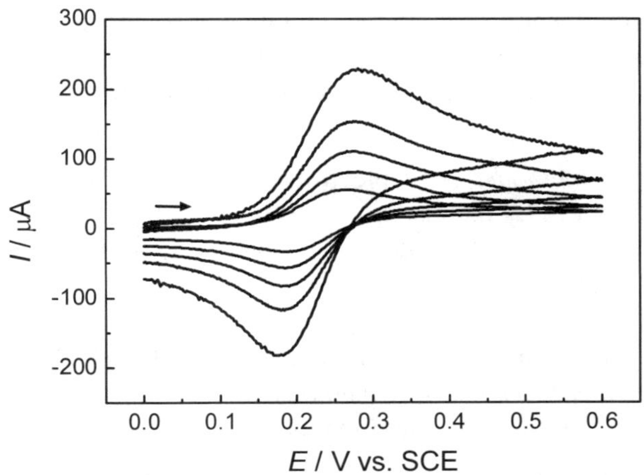

Fig. 11.1. Cyclic voltammograms at carbon film electrode after background subtraction for oxidation of 1 mM $K_4Fe(CN)_6$ in 0.4 M K_2SO_4 electrolyte at scan rates of 20, 50, 100, 200 and 500 mV s^{-1} (adapted from *Electroanalysis*, 13 (2001) 765-769)

Electrode processes can be modelled as electrical equivalent circuits of varying complexity. The simplest electrode process, i.e. an electron transfer, is modelled in Fig. 11.2, where R_{ct} represents the resistance to charge transfer across the electrode-solution interface, C_d the capacity due to charge separation in the interfacial region (in solution and also within the electrode for semiconductors) and R_Ω the resistance of the remainder of the electrochemical cell.

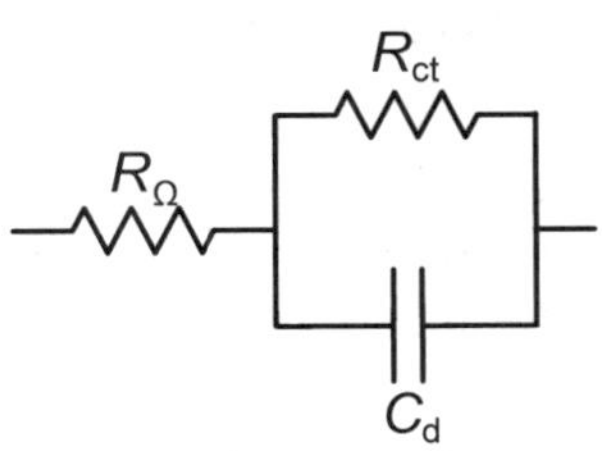

Fig. 11.2. Electrical equivalent of a simple electrode process: R_{ct} charge transfer resistance, C_d capacity of interfacial region and R_Ω the cell resistance

11.6 Electrochemical Techniques and Combination with Non-Electrochemical Techniques

Different types of electrochemical techniques are necessary since usually a complete study of an electrode process requires measurement of kinetic as well as of thermodynamic parameters and details of changes occurring on the electrode surface. Increasing importance of kinetic control of the electron transfer steps of the electrode process can be achieved in the following ways:

- *steady state methods:* microelectrodes by decreasing size; hydrodynamic electrodes by increasing convection,
- *linear sweep and cyclic voltammetry methods*: by increasing applied potential sweep rate (see Fig. 11.1),
- *step and pulse techniques (e.g,. potential step, pulse voltammetry, square wave voltammetry)*: by increasing amplitude and/or frequency,
- *electrochemical impedance methods:* increasing perturbation frequency, registering higher harmonics, etc.

The type of technique chosen will depend on the timescale of the electrode reaction.

Non-electrochemical methods can and should be used for studying electrode surfaces and the interfacial region structure, particularly *in situ* in real time where this is possible. These can include the quartz crystal microbalance (QCM) for monitoring mass changes of the electrode surface as well as viscoelastic effects, atomic force and scanning tunnelling microscopy, infrared reflectance and ellipsometry, etc.

11.7 Applications

A detailed knowledge of the electrode reactions and processes can be used for:

1. tailoring electrode reactions so as to enhance required and inhibit unwanted electrode reactions, perhaps by changing electrode material or developing new electrode materials or modifying the electrode surface,
2. studying complex systems in which many electrode reactions occur simultaneously or consecutively, as in bioelectrochemistry,
3. measuring concentrations of electroactive species, making use of the selectivity of the potential and of the electrode material at or outside equilibrium (as in potentiometric, amperometric, voltammetric, and enzyme sensors).

The vast range of applications also includes electrosynthesis, minimisation of corrosion, sonoelectrochemistry, industrial electrolysis, electroplating, batteries, fuel cells, electrochemical machining, etc.

11.8 Bibliography

There is an extensive bibliography on electrochemistry and its applications. An exhaustive listing can be found in the following internet address: "http://electrochem.cwru.edu/estir/". Texts of physical chemistry usually have a good introduction to electrochemistry. Some electrochemistry texts that go more deeply into aspects of techniques and the practice of electrochemistry are shown next.

11.9 Glossary of Symbols

Table 11.2. Units normally used in electrochemistry.

Symbol	Definition	Unit
a_i	activity of ion i	-
c_i	concentration of ion i	mol cm^{-3} or mol dm^{-3}
C_d	capacity of interfacial region	F cm^{-2}
E	electrode potential	V
$E^{\ominus}$	standard electrode potential	
$E^{\ominus\prime}$	formal potential	
E_{eq}	equilibrium potential	
E_F	Fermi energy	J or eV
$\triangle G^{\ominus}$	standard free energy change	J mol^{-1}
I	Current	A
k	heterogeneous rate constant	cm s^{-1}
k_0	standard rate constant of an electrode reaction	
k_a	rate constant for an anodic reaction	
k_c	rate constant for a cathodic reaction	
k_d	mass transfer coefficient	
n	number of electrons involved in an electrode reaction	-
R_{ct}	charge transfer resistance	Ω
R_{Ω}	cell resistance	Ω
T	Temperature	K
α	electrochemical charge transfer coefficient	-
α_a	anodic	
α_c	cathodic	
γ_i	activity coefficient of species i	-

Table 11.3. Physical constants

F	Faraday constant	9.6485×10^{4} C mol^{-1}
R	gas constant	8.31451 J mol^{-1} K^{-1}

References

1. Brett CMA and Oliveira Brett AM (1993), *Electrochemistry. Principles, methods, and applications*, Oxford University Press, Oxford
2. Bard AJ and LR Faulkner LR (2001), *Electrochemical Methods. Fundamentals and applications*, 2nd Ed., Wiley, NY
3. Gileadi E (1993), *Electrode kinetics for chemists, chemical engineers, and materials scientists*, VCH Publishers Inc., NY
4. Sawyer DT, Sobkowiak A and Roberts JL (1995), *Experimental electrochemistry for chemists*, 2nd Ed., Wiley, NY
5. Kissinger PT and Heinemann WR eds. (1996), *Laboratory techniques in electroanalytical chemistry*, 2nd ed., Dekker, NY

12 Viscoelastic Properties of Macromolecules

Ralf Lucklum[1] and David Soares[2]

[1]Institute for Micro and Sensor Systems, Otto-von-Güericke-University Magdeburg
[2]Institute de Fisica, Universidade de Campinas

12.1 Introduction

An acoustic wave traveling through a material deforms the material, thereby probing its mechanical properties. The classical theory of elasticity deals with the mechanical properties of elastic solids. In accordance with Hook's law, for small deformations stress is always directly proportional to strain. Furthermore, stress is independent of the rate of strain. The classical theory of hydrodynamics deals with properties of viscous liquids. In accordance with Newton's law, stress is always directly proportional to rate of strain but independent of the value of strain itself.

Purely elastic solids and purely viscous liquids are idealizations as well as the two boundaries of system behavior, which combines liquid-like and solid-like characteristics. Those material properties have been called viscoelastic. A viscoelastic material does not maintain a constant deformation under constant stress. It continues to deform slowly with time. On the other hand, when the stress is removed, a viscoelastic material recovers part of its deformation. When a viscoelastic material is constrained at constant deformation, the stress required to hold it, diminishes with time.

When a viscoelastic material is subjected to sinusoidally oscillating stress, the strain is neither in phase with stress as valid for purely elastic solids nor 90° out of phase as valid for purely viscous liquids. Stress and strain are some amount out of phase, which depends on the viscoelastic properties of the body. If both strain and rate of strain are infinitesimal, and if the time-dependent stress-strain relations can be described by linear differential equations with constant coefficients, the system shows linear viscoelastic behavior and the ratio of stress to strain is a function of time or frequency, not a function of stress magnitude [1, 2].

Viscoelastic properties of material should not be mixed up with deviations from purely elastic or viscous behavior due to finite strain (non-

Hookean deformation) or finite strain rates (non-Newtonian flow). The necessity to distinguish between infinitesimal and finite depends on the level of precision of measurement and properties of the material under consideration.

Viscoelastic anomalies are negligible or at least of minor significance for many classical engineering materials. By contrast, the mechanical behavior is dominated by viscoelastic phenomena in macromolecular materials. Due to the extreme importance of those materials for chemical and biochemical sensors, the consequences of viscoelasticity for acoustic sensors will be analyzed in some more detail. A comprehensive analysis of viscoelastic properties of polymers can be found in Ferry's fundamental book [1]. The following text summarizes those statements about the phenomenon of viscoelasticity [1-3] which are important to understand specific features of acoustic-wave based sensors.

12.2 Molecular Background of Viscoelasticity of Polymers

In solids, such as silicon or quartz, all atoms are arranged in a crystal with well defined distances and orientations, giving rise to a long scale order. The interaction forces are strong. In deformation of hard solids, atoms are displaced from equilibrium positions in fields of force with small range. The elastic constants are a result of inter-atomic potentials. The deformation energy is completely stored as potential energy and can be completely recovered. Other phenomena like structural imperfections, which contribute to mechanical properties relevant for acoustic wave devices, typically have distances discontinuously larger than atomic dimensions.

In ordinary liquids, such as water, the molecules are arranged with a rapidly vanishing order. The interaction forces between the molecules are rather small. Viscous flow changes the distribution of molecules surrounding a given molecule with time. Viscosity is a result of the relevant forces between the molecules and the processes of readjustment. The deformation energy is completely dissipated as heat.

Macromolecules cannot be described in a simple way. Each flexible macromolecule pervades an average volume much greater than the atomic dimensions. Furthermore, it continually changes the shape of its contour [1]. The nomenclature of the molecule reflects only the (chemical) constitution of the macromolecule: type, sequence of chain molecules, building blocks, constitutive repeating units, type and number of end groups, as well as type, number, arrangement and distribution of chain links. The configuration describes the specific arrangement of atoms or groups of atoms at a C = C double bond. The spatial arrangement of atoms or groups of atoms as a result of rotation along a single chemical bond is called conformation. Isomers are molecules with an equivalent number of equivalent atoms in a different spatial arrangement [4].

Conformation changes[1] are most important for the understanding of the response of macromolecules to any macroscopic deformation. One has to consider more local relationships as well as long range contour relationships. Rearrangements on a local scale are relatively rapid, whereas rearrangements on a long-range scale are rather slow. Long-range rearrangements require a certain kind and number of cooperative local configuration changes to be realized. In a viscoelastic material a wide and continuous range of time scale exists covering the response of such a system to external stress. From that point of view one can expect a different response of viscoelastic materials to an external perturbation depending on the evaluating experimental method, first of all on the time scale of the applied perturbation.

Another aspect is the morphology of polymers[2]. In principle one can distinguish crystalline and amorphous polymers. A local structure along a polymer chain with enough regularity and symmetry allows a partial ordering in a crystal lattice. The macromolecules can form thin lamellar single crystals, in which the chains run parallel to the thin direction and are folded back and forth many times. Polymers with a high degree of molecular symmetry can form lamellar structures, which are organized into larger morphological units such as fibriles, spherulites or helices. If the molecular weight is not uniform, fractionation occurs with species of lower molecular weight remaining uncrystallized. Side groups, atacticity, or other features, which diminish molecular symmetry, enhance the degree of disorder and amorphous regions within a crystalline polymer become more and more likely to occur. Amorphous polymers are completely disordered; the arrangement of the macromolecules is random.

---

1 Configuration and conformation isomers can be distinguished by the energy barrier between them. The distinction between them is not important as long as the spatial arrangement of atoms is of interest only. However, the lifetime of conformation isomers is much shorter than that of configuration isomers. As a consequence, conformation isomers can be recognized with fast probing methods like acoustic waves. Consequently configuration and conformation isomers can hardly be distinguished by those methods. Although it is not really consistent, many polymer textbooks prefer the words 'configuration changes' to describe rearrangements of the polymer backbone realized by the arrangement of atoms and groups of atoms along the polymer backbone.

2 The word polymers is further used instead of macromolecules, though not limited to classical polymers.

Polymers can be composed of linear chains, chains with small side groups, backbone chains with more or less long side chains, cross-linked backbone chains and complete networks without individual chains. Besides chemical bonds within the macromolecule and between cross-linked chains one has to consider several kinds of physical bonds between different chains and different segments of the chain itself. A special feature of macromolecules is the chain entanglement, which creates a kind of physical network (Fig. 12.1) [1].

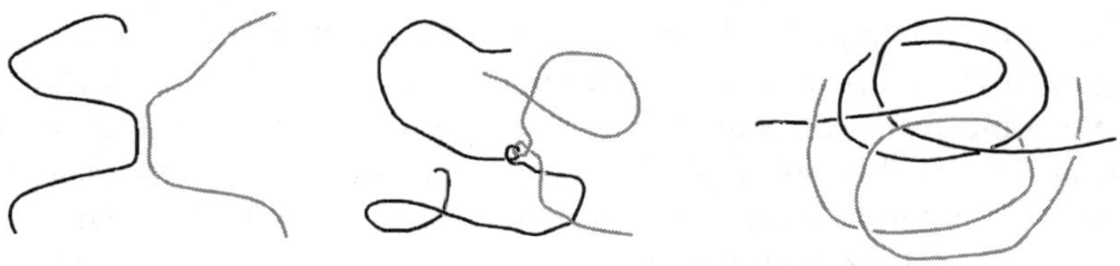

a **b** **c**

Fig. 12.1. Concepts of entanglement coupling: **a** temporary cross-link, **b** local link, and **c** long-range contour loop. Adapted from Ferry [1]

The viscoelastic behavior of polymers is not uniform. It can be illustrated with several examples. The first group considers uncross-linked polymers, which can be treated as viscoelastic liquids. They do not possess any equilibrium compliance and exhibit viscous flow at sufficient times. The viscosity may be very high and hence the material may appear as a rubbery solid. Ferry [1] distinguishes between dilute polymer solutions (I), amorphous polymers of low molecular weight (II), amorphous polymers of high molecular weight (III) and amorphous polymers with high molecular weight with long side chains (IV).

In dilute solutions the polymer molecules are sufficiently separated to move fairly independently without much interaction. Polymers with a low molecular weight up to 10,000 consist of about 100-200 monomer units. The effects of neighboring molecules on viscoelasticity arise from local frictional forces. Viscoelastic properties of polymers above a critical molecular weight of the order of 4,000 to 30,000 reveal a strong coupling to neighbors far more than local frictional forces. The interaction is usually called entanglement coupling. This phenomenon prolongs any molecular rearrangements, which are sufficiently long range to involve regions of a macromolecule separated from each other by one or more entanglement points.

The other group considers viscoelastic solids, which do not exhibit viscous flow and reach or closely approach an equilibrium deformation under a constant stress. Amorphous polymers of high molecular weight below glass transition temperature (V), lightly (VI) and very lightly (VII) cross-linked amorphous polymers, and highly crystalline polymers (VIII) belong to the second group [1]. The chain backbone configuration of an amorphous polymer is highly immobilized below the glass transition temperature. The response to external stress involves primarily very local adjustments. Short-range segmental rearrangements in a network of highly flexible threadlike strands (usually defined as segments between cross-links and entanglements) are rapid whereas long-range rearrangements are profoundly affected by the presence of linkage points. In very lightly cross-linked amorphous polymers one can expect dangling branched structures, which are extensively entangled. Finally, in crystalline materials both ordered and disordered regions will contribute to viscoelastic behavior in their own way [1].

12.3 Shear Modulus, Shear Compliance and Viscosity

In general, the ratio of stress to the corresponding strain is called a modulus and hence, a measure of the strength of a material. The following is restricted to periodical perturbations with a sinusoidal alternation at a frequency f (cycles per second) or ω (radians per second), e.g., shear stress generated from a propagating acoustic wave, and furthermore to linear viscoelastic behavior. Such a periodic experiment at ω is qualitatively equivalent to a transient experiment at time t = $1/\omega$.

The relation between stress, σ, and strain, $\gamma = \gamma^0 \sin \omega t$, can be written as

$$\sigma = \gamma^0 \left(G' \sin \omega t + G'' \cos \omega t \right) \tag{12.1}$$

where the superscript 0 marks the maximum amplitude. The term with $\sin \omega t$ is in phase with γ and the term with $\cos \omega t$ is 90° out of phase with γ. Equation (12.1) thereby defines two frequency-dependent functions, the shear storage modulus, G', and the shear loss modulus, G''. It is instructive to write the stress in an alternative form displaying the amplitude $\sigma^0(\omega)$ of the stress and the phase angle $\delta(\omega)$ between stress and strain:

$$\sigma = \sigma^0 \sin(\omega t + \delta) \tag{12.2}$$

It is evident that each periodic measurement at a given frequency provides simultaneously two independent quantities, either G' and G'' or σ^0/γ^0 and $\tan \delta$. Sinusoidally varying quantities are usually expressed as complex values; then the modulus is also complex:

$$\frac{\underline{\sigma}}{\underline{\gamma}} = \underline{G} = G' + \mathrm{j}G'' \tag{12.3}$$

The data from sinusoidal experiments can also be expressed in terms of a complex compliance:

$$\underline{J} = \frac{\underline{\gamma}}{\underline{\sigma}} = \frac{1}{\underline{G}} = J' - \mathrm{j}J'' \tag{12.4}$$

Note, that although $\underline{J} = 1/\underline{G}$, their individual components are not reciprocally related:

$$J' = \frac{G'}{G'^2 + G''^2} = \frac{1/G'}{1 + \tan^2\delta} \tag{12.5a}$$

$$J'' = \frac{G''}{G'^2 + G''^2} = \frac{1/G''}{1 + 1/\tan^2\delta} \tag{12.5b}$$

or

$$G' = \frac{J'}{J'^2 + J''^2} = \frac{1/J'}{1 + \tan^2\delta} \tag{12.5c}$$

$$G'' = \frac{J''}{J'^2 + J''^2} = \frac{1/J''}{1 + 1/\tan^2\delta} \tag{12.5d}$$

As an alternative to G, the relations can be expressed equally well by a complex viscosity:

$$\underline{\eta} = \eta' - \mathrm{j}\eta'' \tag{12.6}$$

which is frequently used to describe viscoelastic liquids. The ratio of stress in phase with the rate of strain to the rate of strain is η', and η'' is the stress 90° out of phase with the rate of strain to the rate of strain. The individual components are given by:

$$\eta' = \frac{G''}{\omega} \qquad \eta'' = \frac{G'}{\omega} \tag{12.7a,b}$$

The frequency dependence of the quantities in Eqs. (12.3)-(12.7) can be imitated with mechanical models. The two basic elements are the Maxwell element (Fig. 12.2a) and the Voigt element (Fig. 12.2b), both consisting of a spring and a dashpot. For example, when a Maxwell element is subjected to a stress relaxation experiment, the force relaxes exponentially. Written in terms of the modulus $G(t)$ becomes:

$$G(t) \equiv G \exp(-t/\tau) \tag{12.8a}$$

The Maxwell model depicts a viscoelastic liquid since the stress will relax to zero at long times. With a sufficient number of Maxwell elements arranged in parallel a complete relaxation spectrum can be modeled. Each Maxwell element is characterized by a certain relaxation time, τ_i, and a certain shear modulus, G_i:

$$G(t) = \sum G_i \exp(-t/\tau_i) \tag{12.8b}$$

The Voigt model represents a viscoelastic solid, however, both models are equivalent with an appropriate assignment of parameters. To extend the Voigt model for a set of relaxation times, an equivalent number of Voigt elements must be arranged in series.

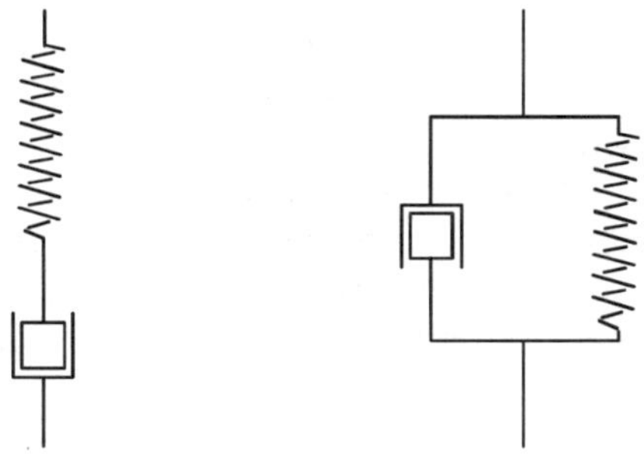

a **b**

Fig. 12.2. **a** The Maxwell element and **b** the Voigt element

For a liquid, all viscosities in the Maxwell model must be finite and one spring in the Voigt model must be zero. For a solid, one viscosity in the Maxwell model must be infinite and all springs in the Voigt model must be nonzero [1, 2].

Bulk compression and dilatation may be introduced from a propagating acoustic wave with displacement components normal to the surface of the viscoelastic material, e.g., from a Raleigh-wave propagating on SAW-devices. For this component, the bulk modulus, K, or the compliance, B, must be applied to describe the viscoelastic properties of the material. K and B are complex values in analogy to Eqs. (12.3) and (12.4). For a viscoelastic solid at equilibrium, the equilibrium Young's modulus, E_e, is related to the (equilibrium) shear and bulk moduli as follows [1]:

$$E_e = \frac{9G_e K_e}{G_e + 3K_e} = 2G_e(1+\mu_e) \tag{12.9}$$

Here, $\mu = -\gamma_{22}/\gamma_{11}$ is the relation between axial extension and lateral contraction for a perfectly elastic solid and is called Poisson's ratio with the typical notation of indices known from "Mechanics of Materials" textbooks. For viscoelastic solids the glasslike modulus G_g and K_g may replace G_e and K_e, respectively.

The shear storage modulus is a measure of the energy stored and recovered in each cycle of a periodic deformation. The frequency dependence is usually analyzed with logarithmic scales of the radian frequency, ω.

At low frequencies, $G'(\omega)$ increases rapidly for all viscoelastic liquids by several orders of magnitude. In molecular terms it corresponds to random average configurations of the macromolecular coils at low frequencies and the resumption of constraints with increasing frequency. The magnitude depends on what contour rearrangements can take place within the period of oscillatory perturbation. With increasing frequency the time necessary for long-range configuration rearrangements exceeds more and more the period of oscillation.

The increase of G' is usually associated with the transition from rubbery-like to glassy-like consistency. Note, that this is not a transition in any thermodynamic sense. It reflects a change in the response of the system to experiments with different frequencies. Even if the magnitude of the shear modulus changes from rubbery-like to glassy-like, there is no change in the thermodynamic state of the material as it occurs when the glass transition temperature, T_g, is traversed in a thermal experiment. In the latter case the system is in equilibrium above T_g and not in equilibrium below T_g.

The rise in G' occurs in a single stage for polymers with low molecular weight (II). The transition can be separated into two stages for polymers with high molecular weight (III and IV). The main step reflects the relative motion of chain segments between the entanglement loci and corresponds to the glass-rubber transition. With decreasing frequency, the stress gradually falls as the distortion of the chain backbone adjusts itself through Brownian motion, first of the relative position segments near each other, then farther and farther apart along the backbone contour, requiring more and more mutual cooperation and therefore more and more time. In the terminal zone the properties are dominated by the entanglement network or equivalent topological restraints, which strongly inhibit long-range configurational motions. Here, molecular weight or molecular weight distribution strongly influences the viscoelastic behavior. The terminal relaxation time is much longer than it would be in the absence of topological restraints and increases rapidly with molecular weight. The terminal zone of the storage modulus is displaced toward lower frequencies with increasing molecular weight. As a consequence, a plateau zone appears between the transition zone and the terminal zone, where *G* changes only slightly with frequency ($G' \approx 10^5$ Pa).

At very high frequencies *G*' approaches a limiting value of the order of 10^9 Pa. This value is G_g and represents the rigidity of the polymer in the absence of backbone rearrangements within the interval of the experiment. The corresponding region is called the glassy zone.

Viscoelastic solids, except lightly cross-linked amorphous polymers (VI), do not show such pronounced frequency dependence. The glassy polymer (V) exhibits very little stress relaxation over many decades of frequency, since no backbone contour changes occur ($G' \approx 10^9$ Pa), whereas the highly crystalline polymer (VIII) shows some relaxation only at very long times ($G' \approx 10^8$-10^9 Pa). For cross-linked rubbers (VI, VII) *G*' approaches a limiting value of about 10^5-10^6 Pa at low frequencies, which is characteristic for rubber-like elasticity.

The shear loss modulus is a measure of the energy dissipated or lost as heat in each cycle of sinusoidal deformation. The common feature of $G''(\omega)$ can be stated qualitatively as follows.

In frequency regions where $G'(\omega)$ changes slowly, the behavior of polymers is more or less nearly elastic. In such regions $G''(\omega)$ tends to be considerably less than *G*'. Hence comparatively little energy is dissipated. Consequently the curves of the glassy polymer (V) and crystalline polymer (VIII) are relatively flat throughout the entire frequency range ($G'' \approx 10^7$ Pa). The flat parts of the *G*'-curves at frequencies below the transition zone for amorphous polymers (III) and those with long side groups (IV) or cross-linking (VI, VII), which exhibit entanglement coupling, is also accompanied by a plateau or a minimum in *G*'' ($\approx 10^3$-10^5 Pa). At very high frequencies $G''(\omega)$ does not approach zero, although one can expect perfect elastic behavior from the mechanical model. Atomic adjustments are still capable of dissipating energy within the period of deformation, although losses due to backbone configuration changes diminish (up to $\approx 10^8$ Pa).

At frequencies where $G'(\omega)$ changes rapidly, maxima in *G*'' appear and $G' \approx G''$. The relaxation time of configuration rearrangements corresponds to the frequency of the periodic mechanical perturbation and the energy of deformation can be efficiently dissipated and used to realize them.

At very low frequencies, *G*'' for viscoelastic liquids is directly proportional to ω with a slope approaching 1 on a logarithmic scale. This is the

Newtonian steady-flow viscosity, η_0. Only for a simple Newtonian liquid, $G'' = \omega\eta_0$ over the entire frequency range. The in-phase component of the complex dynamic viscosity, η' in Eq. (12.6), is most useful for viscoelastic liquids. From Eq. (12.7), in regions where G'' is flat, η' is inversely proportional to frequency, whereas when G'' rises steeply η' may flatten out, as particularly for the cross-linked rubber (VI, VII). However, η' decreases monotonically with increasing frequency, altogether by many orders of magnitude in the entire frequency range. For the dilute solution (I), the steady-flow viscosity, η_0, is about eight times that of the solvent; at high frequency it approaches a limiting value slightly higher than that of the solvent.

A very useful dimensionless parameter is the loss tangent. It is a measure of the ratio of energy lost to energy stored in a cyclic deformation:

$$\tan\delta = \frac{G''}{G'} = \frac{J''}{J'} \tag{12.10}$$

The loss tangent reveals several characteristics. For the dilute solution (I) $\tan\delta$ is very high. Both solvent and solute contribute to G'' but only the solute contributes to G'. The minimum is much more pronounced than the transition in η'. All uncross-linked polymers have a large loss factor at low frequency and $\tan\delta$ becomes inversely proportional to the frequency. The glassy polymer (III) and the crystalline polymer (VIII) have values in the neighborhood of 0.1. All amorphous polymers, whether cross-linked or not, have values in the transition zone between the glasslike and rubberlike consistency of about 1. The loss tangent goes through a pronounced maximum, which is on the right of the maximum in J'' and on the left of the maximum in G''. Each of the three functions is a measure of energy dissipation, but the frequency region in which the "loss" occurs, depends on the choice of function [1].

The loss tangent determines macroscopic physical properties such as the damping of a free vibration or attenuation of a propagating wave.

12.4 The Temperature-Frequency Equivalence

In Sect. 12.3 the viscoelastic functions have been discussed in terms of time and frequency. Just as the dependence on time or frequency, the dependence of viscoelastic functions on temperature is most spectacular in the transition range.

The glass transition is the point or narrow region on the temperature scale, where the thermal expansion coefficient, α, undergoes a discontinuity and below which configuration rearrangements of the polymer backbones are extremely slow. It corresponds to a change in the slope in a plot of the specific volume against temperature, as illustrated schematically in Fig. 12.3. Above the glass transition temperature, T_g, α_f has a magnitude characteristic for liquids in the order of 10^{-3} deg^{-1}. Below T_g, α_g is smaller by a factor of one-half to one-third.

From the relaxation times point of view, $\omega\tau_i >> 1$ in the glassy state at low temperatures. Practically no configuration changes occur within the period of deformation. With increasing temperature τ_i decreases and $\omega\tau_i \cong 1$ in the transition zone around T_g. The configurational modes of motion within the entanglement coupling points become fast enough to occur within the period of deformation. In the rubbery state at high temperatures τ_i is further decreased, hence $\omega\tau_i << 1$, and configuration rearrangements within the entanglements are fast enough to keep the material in equilibrium.

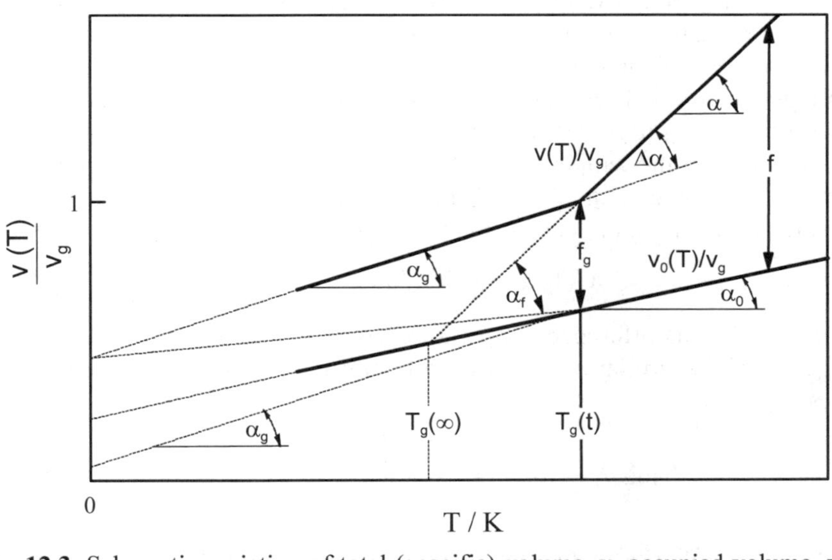

Fig. 12.3. Schematic variation of total (specific) volume, v, occupied volume, v_0, and free volume, v_f, (relative to specific volume, v_g, at the glass transition temperature, T_g) with temperature, T. $f = v_f/v_g$ is the fractional free volume, α is the thermal expansion coefficient. Adapted from Ferry [1] and Kovacs [8]

This behavior can be readily understood in terms of the free volume concept. In this concept the total volume per gram, v, is the sum of the occupied volume, v_0, and the free volume, v_f. The occupied volume includes not only the volume of the molecules as represented by their Van-der-Waals radii; it also includes the volume associated with vibrational motions. The occupied volume increases with temperature, given by the thermal expansion coefficient, α_0. It cannot be identified with the Van-der-Waals co-volume and hence can be estimated only indirectly.

The free volume can be understood as holes of the order of monomeric dimensions or smaller voids associated with packing irregularities. The increased thermal expansion coefficient above T_g represents, primarily, the creation of free volume with rising temperature. At temperatures high enough so that Brownian motion is rapid in a viscoelastic liquid or soft solid, a lowering of temperature is accompanied by a collapse of free volume as the molecular adjustments take place within the time scale of cooling. At lower temperatures, the adjustments are slower and the collapse can no longer occur within the time scale of cooling. The residual contraction is of solid-like character, and α_g is only a little or possibly no larger than α_0.

It is inherent in the above discussion that T_g must depend on the time scale of the volumetric measurement. A slower measurement pattern leads to a lower value of T_g. The appearance of the glass transition results from the reduction of molecular mobility as the temperature falls, slowing the collapse of free volume. The mobility at any temperature depends primarily on the free volume remaining.

The shift on the temperature scale as a result of the increased measuring frequency is a general phenomenon and can be described as the time-temperature correspondence principle of polymer relaxation:

$$G(T_0,t) = G(T,t/a_T) \tag{12.11}$$

T_0 may act as reference temperature, a_T, is the shift factor. The function $\log a_T$ exhibits similar behavior for all amorphous polymers and can be expressed as:

$$\log a_T = \log\frac{\omega_0}{\omega} = \frac{-C_1(T-T_g)}{C_2+T-T_g} \tag{12.12a}$$

Equation (12.12a) is known as the WLF equation [5]. It is a scaled temperature-frequency relation of characteristic relaxation processes. C_1 and C_2 are originally thought to be universal constants ($C_1 = 17.4$, $C_2 = 51{,}6$ K). Equation (12.12a) can be rearranged using $C_1 = \log(\Omega/\omega_0)$ and $C_2 = T_0 - T_\infty$:

$$\left(T - T_\infty\right)\log\left(\frac{\Omega}{\omega}\right) = \left(T_0 - T_\infty\right)\log\left(\frac{\Omega}{\omega_0}\right) \tag{12.12b}$$

Equation (12.12b) defines a set of hyperbolas with common asymptotes at T_∞ (the so-called Vogel temperature) and $\log \Omega$ ($\Omega \approx 10^{12} \ldots 10^{15}$ Hz for simple glass formers). Figure 12.4 depicts Eq. (12.12b). The set of hyperbolas is parameterised by one value, T_0 at a given mobility ($\log \omega_0 < \log \Omega$) or ω_0 at a given temperature T ($T_0 > T_\infty$). C_1 and C_2 depend on these reference points. For $\omega/(2\pi) \approx 10^{-2}$ Hz $T_0 \approx T_\mathrm{g}$ and C_1 can be approximated to be in the range of 10 to 20, whereas C_2 varies quite widely (e.g. PIB: $C_1 = 16{,}6$ and $C_2 = 104$ K) [6]. At low temperatures, the glass transition curve in Fig. 12.4 is very steep and effects are of thermodynamic nature. At high frequencies the curve is rather flat, effects are of kinetic nature.

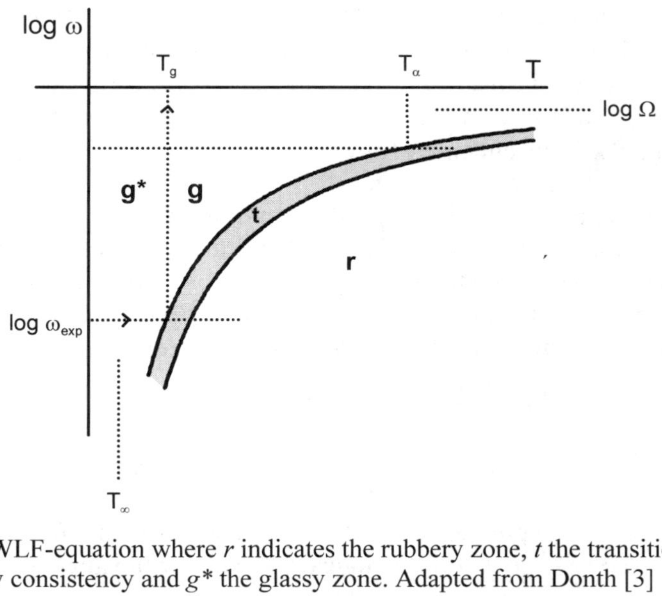

Fig. 12.4. WLF-equation where *r* indicates the rubbery zone, *t* the transition range, *g* the glassy consistency and *g** the glassy zone. Adapted from Donth [3]

Finally, a distinction has been made in Fig. 12.4 between glassy zone and glasslike consistency. The glassy consistency is that part of the glassy zone, where $\tau_i >> t_{exp}$. The mobility relevant for the glass transition is characterized by relaxation times much larger than the experimental time. The glassy zone is characterized by $\omega_{exp} >> \omega_i >> 1/t_\mathrm{exp}$: The experimental time is large enough to realize molecular rearrangements; however, the probing frequency is much too high to respond to the perturbation. Performing a dynamic experiment in a large temperature range including T_g, the border between glassy zone and glassy consistency can be recognized if a change in certain thermal, mechanical or geometric properties have an influence on the signal [1].

The time-frequency equivalence principle is nothing else than a conclusion from the qualitatively similar behavior of experimental findings during the glass transition, e.g., the value of the shear modulus. The response keeps its characteristics when shifted along the glass transition curve in Fig. 12.4 due to a change in temperature in a very broad frequency range of several orders of magnitude. It indicates a basically similar relaxation process, independent of whether it takes place at low temperatures at a mHz frequency range or at high temperatures in the MHz frequency range. It is therefore common-sense to reduce the response curves at different temperatures, which is technically speaking a temperature dependent shift along the log ω-axis, the so-called master curve construction. The master curve is, in the optimal case, a universal curve with a reduced abscissa log $(a_T\omega)$. The variable log $(a_T\omega)$ describes the dependence on log ω in the isotherm case (T = const), and on T in the isochronous case (ω = const). After some "equalization" of the hyperbola one gets similar curves for the log ω- and the T-dependence [1, 6].

The absorption of vapor molecules, presumed to be of low molecular weight in comparison with the polymer and molecularly dispersed, causes dilution of the polymer. The effect of the diluent on polymer viscoelastic properties can be understood as a generation of additional free volume in proportion to the volume fraction, V. From that point of view solvent absorption has an effect similar to a temperature increase:

$$f(T) = f_0 + \alpha_f (T - T_0) \tag{12.13}$$

$$f(T, V_1) = f_2(T) + \beta' V_1 \tag{12.14}$$

In the former expressions, f is the fractional free volume (free volume related to the volume at T_g), the index 1 stands for the analyte, index 2 for the polymer and β' is a parameter relating the volume fraction of the analyte to the free volume and is marginally smaller than the fractional free volume of the analyte as a liquid. At low analyte concentrations ($V_1 << 1$) the contributions to the free volume of polymer and analyte are additive, approximately. The difference arises from the fact that the increase in volume from temperature increase directly influences the modulus, whereas solvent absorption influences the modulus in relation to the created additional free volume. Temperature effect, Eq. (12.13), and solvent absorption effect, Eq. (12.14), can be combined and result in a modified WLF-equation [7]:

$$\log a_T = -C_1 \left(\frac{\alpha_f (T - T_0) + \beta' V_1}{F_0 + \alpha_f (T - T_0) + \beta' V_1} \right) \tag{12.15}$$

Analog to the temperature-frequency relation, changes in viscoelastic properties with vapor absorption can be expressed as a translation a_T in the polymer relaxation time or the apparent probing frequency. As the probing frequency shifts the activation of a certain relaxation process to higher temperatures, solvent absorption results in the opposite shift. Hence, relaxation processes, which are too slow to respond to high frequency mechanical perturbations, may be activated after solvent absorption.

12.5 Conclusions

The acoustic wave propagating in a coating is a mechanical perturbation of the film. In case of a viscoelastic film, the polymer chain segments try to relax back to equilibrium conditions. As long as the characteristic relaxation time, τ, is much longer than the period of the oscillation, the relaxation is "frozen". Molecular rearrangements on a long-range scale are severely restricted. Distortions generated by external stress, being of a rather high energy, result in the high storage modulus. This is the glassy state. In this state the material is rigid. The shear storage modulus is about 10^9 Pa and decreases slowly with temperature. The shear loss modulus is significantly lower but increases with temperature.

The rubbery state is characterized by a deformation on a time scale larger than the relaxation time. All configurational modes of motion within the entanglement coupling points can freely occur. Distortions generated by the external stress, being of a much lower energy and G' is in the order of 10^6 Pa and $G'' < G'$. Only at still higher temperatures more and more chains can escape from topological restraints and the material is capable to flow. The temperature of the transition from glassy to rubbery consistency is the glass transition temperature. The so-called dynamic glass transition, T_α, must be applied for acoustic measurements; at this temperature $\omega\tau \approx 1$. T_α at probing frequencies, typically for acoustic devices, is a few tens of degrees higher than T_g, the static glass transition temperature, which is known, e. g., from differential scanning calorimeter (DSC) measurements.

Changes in the intrinsic polymer properties do not appear directly as changes in the electrical properties of the composite resonator. They become effective via acoustic properties of the coating through the effective surface impedance experienced by the resonator at its interface to the coating. Performing a thermal experiment with a reasonably slow cooling rate and further assuming an ideal viscoelastic film one can expect a major response of an acoustic wave device around the dynamic glass transition temperature. However, due to the transduction mechanism, a strong response from acoustic devices can also be generated from rather small material property changes within the rubbery state. They usually have an acoustic origin. By contrast, changes within the glassy state usually do not appear as significant sensor responses of quartz crystal resonators but contribute to the sensor signal of surface acoustic wave devices.

The change in the thermal expansion coefficient around T_g can hardly be recognized with an acoustic wave device, because the thickness-density product does not change. Effects at this temperature are mostly generated from deviations from the ideal behavior of the viscoelastic film, e.g., residual or thermally generated static stress. As discussed above, the different time scales in the experiment (acoustic wave frequency on the one hand and cooling/heating rate on the other hand; the same holds for solvent penetration, oxidation/reduction etc. in other kinds of experiment) allow for relaxation processes to take place and to be monitored, although they seem to be frozen on the time scale of the acoustic wave.

In contrast to the more or less broad temperature range of the glass transition, the melting of the highly ordered regions in crystalline polymers or phase transitions in liquid crystals are not subjected to the time-frequency equivalence. They are characterized by a sharp change in G', usually recognized as a sharp frequency change at a temperature independent of the probing frequency and therefore similar to textbook values.

References

1. Ferry JD (1980) *Viscoelastic Properties of Polymers*, John Wiley & Sons, New York
2. Aklonis JJ, MacKnight WJ (1983) *Introduction to Polymer Viscosity*, John Wiley & Sons, New York
3. Donth E (1992) Relaxation and Thermodynamics in Polymers. Glass Transition, Akademie-Verlag, Berlin
4. Elias Hans-Georg(1997) *Introduction to Polymer Science*, VCH, Weinheim
5. Williams ML, Landel RF, Ferry JD (1955) J. Amer. Chem. Soc. 77, 3701
6. McCrum NG, Reed BE, Williams G (1967) *Anelastic and Dielectric Effects in Polymeric Solids*, John Wiley & Sons, New York

7. Martin SJ, Frye GC, Senturia SD (1994) "Dynamics and response of polymer-coated surface acoustic wave devices: Effect of viscoelastic properties and film resonance", Anal. Chem. 66 , 2201-2219
8. Kovacs A (1964) Adv. Polym. Sci., 3 , 394

13 Combination of Quartz Crystal Microbalance with other Techniques

Ernesto Calvo[1] and Kay Kanazawa[2]

[1]Facultad de Ciencias Exactas y Naturales, Universidad de Buenos Aires
[2]Department of Chemical Engineering, Stanford University

13.1 Introduction

As implied by its acronym, the quartz crystal microbalance was first used to determine the mass of material deposited on its surface. A heuristic description of the linear relation between the change in its resonant frequency Δf, from its unloaded resonant frequency f_0, and the mass density, m', was recognized by Sauerbrey [1] and led to the now standard use of the QCM to measure mass deposition. This Sauerbrey relation is described by (Eq. (4.10)):

$$\Delta f = -\frac{2 f_0^2}{\sqrt{\rho_Q \mu_Q}} m' \tag{13.1}$$

The coefficient preceding m' is a fixed quantity, depending only on the parameters of the unloaded quartz. Using this relation, one can determine the mass density (kg/m^2) deposited onto the QCM surface. Although the range of validity of Eq. (13.1) has limits, it has created a whole industry to measure deposition. From the single measurable of Δf, one can measure the single parameter m'.

As shown in Chaps. 4 and 16, it is possible to extract more information from the QCM. For example, when m' exceeds certain values, Eq. (13.1) is no longer satisfied. A non-linear behavior is observed and the shape of the non-linearity can be used to extract information on elastic films concerning the shear modulus, μ_F, of the film (Chap. 16). This was put on a quantitative basis by Lu and Lewis in 1974 [2]. More recently, there are active studies on the determination of many other parameters using the QCM. For example, when there is a viscoelastic film on the QCM and the quartz/film is immersed in a liquid, then there are a number of parameters which are involved in QCM measurements. In addition to the quartz parameters, there is the density of the film ρ_F, the shear modulus of the film μ_F, the viscosity of the film η_F, the thickness of the film d_F, the density of the liquid ρ_L and the viscosity of the liquid η_L. It is clear that there are a large number of parameters that influence the behavior of the QCM. Most of the measurables to date have focused on two values, the change in frequency and the change in dissipation as the QCM is loaded. It is not possible to deconvolute these two variables into a determination of the materials' parameters (see Chap. 16). Therefore, it is not surprising that additional measurements in conjunction with the QCM measurements are being undertaken to increase the number of measureables.

13.2 Electrochemical Quartz Crystal Microbalance and other Techniques

One of the earliest of these hyphenated techniques is the EQCM, or electrochemical quartz crystal microbalance [3, 4]. Descriptions of EQCM works are to be found in this Chapter and in other books and publications [5]. One of the metal electrodes on the QCM serves as a working electrode which is interfaced to a conducting solution containing electroactive moieties. For example, metal cations may be dissolved in an acidic solution. By applying a negative potential to the working electrode, the cations can be deposited onto the electrode as metal atoms. The deposited mass is sensed using the QCM. In addition, the current during the deposition is also monitored and by integration, the charge per unit area Q' can be measured. If it is known from the chemistry that all of the charge is used for the deposition, the charge-to-mass ratio can then be determined. Or if some of the current goes into other reactions (for example the creation of hydrogen molecules), then the fraction of the current used to deposit the metal atoms (the Faradayc efficiency) can be measured.

There are a host of other techniques used by the electrochemical community to study the details of the reaction at the electrode [6]. For example, the kinetic behavior of the reaction can be investigated using impedance spectroscopy. In this case, a small ac voltage is superimposed onto the working electrode potential and the phase and magnitude of the resulting current are recorded. The ratio of the voltage to the current defines the electrochemical impedance and its behavior as a function of frequency can be used to study the reaction kinetics. In addition, the phase and magnitude of the mass deposited during this ac stimulation adds additional information to the study of the reaction. Another very useful technique for reaction studies has been the study of reactions under controlled flow conditions. These can be obtained using the so-called rotating disk electrode (RDE) and rotating ring disc electrode (RRDE) geometries. Both of these techniques have also been adapted for the QCM [7].

As noted above, one of the materials parameters that affects the QCM behavior is the film thickness d_F. An independent method of measuring the film thickness during deposition would be extremely useful in eliminating one of the unknown variables. The importance of this thickness determination is underscored by the fact that one of the favored techniques for deconvolving the materials parameters from the measurements is to try to fit the film behavior for a set of film thicknesses. Of course, it is assumed that the film is homogeneous and has properties unchanging with thickness. For transparent films, such as many polymer films, optical interferometry can be used for thickness determination for films which are relatively thick. Examples of these types of measurements are to be found in Forzani et al., [8] and Hinsberg et al. [9]. For very thin films in the Angstrom to nanometer scale, the technique of surface plasmon resonance (SPR) can be used. In this method, a laser probes the film/noble metal interface. This radiation generates a resonance of the free electrons in the metal, called the plasmon resonance, creating an evanescent wave which probes the film. In reflectance, θ-θ type geometry is used with the angle of incidence and reflectance being the same. Depending upon the film thickness, a minimum in reflectance is observed at a particular angle. Changes in the film thickness are tracked by tracking the angular position of the minimum.

Hinsberg has found it useful for the purposes of chemical pathway determination of tracking the IR spectrum. This is effective only when there is a window in the neighborhood of the peak(s) of interest. It has been possible to trace the evolution of absorption bands in the film during film swelling as well as the changes in the film species and solution species during dissolution by using flow cell techniques. For example, one peak in the dry film resulting from acid groups is seen to decrease in intensity during swelling, consistent with conjectured deprotonation. An exploded view of the cell used for both these IR measurements is shown in Hinsberg et al. [9].

The QCM has also been coupled with surface probe microscopes. Some beautiful work was done by Krim, Borovsky et al. using scanning tunnelling microscopy (STM) at low temperatures [10], showing the amplitudes of motion on the surface of a QCM. This has been extended to the study of the dynamics of contact studies on the surface by Borovsky et al. [11]. In a similar vein, Bailey et al. have used the atomic force microscopy (AFM) to study the surface displacements of the QCM under liquid immersion [12].

Undoubtedly, there are more hyphenated techniques which have not been discussed. For example, the use of calorimetry with the QCM has been demonstrated by Allan Smith [13]. Various endoergic and exoergic phenomena are then studied simultaneously with the QCM measurements, providing a great deal of additional information on the state of the films. It is clear that these techniques, used in conjunction with the QCM, will play an important role in the determination of film parameters, film chemistry and film kinetics.

We shall discuss in some detail the combination of EQCM with optical techniques and electrochemical microscopy.

Ellipsometry is a powerful tool to characterize thin films and is based on the analysis of intensity and phase shift of light externally reflected on a sample surface. Reflection of linearly polarized light from a surface generally produces elliptically polarized light, because the parallel and perpendicular components are reflected with different efficiencies and phase shift.

Film thickness, refractive index, and extinction coefficient can be assessed with ellipsometry using expressions derived from Fresnel equations, that relate the changes of elliptically polarized light parameters measured after incidence on the sample surface [14-16]. Other related optical methods such as surface plasmon resonance and scanning angle reflectometry have also been used for the study of layered films like thin protein films.

Ellipsometry has been extensively employed to study the growth of oxide films on metals, polymer films, multilayers self-assembled composed of linear and hyperbranched polycations and polyanions, etc. Except in few cases where spectroscopic ellipsometry was employed or ellipsometric measurements were supported by XRR, the value of a complex refractive index has been assumed on the basis of literature data to extract the film thickness values. It should be noticed, however, that film optical properties may change during film growth.

Gottesfeld et al. have shown that the combined information of ellipsometric film thickness and QCM film mass makes it possible to assess the apparent mass density of the GOx-polypyrrole films during film growth [15].

Forzani et. al. combined ellipsometry and QCM under “ex-situ” and “in-situ” conditions to study self-assembled multilayers comprised of alternated layers of the enzyme glucose oxidase (GOx) and an osmium complex attached to poly(allylamine) PAH-Os, deposited on a gold surface primed with mercapto propanesulfonate [8]. The ellipsometric parameters of the thiol film on gold in the first layer were analyzed in terms of an anisotropic single-layer model since the strong vertical interaction of the S-Au bond differs from the lateral Van der Waals interaction of thiol molecules. For the subsequent $(PAH\text{-}Os)_n(GOx)_n$ multilayers on gold, a two-

two-layer model with the anisotropic thiol film and the isotropic enzyme/polyelectrolyte film yielded identical results than an isotropic one-layer model with the substrate parameters measured after thiol adsorption to offset any effect due to the Au-S bond. Film thickness and complex refractive index for each adsorbed layer in protein-polyelectrolyte multilayers were determined for dry films and for films in contact with water, revealing the importance of water content control in these self-assembled structures. The values of complex refractive index determined from "in situ" experiments suggested that the enzyme kept its native state after the adsorption step.

Plots of acoustic mass vs. ellipsometry thickness obtained for different enzyme/polycation self-assembled multilayers built up under similar conditions yield coincident results and, therefore, it is possible to estimate the average apparent film density [8]. It should be noticed that the film thicknesses derived from acoustic and optical measurements do not coincide since water molecules which are not polarized in the visible spectral region are not detected in reflectance methods, unlike the shear wave that senses water in the film as a coupled mass. The acoustic contrast between the film and liquid depends on the difference in shear modulus (i.e. 10^6 Pa for the film and 10^4 Pa for water) and this is more pronounced than the optical contrast given by the difference in refractive index of the hydrated film and water (ca. 1.43 for the protein film and 1.33 for water). Therefore, the most highly hydrated outer regions of the film cannot be distinguished from the aqueous liquid electrolyte by the optical methods but can be sensed by the acoustic wave.

An ellipsometric areal mass density, Δm_e can be derived from the optical film thickness (d_f) and film refractive index (n_f) through the equation proposed by Feijter et al.:

$$\Delta m_e = d_f \frac{n_f - n_0}{\frac{\partial n}{\partial c}} \tag{13.2}$$

where n_o is the electrolyte solution refractive index, n_f the film refractive index, and $\partial n/\partial c$ is the refractive index increment of the adsorbed substance measured.

Höök and Kasemo have represented the protein film with an effective hydrodynamic thickness, d_{eff} and an effective density, ρ_{eff} [17]. The effective thickness can be expressed in terms of the Sauerbrey mass, Δm_{QCM}, the ellipsometric mass, Δm_e, and the respective densities of the dry protein film, $\rho_{dry} = 2.0$ g cm^{-3} and water, ρ_{water} as follows:

$$d_{eff} = \frac{\Delta m_{QCM}}{\rho_{eff}} = \frac{\Delta m_{QCM}}{\rho_{dry}\dfrac{\Delta m_e}{\Delta m_{QCM}} + \rho_{water}\left(1 - \dfrac{\Delta m_e}{\Delta m_{QCM}}\right)} \qquad (13.3)$$

The effective thickness results always larger than the ellipsometric thickness and smaller than the QCM thickness. In the few first layers the effect of the "hairy" external film/water structure, which cannot be seen by ellipsometry, becomes dominant and as the film grows, less important is the effect of the interface with the external electrolyte.

Another related optical technique that has been employed in combination with the quartz crystal microbalance is the surface plasmon resonance (SPR). As mentioned, surface plasmons are collective oscillations of the free electrons at the metal surface, which under appropriate conditions can resonate with light with absorption at certain resonant frequencies. This resonant condition is extremely sensitive to the optical properties of the media (metal film, adsorption film or monomolecular layer and liquid in contact), namely the refractive index of these media. Changes in refractive index due to the adsorption of molecules or conformational changes in the adsorbed molecules result in changes in the resonant condition which can be accurately measured.

When the incident light reaches an appropriate angle, the reflection decreases sharply to a minimum, corresponding to the excitation of surface plasmon waves in the film. The total internal reflection has been detected with a photodetector as a function of incident angle by rotating both the prism and the photodetector. The angular resolution achieved by this rotating prism approach is typically 10^{-2}–10^{-3} degrees! limited by errors in the angular position and noise in the intensity of the reflected light. For comparing different SPR detection techniques, the SPR resolution is often described in terms of the smallest detectable change in the refractive index of an analyte in refractive index units. The above angular resolution corresponds to 10^{-5}–10^{-6} refractive index units at a wavelength of 632 nm. For higher angular resolution, a large distance between the prism and the photo detector is required which makes the setup not only bulky but also more susceptible to mechanical noise and thermal drift. The response time is slow because of the mechanical movements in the setup.

Both surface plasmon resonance (SPR) and quartz crystal microbalance (QCM) involve resonance phenomena that are perturbed by changes in the characteristics of an interfacial layer. These two techniques can be used to measure film thickness with resolution on the angstrom-to-nanometer level. SPR is a noninvasive optical-measurement technique that probes the thickness and dielectric constant of thin films at a noble-metal surface. Thickness resolution is typically on the angstrom level, depending on the optical contrast between the film and the solvent, and films up to several hundred nanometers thick can be studied using this technique.

It is worth mentioning the work of Johannsmann et al. [18] who measured the second-order stress birefringence in a 2 μm film of poly-methylmethacrylate sandwiched between the top electrode of a quartz resonator and a second, semitransparent gold layer. The gold–film–gold system formed an optical Fabry–Perot resonator with characteristic dips of reflectivity at certain angles of incidence.

By combining SPR and QCM, Bailey et al. have been able to take advantage of the strengths of each technique to monitor "in situ" the interfacial behavior of thin organic films including perfluoro-poly-ether lubricant, electropolymerization of pyrrole and desorption of 1-octadecanethiol from a gold surface [19].

The stability of SPR measurements is generally quite good and sensitivity to environmental factors is minimal. However, there must be optical contrast between an adsorbed film and the solvent for the film to be seen by SPR. In addition, the refractive index of the film must be known regardless of whether a simple measurement is to provide the thickness and mass of an adsorbed layer.

Both SPR and QCM can measure the properties of a film adsorbed on a metallic electrode in contact with a liquid electrolyte. The system that utilizes these two techniques can be also under electrochemical control in a measurement cell fitted with counter and reference electrodes. This addition opens the door to the study of a vast number of systems redox systems and chemically modified electrodes. Examples of SPR-QCM combination are recent publications by Kasemo et al. who combined SPR and QCM to study variations in coupled water, viscoelastic properties, and film thickness of a protein film during adsorption and cross-linking [17]; and Willner et al., who reported the swelling of acrylamido-phenylboronic acid-acrylamide hydrogels upon interaction with glucose [20].

There are many benefits to the combination of optical techniques and EQCM measurement. Using optical techniques and QCM independently to monitor interfacial processes in situ, two measures of a process can be obtained, which rely on fundamentally different principles of physics. By carefully considering the nature of each measurement, our ability to interpret the results obtained from either technique is greatly enhanced and the weaknesses of assumptions inherent to data analysis become apparent, and differences measured by each technique can often be reconciled.

While in conventional EQCM the current that flows through the external circuit or the electrode potential is the resultant of integration at the whole electrode surface, space resolved electrochemistry can be achieved with the aid of the scanning electrochemical microscope (SECM). The SECM, introduced in 1986 by A.J. Bard, is a scanning probe technique where the probe signal is generated by an electrochemical reaction at the tip of an ultra-microelectrode [21, 22]. Several electrochemical scanning probe techniques have been described: Localized electrochemical impedance spectroscopy (LEIS) [23], scanning reference electrode (SRET) and scanning vibrating electrode (SVET) techniques [24], as well as the scanning Kelvin probe technique [25], the approach of the scanning droplet cell [26, 27] and scanning electrochemical microscopy (SECM) [28, 29], and the alternating current scanning electrochemical microscopy (AC-SECM) for measuring local interfacial impedance properties with high lateral resolution [30]. These techniques can map electrically or chemically a surface sample, depending on whether a local potential, current or impedance is recorded as a function of sample surface x-y position, or alternatively a local surface concentration is mapped. SECM can be used to study the flux of ions at membranes, polymer modified surfaces or liquid-liquid interfaces.

The scanning electrochemical microscope (SECM) can work in either potentiostatic [31] or amperometric modes [32]. In amperometric mode, a bipotentiostat controls the tip potential and in some cases the conductive substrate. The microelectrode (1-25 μm Pt or carbon) sealed at the moving tip is positioned by x,y,z-stepper motors and piezo controllers and rasters the sample surface.

In a redox electrolyte solution and far from interference by the sample surface, the microelectrode current $i_{T,\infty}$ at a disc ultra-microelectrode (UME) for the reaction:

$$O \pm e \rightarrow R \tag{13.4}$$

is given by:

$$i_{T,\infty} = 4nFDC^{*}r_{o} \tag{13.5}$$

where r_o is the radius of the tip ultra microelectrode, n is the number of electrons exchanged in the electrochemical reaction, D is the diffusion coefficient of the reacting species and C^* its analytical concentration.

When the ultra microelectrode tip is brought close to the sample substrate within few tip radii, the insulating surface blocks the diffusion of O to the tip and the UME current decreases. If the surface can regenerate O, because it is a metal where O and R can be exchanged through surface electrons, or if there is a surface catalytic process that can regenerate O from R then an enhancement in the UME current is observed.

It is important to record the tip current as a function of tip-to-substrate distance (approach curves) which in the first case show a decrease and in the second one, a current increase as the tip gets closer to the substrate.

Two operation modes are possible [33]: *feedback mode*, when regeneration of *O* takes place and *collection mode* when the tip close to the substrate surface is poised at a potential where the electroactive species *R* is detected at the UME tip. In addition, the SECM can operate at *constant current* with feedback at the z-piezo and recording the error signal of the feedback loop or at *constant height* recording the current or potential at the tip. In the constant-height mode of the SECM, the UME tip is slowly approached in the z-direction towards the sample surface, and the approach is stopped when a predefined decrease (negative feedback) or increase (positive feedback) in the diffusion limited current is detected. This z-height is kept constant during the following scanning leading to a convolution between the tip-to-sample distance and the surface activity which are both influencing the measured current through the microelectrode.

An optical method to control the constant tip-sample distance of the SECM, based on the detection of shear forces occurring between the vibrating tip and the sample surface [34] has been reported. By focusing a laser beam on the very end of the microtip which is vibrating in its resonance frequency a Fresnel diffraction pattern is generated which is focused on a split photodiode (SPD). The difference current is amplified with respect to the vibration frequency using a lock-in amplifier [35-38]. Thus, a constant distance between any tip and sample can be kept, allowing the use of non-electrode tips in SECM.

The electrochemical quartz crystal microbalance (EQCM) and the scanning electrochemical microscope (SECM) have been used independently to study the redox behavior of thin films on electrodes [39-40]. Each technique has advantages and limitations, but if combined into one instrument they can provide complementary in situ valuable information on the electrochemical and rheological behavior of polymer films and overcome the limitation of the individual techniques.

Positioning the tip of a microelectrode close to the film's surface allows us to measure concentration changes of redox-active species crossing the film/ solution interface, so we are able to identify fluxes of counter ions during oxidation-reduction cycles and to assess their influence on the film's quartz impedance parameters measured by the quartz crystal microbalance.

Several reports on the combination of an EQCM with a SECM have been published [41-44]. Hillier and Ward [41] used a single-potential control between substrate and microelectrode, where the tip acted as the counter electrode to study the radial sensitivity, mapping different quartz crystals. Cliffel and Bard [42] described a SECM/EQCM instrument under bipotentiostatic control, which was employed for studies of silver and C60 films. The EQCM in that case, however, only measured frequency changes and is thus not capable of distinguishing between acoustically thin and viscoelastic films or cases where mass and solution properties change simultaneously. Shin and Jeon [43] described the use of frequency changes with the EQCM as a method of calibrating the tip-substrate distance in the SECM. Gollas et al. reported a novel combination of an electrochemical quartz crystal microbalance (EQCM) and a scanning electrochemical microscope (SECM) with rapid in situ measurement of the modulus of the quartz crystal's transfer function [44]. This technique based on a complex voltage divider method [45-46] allows data analysis in the complex plane for the Butterworth-Van Dyke (BVD) equivalent circuit and yields the real and the imaginary components R (damping resistance) and X_L (reactive inductance) of the crystal's electro acoustic impedance around its resonant frequency of 10 MHz. The authors tested the method by cyclic voltammetry for copper plating/stripping experiments from copper(I) chloride solution of high concentration in 1 M HCl recording simultaneously four parameters, X_L, R, the substrate, and the tip current. The amperometric response of the SECM tip positioned closely to the substrate reflects the concentration changes of electroactive ions in the diffusion layer of the substrate electrode.

References

1. Sauerbrey G (1959) Z. Phys.155:202
2. Lu C., Lewis O (1972) J. Appl. Phys. 43:4385.
3. Reed C.E. et al. (1985) Anal. Chem. 57:1770.
4. Bruckenstein S. and Shay M (1985) Electrochim. Acta 30:1295.
5. Calvo E.J. et al. *Layered protein films: quartz crystal resonator frequency and admittance analysis.* In Rusling J (Ed) (2003) Biomolecular Films: Design, Function and Applications, Marcel Dekker, Inc.
6. Bard A.J. and Faulkner L.R (2001) Electrochemical Methods. Fundamentals and Applications, 2nd. Ed. John Wiley and Sons, New York.
7. Calvo E.J. and Etchenique R.A. *Kinetic applications of the electrochemical quartz crystal microbalance (EQCM).* In Compton R.G. and Hancock G. (Eds) (1999) Comprehensive Chemical Kinetics, Ch. 12, pp.461-487. Elsevier, Amsterdam.
8. Forzani E.S. et al. (2002) Langmuir 18:4020.
9. Hinsberg W.D., Kanazawa K.K. (2002) Anal. Chem. 74(1):125.
10. Borovsky, B. et al. (2000) J. Appl. Phys. 88:4017.
11. Borovsky, B. et al.(2001) J. Appl. Phys. 90: 6391

12. Bailey, L.E. et al. (2001) Langmuir, 17: 8145
13. Korobov M. V. et al. (1999) J. Phys. Chem. B. 103(8):1339
14. Azzam R. M. A. and Bashara, N. H. (1977) Ellipsometry and polarized light. North-Holland, Amsterdam.
15. Gottesfeld, S. et al. *Recent Applications of Ellipsometry and Spectroellipsometry in Electrochemical Systems*. In Rubinstein, I. (1995). Physical Electro-chemistry: Principles, methods and applications. Marcel Dekker. New York.
16. Arwin, H (2000) Thin Solid Films. 48:377
17. Hook F. et al. (2001) Anal. Chem. 73:5796.
18. Domack A. and Johannsmann D. (1998) J. Appl. Phys. 83 (3):1286.
19. Bailey L.E. et al. (2002) Langmuir 18(2):479.
20. Gabai R. et al. (2001) Phys. Chem. B. 105(34):8196.
21. Liu H.Y.et al.(1986) Am. Chem. Soc. 108:3838
22. Bard A.J. et al.(1989) Anal. Chem., 61:1221
23. Zhou F. et al. (1997) J. Electrochem. Soc. 144:1957.
24. De Wit J. W. H. et al. (1989), Electrochemical Methods In Corrosion Research VI, Pt. 1 & 2, 289(2):69.
25. Stratmann M. et al. (1994) Electrochim. Acta 39:1207.
26. Lohrengel M.M. et al. (2000) J. Anal. Chem. 367:334.
27. Suter T. and Böhni (1997) Electrochim. Acta 42:3275.
28. MacPherson J. V. et al. (Eds) (2001) Scanning Electrochemical Microscopy, Marcell Dekker Inc. New York.
29. Horrocks R. H. et al. (1993) Anal. Chem. 65:3605.
30. Ballesteros B. et al.(2002) Electrochemistry Comm., 4:134
31. Horrocks B.R. et al (1993) Anal. Chem. 65:1213
32. Wei C. et al.(1995) Anal. Chem. 67:34
33. Kwak J. and Bard A.J (1989) Anal. Chem. 61:1221
34. Ludwig, C. et al.(1995) Rev. Sci. Instr. 66: 2857-2860.
35. Shi, L.F. et al. (1998) J. Electrochem. Soc. 145: 2011
36. Kranz C. et al. (1996) Adv. Mater. 8:634
37. Hengstenberg A. et al. (2000) Chemistry - A European Journal. 6:1547
38. Hengstenberg A. et al. (2001) Angew. Chem. Int. Ed. Engl. 40:905
39. Bard, A. J. et al (1994) Electroanalytical Chemistry. 18:243
40. Mirkin, M. V. (1996) Anal. Chem. 68:177A. (1995) 91:4083
41. Hillier, A. C. and Ward, M. D. (1992) Anal. Chem. 64:2539
42. Cliffel, D. E. and Bard, A. J (1998) Anal. Chem. 70:1993
43. Shin, M. and Jeon, I. C. (1998) Bull. Korean Chem. Soc.19:1227
44. Gollas B. et al. (2000) Anal. Chem. 72:349
45. Calvo, E. J. et al. (1995) J. Chem. Soc., Faraday Trans. 91:4083.
46. Calvo, E. J. et al. (1997) Faraday Discuss. Chem. Soc. 107:141

14 Ultrasonic Systems for Non-Destructive Testing Using Piezoelectric Transducers

Antonio Ramos and José Luis San Emeterio

Departamento de Señales, Sistemas y Tecnologías Ultrasónicas, Instituto de Acústica (CSIC)

14.1 Generalities about Ultrasonic NDT

In very distinct areas of industry, the quality control of the fabrication products, as well as of the equipment used to fabricate them, is performed by means of Non-Destructive Testing (NDT) techniques. This control procedure acquires still more relevance when aspects of public security are present such as in the nuclear and aeronautics sectors.

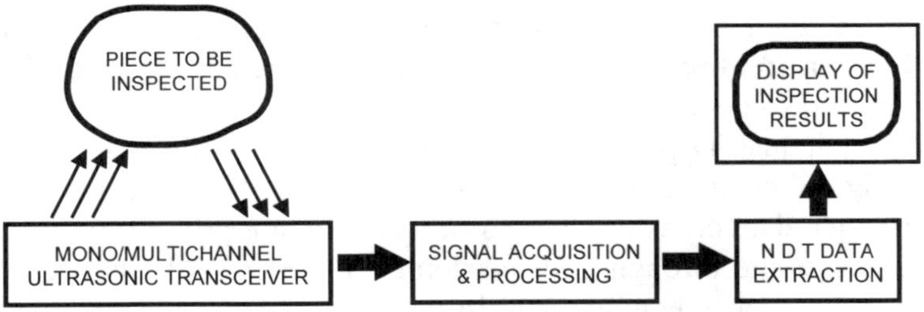

Fig. 14.1. Functional blocks scheme of an ultrasonic NDT process.

In many NDT applications, high frequency ultrasonic waves are employed. In most cases they are produced and detected by piezoelectric transducers operated in a pulsed mode. The ultrasonic radiation of these transducers is used as a tool to visualize internal parts. The propagation of these pulsed elastic waves can provide information about the internal structure of the medium by means of the analysis of reflections, scattering and attenuation in the mechanical discontinuities encountered along the ultrasonic path. Figure 14.1 shows a very general block diagram of NDT systems based on ultrasonic waves. It includes one or more ultrasonic transceivers depending on the number of piezoelectric channels involved.

Different acoustic propagation modes can be used but some of them are more frequent in NDT. Fluids propagate longitudinal waves. Solids also propagate shear or transversal waves, which are useful for testing purposes [1]. Other secondary modes are those corresponding to Rayleigh waves (over solid surfaces), the lateral or creeping waves, the Love shear waves, and the Lamb waves [2]. In laminar and tubular industrial pieces, other complex waves can propagate, thus interfering with the basic transversal and longitudinal modes [1]. Acoustic propagation, originated by lateral vibrations of the piezoelectric vibrators, has also been detected; they are known as head waves [3-5].

Longitudinal waves propagate in general with a higher velocity than transversal waves and they are the most commonly used in NDT applications.

14.1.1 Some Requirements for the Ultrasonic Responses in NDT Applications

Regardless of the acoustic propagation mode, a good efficiency in pulsed regime is required for the piezoelectric responses in ultrasonic NDT because that makes it possible to obtain high amplitudes in the collected ultrasonic pulses, thus improving the sensitivity of the NDT processes. In addition, a good bandwidth in the testing signals (short time duration) would permit to improve the time-resolution related to the axial discrimination in ultrasonic testing.

NDT of industrial pieces can be performed using different procedures depending on the geometry of the tested structure, the constructive materials and other characteristics. The two most common procedures are known as pulse-echo and through-transmission (T/T) testing. The first one is the most widely used in conventional inspections whereas the T/T method is applied in more specific cases, for instance in the inspection of large pieces of advanced materials (space, aircraft and nuclear materials) and in general when high acoustic attenuation in the testing pulses is produced through the propagation medium.

In most applications, a time-domain analysis of the received signals is performed. Thus, it is necessary to obtain a good time resolution by using compact ultrasonic pulses. In fact, NDT systems should be able to generate ultrasonic pulses as short as possible. In those aspects that depend on the piezoelectric transducers, a shortening in the ultrasonic pulses can be attained in two ways: selecting a high oscillation frequency (which increases the acoustic attenuation in the medium as a secondary effect) or increasing the transducer bandwidth (with notable internal losses in the transducer).

An undesired consequence of both ways is a minor global efficiency, which produces worse signal-to-noise ratios in the testing signals. Therefore, the design of piezoelectric transducers adequate to NDT applications is mainly a question of trade-off between these opposite requirements: high amplitude and low length of the ultrasonic pulses.

In applications requiring a frequency-domain analysis, a large bandwidth should be achieved in testing signals. Simultaneously, it is important to assure that enough ultrasonic energy distributed in many frequencies was supplied into the testing piece, in order to resolve flaw indications from electronic and acoustic noises induced in the ultrasonic signals.

An alternative way for designing broadband transduction systems, with a minor disadvantage in efficiency, is to modify the working frequency band by means of electrical methods, in such a way that the final ultrasonic band, balancing electronic and transducer contributions, is adequately broad and smooth. There are some ways to achieve this objective [6-8]. In particular, a notable enlargement of both, amplitude and bandwidth of the testing signals, can be obtained by using matching and tuning circuits properly optimised for specific transducers and electronic transceivers [9, 10].

In next sections, some cases of improvement in global NDT emitter-receiver performance will be analysed. Other improvements of NDT results, related to lateral resolution and operation time, by means of multi-channel equipment, will also be described, including the consideration of their technological implications.

14.2 Through-Transmission and Pulse-Echo Piezoelectric Configurations in NDT Ultrasonic Transceivers

Each ultrasonic channel in Fig. 14.1 usually has one of these two configurations: through-transmission (T/T) or pulse-echo. The general schemes of these testing procedures are respectively detailed in Figs. 14.2 and 14.3. In both cases, a selective damping block appears in cascade with the basic generation section. This block notably reduces certain oscillations induced in driving pulses when the emission matching circuit includes an inductive tuning [8].

In pulse-echo testing (Fig. 14.3), both piezoelectric stages share the same matching circuits (damping and tuning) because only one transducer is working successively as an emitter and as a receiver. A decoupling circuit protects the signal amplifier circuit from high-voltage impulses during driving instants.

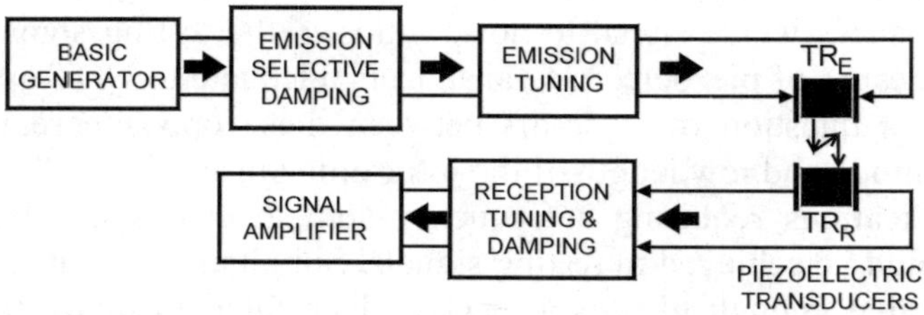

Fig. 14.2. Ultrasonic NDT inspection using a through-transmission procedure

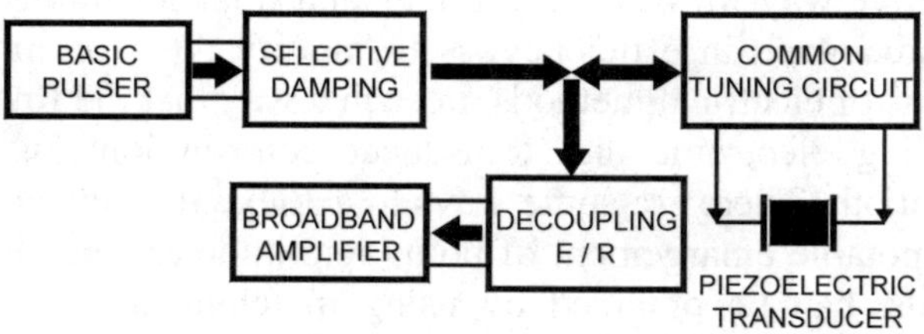

Fig. 14.3. Ultrasonic NDT inspection by means of a pulse-echo procedure

The main difference between both dispositions is that matching circuits of each stage are independent in through-transmission mode (scheme in Fig. 14.2).

Although the pulse-echo scheme is the most frequently used in NDT, an example of the T/T usefulness is the testing of aircraft manufacturing where high-speed is required during the inspection of large laminates of composite material [10]. Moreover, this alternative is the only possibility when inspecting pieces with strong acoustic attenuation.

Ultrasonic transceivers adequate to implement the testing procedures of Figs. 14.2 and 14.3 can be represented by a unique equivalent circuital diagram, as shown in Fig. 14.4. This diagram includes specific circuits for the pulsed transmitter and the input impedance in reception (parallel circuit: $L^R{}_{Tun} // C_R // R_R$).

In Fig. 14.4, two electrically independent stages were assumed, but this scheme may also represent the pulse-echo case if the amplitude of the received echo-graphic signals is always below a predetermined voltage threshold (between ± 0.6 V or ± 5 V, depending on NDT equipment). As this condition is often fulfilled, the representation in Fig. 14.4 describes the behaviour of most NDT piezoelectric transceivers.

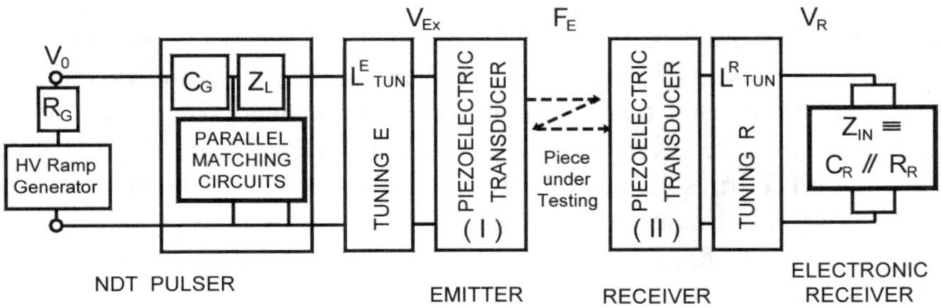

Fig. 14.4. Schematic representation of a mono-channel NDT piezoelectric transceiver

14.3 Analysis in the Frequency and Time Domains of Ultrasonic Transceivers in Non-Destructive Testing Processes

The frequency response of the transmitting stage notably influences the received testing signal V_R in Fig. 14.4, because its behaviour can be expressed as a product of the pulser response $V_{Ex}(s)$ with successive transfer functions (TF) in cascade related to: linear emission and reception in transducers I, II, respectively, ultrasonic propagation (Pr) and inductive tuning in reception (L_{tun}).

$$V_R(s) = V_{Ex}(s) \cdot TF_I(s) \cdot TF_{\mathrm{Pr}}(s) \cdot TF_{II}(s) \cdot TF_{Ltun}(s) \qquad (14.1)$$

Figure 14.4 includes an equivalent block diagram for the pulser electronics on which the response $V_{Ex}(s)$ depends. This block contains: a high-voltage ramp generator (with output impedance R_G) of amplitude V_0 in series with a capacitor C_G and in parallel with a generic matching and tuning network. The value of the series impedance $[R_G + Z_L]$ is usually very low in efficient driving cases [8]. In many NDT equipments, there are non-linear components in the matching circuits altering the trailing edge in the excitation pulse V_{Ex}, but their effects may be neglected when analysing specifically the first main part of the spike V_{Ex} in Laplace domain:

$$V_{Ex}(s) = -C_G V_o / \left[(1+s\tau)\left(s(C_T + C_G) + (1/R_\alpha) + (1/sL^E_{TUN})\right)\right] \qquad (14.2)$$

In this expression, a parallel circuit R_T-C_T has been assumed to be the equivalent pulser load related to the transducer; R_α is the equivalent parallel of R_T and the damping resistor in the matching circuit, and τ is the fall time of the real exponentially decaying ramp from a V_o value. This expression is also useful for describing the complete spikes in the frequency domain when they are relatively wide, which occurs in NDT applications developed in the low MHz range. As an example, Fig. 14.5 shows some effects of matching and tuning circuits on the frequency responses of the pulser and of the received signal V_R in a T/T system formed by two low-frequency NDT transducers (1 MHz) in contact with a plate of methacrylate 20 mm thick. In these cases, the applied driving V_{EX} were high-voltage spikes: a delta of Dirac $\delta(t)$ in case **a** and distinct broader pulses V_{SP} in **b** and **c**.

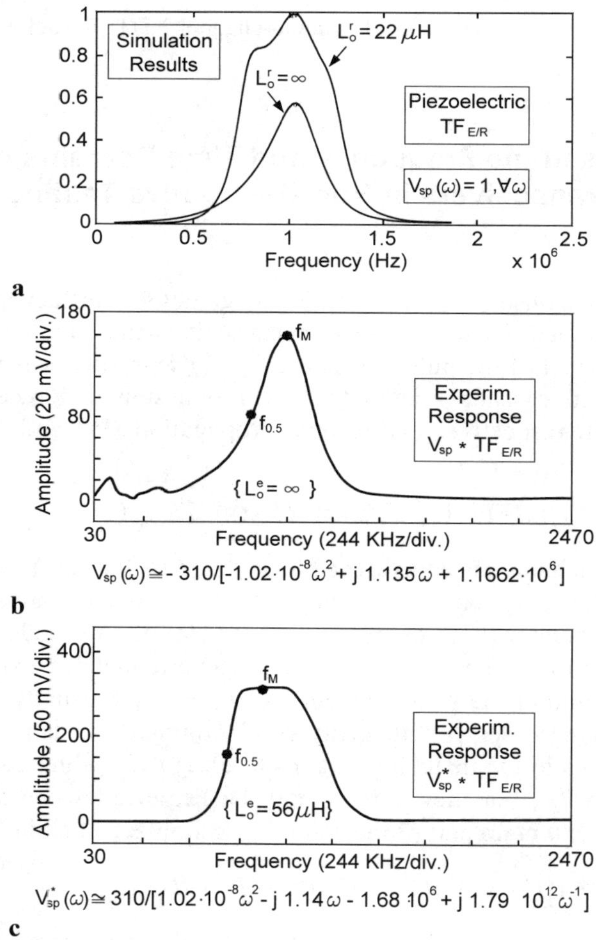

Fig. 14.5. a Simulated and experimental results related to parallel tuning (in emission L_0^e and reception L_0^r) effects on transfer functions; **b** and **c** frequency responses in NDT of a plastic plate by means of T/T testing: **a** $L_0^e = \infty$, **b** $L_0^r = \infty$, and **c** $L_0^r = 22$ µH

In the remaining cases, when the previous approach of 14.2 does not apply, the transceiver analysis can be performed by using an intermediate time-simulation step based on circuital analysis programs, like Spice. As an example of this option, the simulated pulser response in time and its experimental confirmation, obtained for the NDT emitter probe of Fig. 14.5 driven with a narrow spike and emitting in carbon fibre reinforced polymer (CFRP) composite material [11], are shown in Fig. 14.6. The force pulse emitted under this specific driving is shown in Fig. 14.7. From these results, the corresponding frequency responses could be obtained by FFT transforms.

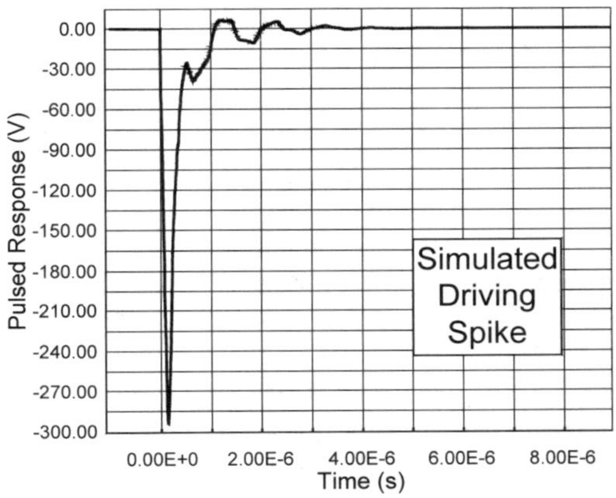

a

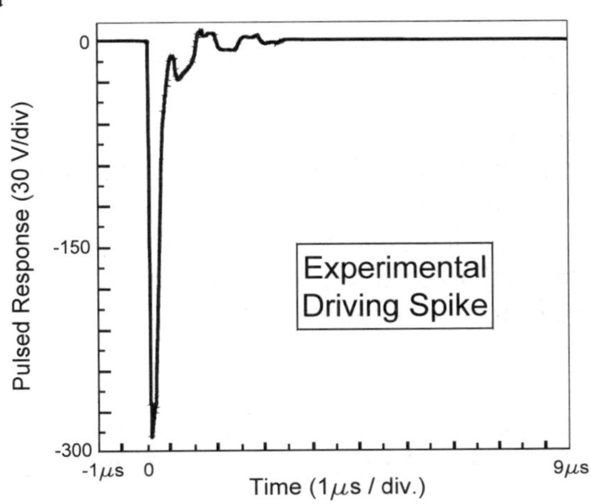

b

Fig. 14.6. Waveforms of the driving spikes with a particular pulser setting, when loading with a broadband piezoelectric NDT probe: **a** simulation result, **b** measured pulse

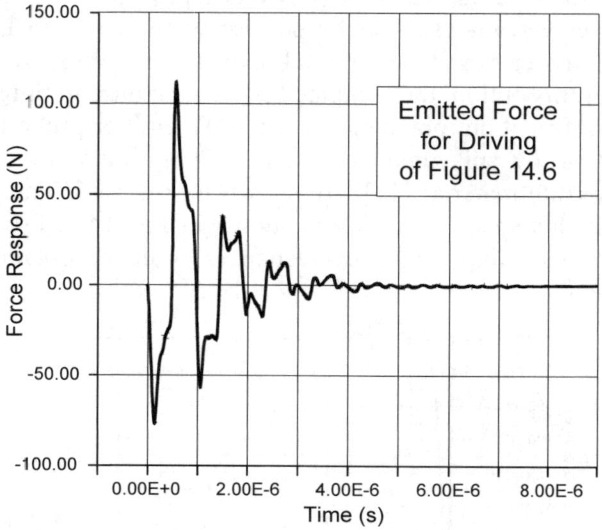

Fig. 14.7. Simulated force pulse in the probe surface F_E for driving conditions of Fig. 14.6.

The matching circuits in practical NDT ultrasonic channels are usually manually adjusted in order to find the best sensitivity and a good frequency response. Procedures to improve transceiver responses by electrical matching have been applied to pulse-echo [9, 14] and through-transmission (T/T) testing [23]; in the first case, a trade-off reasonably well fitted to the emitter and the receiver has to be achieved because impedances presented from the transmitter and the receiver electronics are very different in many cases.

On the other hand, with T/T schemes the matching circuits can be calculated with different criteria for the emitting and receiving phases. This possibility offers some advantages to better adjust these circuits, separately in each E/R stage, in order to optimise both E/R frequency bands composing a compact time pulse for testing purposes (see Fig. 14.5c).

14.4 Multi-Channel Schemes in Ultrasonic NDT Applications for High Resolution and Fast Operation

In ultrasonic NDT using high-resolution imaging or high-speed scanning of piezoelectric transducers, multi-channel schemes are adopted to perform an electronic control of the acoustic aperture. These control procedures include Rapid Scanning and Steering of ultrasonic beams, Variable Focusing of Beams and Synthetic Aperture Focusing [15, 16].

14.4.1 Parallel Multi-Channel Control of Pulse-Echo Transceivers for Beam Focusing and Scanning Purposes

By means of a dynamic focusing effect in each of the scan lines emitted from multi-element piezoelectric transducers (arrays), the lateral resolution can be improved through an artificial reduction of the beam-width. When this variable focusing has to be made quickly, electronic methods must be used; they operate with a number of channels working in parallel equal to the number of elements used as ultrasonic aperture. This procedure (see Fig. 14.8) can produce a focal distribution in the transmitted wavefronts, varying in successive time intervals [9] by means of modifications in the driving delays associated with the aperture elements. This figure includes an expression to calculate these delay τ_{ij} depending on the focal depth. Figure 14.9 shows the block diagram of an electronic implementation for a focusing / steering procedure of this type.

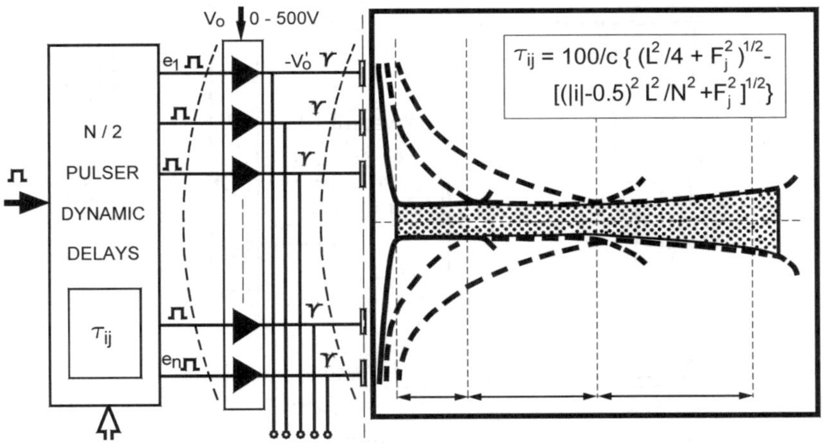

Fig. 14.8. Time-Delay driving procedure for electronic focusing of transmitted wavefronts in an array ultrasonic aperture. In the expression for τ_{ij}: F_j is the focal depth, i is the element number, N the total number of elements, and L the aperture size.

Figure 14.10 shows a snapshot of the pulsed wavefront emitted by an (4 MHz, 32 elements) ultrasonic array aperture, electronically focused at 20 mm in depth. The acoustic pressure P_j produced in the area radiated by the array is modified by the elements delays because they alter the global impulse response following the expression detailed in the same figure, in which $h_i(\mathbf{x}, t)$ is the spatial impulse response associated with the rectangular array elements [13], and $v(t)$ is the velocity in the radiating surface. The depicted transient wavefront was detected when it propagates around the focal zone. It can be appreciated how, at the focal zone, the beam-width has been drastically reduced from the unfocused beam-width (approximately of the same size than the acoustic aperture).

In a similar way, this type of beam focusing/steering process can be performed also in reception, by using the scheme detailed in Fig. 14.11, which permits to properly delay the echo-signals received in all the aperture ultrasonic channels.

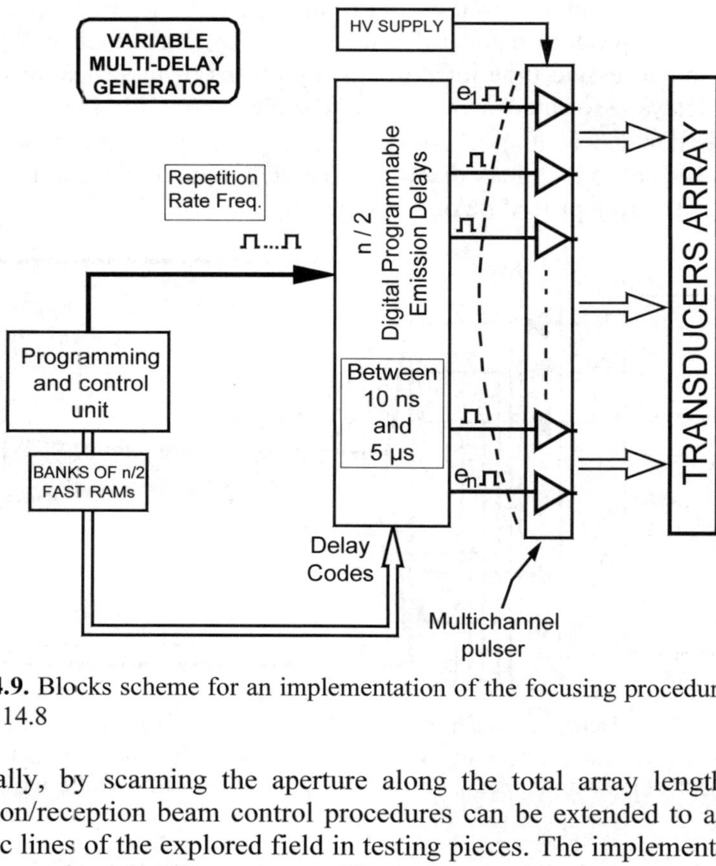

Fig. 14.9. Blocks scheme for an implementation of the focusing procedure shown in Fig. 14.8

Finally, by scanning the aperture along the total array length, these emission/reception beam control procedures can be extended to all echographic lines of the explored field in testing pieces. The implementation of these procedures implies some specific design requirements:

1. The first one is due to the short time differences between two consecutive driving pulses (as short as a few nanoseconds). For this rea-

son it is not possible to share one electronic transceiver for all the aperture transducers and further very accurate designs in the delayed transceiver are required [12].

2. In addition, the number of pulsers/receivers may range between half of the aperture elements and the total number of array elements. The number of elements of a linear array is usually about one hundred. This number increases in a quadratic way when 2-D arrays are used for bidimensional focusing. As a consequence, the basic pulser and reception units must be as simple as possible.

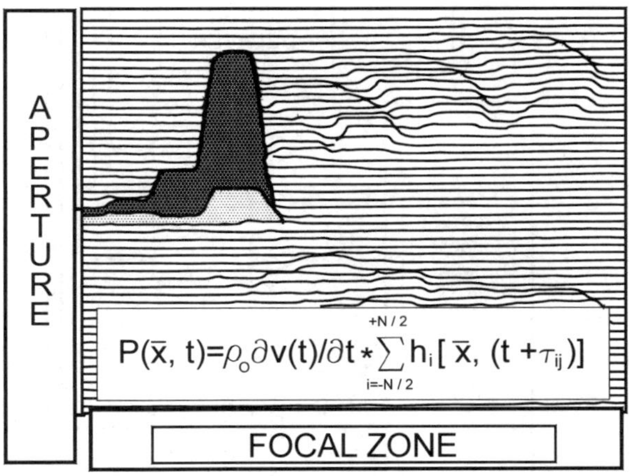

Fig. 14.10. Snapshot of the acoustic pressure field produced by a 4 MHz ultrasonic array electronically focused at 20 mm, when it propagates around the focal zone.

14.4.2 Electronic Sequential Scanning of Ultrasonic Beams for Fast Operation in NDT

There are some NDT applications requiring fast methods to make a sequential scanning of the acoustic beam during the ultrasonic examination. For instance, many ultrasonic NDT procedures employed in aircraft and nuclear areas, with mechanical scanning of a single transducer, need a drastic increase in inspection speed because the testing of these structures, normally large in size, is very time-consuming. In manual testing with pieces of complex geometry, this aspect is specially relevant.

Alternative solutions to these problems, by using an electronic scanning of multielement ultrasonic probes, are discussed in the following sections.

Individual pulsers/receivers for each NDT channel have been reported in references [19, 20]. Such full-parallel systems are bulky and very expensive.

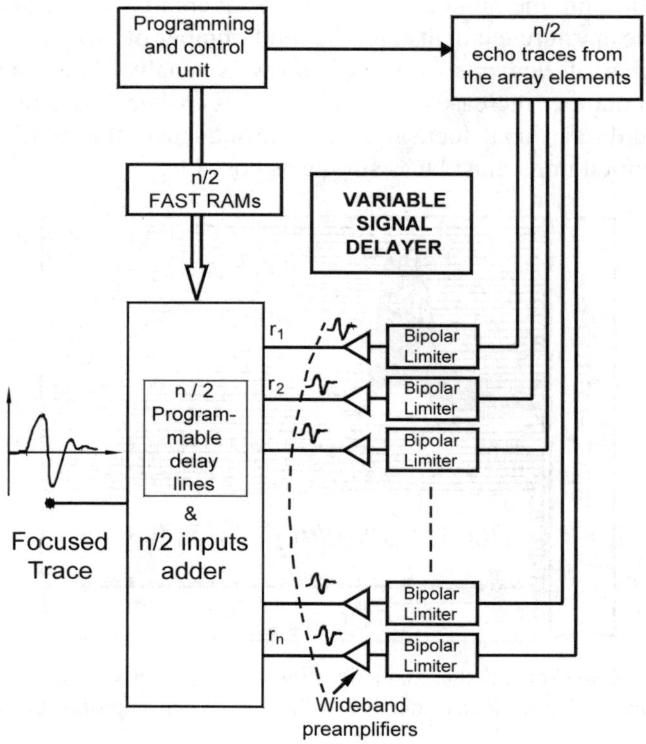

Fig. 14.11. Electronic system for dynamic focusing of received wavefronts in ultrasonic arrays by properly delaying the echo-signals in the different channels of the aperture

The use of commercial multiplexers for these purposes presents difficulties because of the high voltage in driving pulses (up to 400 V) and the low electrical impedance in the HF piezoelectric transducers.

Some acoustical imaging applications [21, 22], describe multiple switches for sequential acquisition of ultrasonic signals from the inspected medium, but with limitations for low impedance transducers and high-voltage drivers.

Figure 14.12 shows the block diagram of a NDT system based on a high-voltage analog Multiplexer/De-multiplexer device designed to fulfil the requirements commented [17, 18].

This high-voltage analogue switch network, with a low cross-coupling between channels, is able to control multiple short electrical pulses, functioning as a de-multiplexer during driving processes and as a multiplexer for the signals coming back from the tested material.

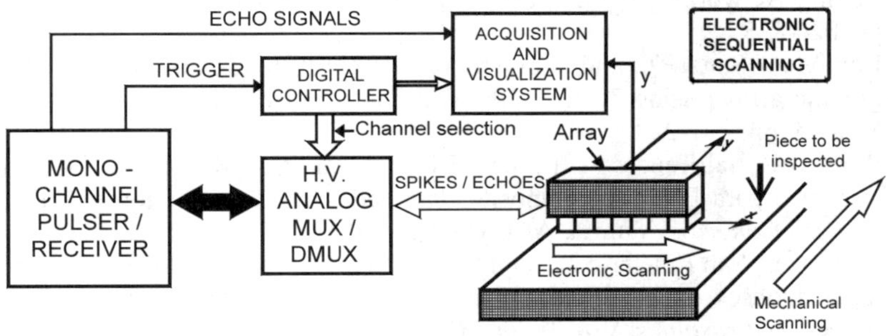

Fig. 14.12. Blocks scheme of a NDT system for rapid-scan inspection, based on a digitally-controlled high-voltage Mux/DMux

A rapid scanning of the ultrasonic beam, along *x* axis, is made by fast electronic switching of the probe elements. While the probe is moving along the *y* axis, this device sequentially switches the probe elements with a very high speed, which facilitates to obtain C-scan displays at different testing depths. Therefore, inspection times can be drastically decreased by using only one electronic pulser / receiver and the described Mux-Dmux device.

References

1. Krautkramer J and Krautkramer H (1983), *Ultrasonic Testing of Materials,* Berlin: Springer-Verlag
2. Silk MG (1984), Ultrasonic Transducers for Nondestructive Testing, Bristol: Adam Hilger
3. Uberall H (1973), *Physical Acoustics Vol. IX* (W. P. Mason, Acad. Press), New York, pp. 1-20
4. Hayman AJ and Weight JP (1979), "Transmission and reception of short ultrasonic pulses by circular and square transducers", *J. Acoust. Soc. Am.*, 66(4), pp. 945-951
5. Riera E, Ramos A, Rodríguez F (1994), "Temporal evolution of transient transverse beam profiles in near field zones", *Ultrasonics*, Vol. 32, nº 1, pp. 47-56
6. Lakestani F, Baboux JC and Fleischmann P (July 1975), "Broadening the bandwidth of piezoelectric transducers by means of transmission lines", *Ultrasonics*, Vol. 13, pp. 176-180

7. Kazys R and Lukosevicius A (May 1977), "Optimization of the piezoelectric transducer response by means of electrical correcting circuits", *Ultrasonics*, Vol. 15, pp. 111-116
8. Ramos A, San Emeterio JL, and Sanz PT (2000) "Improvement in Transient Piezoelectric Responses of NDE Transceivers Using Selective Damping and Tuning Networks", *IEEE Trans. Ultrason., Ferroelect., Freq. Cont*, Vol. 47, pp. 826-835
9. Ramos A, Sanz PT, and Montero FR (1987) "Broad-band driving of echographic arrays using 10 ns - 500 V efficient pulse generators", *Ultrasonics*, Vol. 25, pp. 221-228
10. Ramos A , San Emeterio JL, Sanz PT, and García J (May 1999) "Sistema ultrasónico para END de composites laminados con núcleos nomex", IX *NDE Spanish Congress*, Vitoria, AEEND, pp. 43-51
11. Ramos A , San Emeterio JL, and Sanz PT (2000) "Dependence of pulser driving responses on electrical and motional characteristics of NDE ultrasonic probes", *Ultrasonics*, Vol. 38, pp. 553-558
12. Certo M, Dotti D and Vidali P (July 1984) "A programmable pulse generator for piezoelectric multielement transducers", *Ultrasonics*, pp. 163-166
13. San Emeterio JL, Gómez-Ullate L (1992) "Diffraction impulse response of rectangular transducers", *Journal of the Acoustic Society of America*, Vol 92, pp. 651-662
14. Rhyne TL (1996), "Computer optimization of transducer transfer functions using constraints on bandwidth, ripple, and loss", *IEEE Trans. Ultrason. Ferroelect. Freq. Cont.*, Vol. 43, No 6, pp. 1136-1149
15. Howard PJ and Chiao RY (1995), "Ultrasonic maximum aperture SAFT imaging", *Review of Progress in Quantitative Nondestructive Evaluation (QNDE)*, NewYork: Plenum Press, Vol. 14, pp. 901-908
16. Ullate LG, Ramos A and San Emeterio JL (1994), "Analysis of the ultrasonic field radiated by time-delay cylindrically focused linear arrays", *IEEE Trans. Ultrason. Ferroelect. Freq. Cont.,* Vol. 41, No 5, pp. 749-760
17. Ramos A, Montero de Espinosa FR, Sanz PT, and Torregrosa JM (1993), "A 5 MHz high-voltage demultiplexed ultrasonic array system for rapid-scan testing of advanced materials", *Sensors and Actuators* A, No. 37-38, pp. 385-390
18. Ramos A, Sanz PT (1996) "Fast bidirectional analog switching system for HF pulses of high instantaneous power ", *USA Patent* No 5,592,031
19. Addison RC, Marsh KA, Richardson JM and Ruokancas CC (1982) "NDE Imaging with multielement arrays", *Acoustical Imaging*, Vol. 12, London: Plenum Press, pp. 643-663
20. Vos HCL, Schepers R and Vocel JA (1988), "An ultrasonic circular array transducer for pipeline and borehole inspection", *IEEE Ultrasonics Symp.*, Chicago, pp. 659-662
21. Gelly JF and Maerfeld C (1981) "Properties of a 2D multiplexed array for acoustic imaging,", *Proc. IEEE Ultrasonics Symp.*, pp. 685-689
22. Peterson DK, Bennett SD and Kino GS (1982) "Real-time NDE of flaws using a digital acoustic imaging system", *Materials Evaluation*, No. 40, pp. 1256-1262

23. Ramos A, San Emeterio JL and Sanz PT (1997) "Electrical matching effects on the piezoelectric transduction performance of a through transmission pulsed process", *Ferroelectrics,* Vol. 202, pp.71-80

15 Ultrasonic Hyperthermia

Gilberto González, Lorenzo Leija and Roberto Muñoz

Departamento de Ingeniería Eléctrica, Centro de Investigación y Estudios Avanzados (CINVESTAV)

15.1 Introduction

In the beginning, Ultrasonics in the field of medicine was oriented to applications in therapy rather than in diagnosis, using its heating and dissociation effects on biological tissues.

We call ultrasonic therapy to the use of the high-intensity ultrasounds to induce changes in the state of the tissue by means of its thermal and other effects.

Hyperthermia is a relatively new cancer treatment. Hyperthermia effects are obtained by increasing the temperature range in the tumor target to 42-45 ºC. In a solid cancerous tumor, when the rate of temperature increase of the body exceeds the ability of the regulation system to dissipate the heat, we obtain the death of the cells

Hyperthermia systems need highly accurate control of the changes in the focalization of the field resulting from the changes in the properties of tissue after heating, and in the temperature increase in the zone treated. It is easier to focus energy with ultrasound than with other hyperthermia techniques so that tumors located deeply within the body can be treated. The main problem is the accuracy in the method of directing, measuring and controlling the heat.

15.2 Ultrasonic Fields

We are only to underline some basic issues related to therapy applications. A detailed analysis of the problem can be found in [1]

The general equation for wave propagation in liquids and isotropic solids is:

$$\nabla^2\phi - \frac{1}{V^2}\frac{\partial^2\phi}{\partial t^2} = 0 \tag{15.1}$$

where ϕ is the potential from which the displacement u of particles under a longitudinal wave in an isotropic medium can be derived, and V is the longitudinal wave velocity.

For a wave, in spherical coordinates, whose components vary as $\exp(j\omega t)$, where $k = \omega/V$ we can write Eq. (15.1) as

$$-\frac{1}{R^2}\frac{\partial}{\partial R}\left(R^2\frac{\partial\phi}{\partial R}\right)+k^2\phi=0 \tag{15.2}$$

the solution is $\phi = \dfrac{Ae^{\pm jkR}}{R}$, or more generally $\phi = \dfrac{f(t \pm R/V)}{R}$

For a cylindrical planar piston transducer, considering it is placed in a rigid baffle (that is, displacement at the face $u_z(r,0) = u_0$ uniform over its radius and zero outside) and the piston transducer radiating into a liquid, Fig. 15.1, the amplitude of a wave at point z on the axis is

$$\phi(0,z)=-u_0\int_{r'=0}^{a}\frac{e^{-jkR}}{R}r'dr' \tag{15.3}$$

that can be solved as

$$\varphi(0,z) = \frac{2}{k}\,u_z(0)\sin\frac{ka}{2} \tag{15.4}$$

Then, if $z^2 >> a^2$ the paraxial approximation (small angles from the axis) can be used. Introducing the Fresnel normalized parameter $S = z\lambda/a^2$, two regions are defined:

1. Fraunhofer zone, when $z >> a^2/\lambda$ or S >> 1.
2. Fresnel zone, when $z >> a^2/\lambda$ or S >> 1.

In the Fraunhofer zone $I(z)$ decreases with $1/z^2$. In the Fresnel zone there are values of z where ϕ is maximum on the axis for $z \approx a^2/[(2m+1)\lambda]$ and $ka >> (2m+1)\,\pi$.

Figure 15.2 shows both zones for a circular planar piston transducer. The values of *x*, the distance *z*, and *a*/λ are the conditioning factors for the number of maxima and minima along the ultrasonic beam. Frequency of peaks increases when *x* decreases.

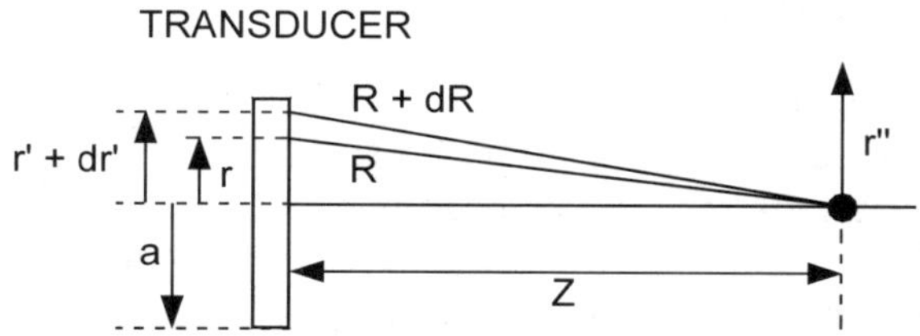

Fig. 15.1. Scheme for the study of the diffraction theory for the planar piston transducer

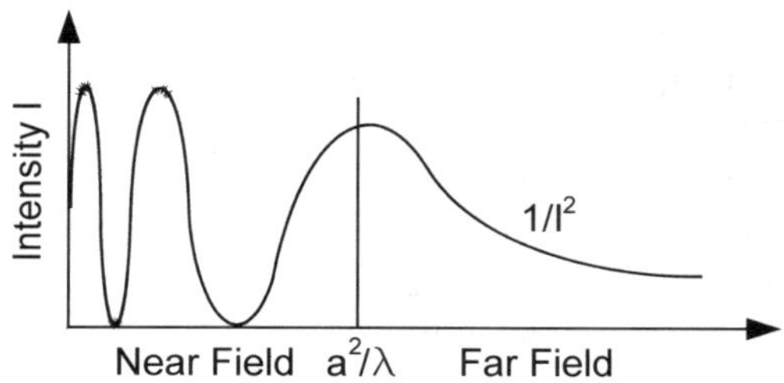

Fig. 15.2. Distribution of intensity along the axis of a transducer

The width of the beam also changes along z, narrowing towards the last axial maximum, being about one-quarter of the diameter of the transducer (-3dB beam diameter) at the last axial maximum.

The ultrasonic field is diverging after the last axial maximum and, at large distances ($z >> a^2/\lambda$ or $S >> 1$), the intensity follows the inverse square law, $I(x) \sim 1/x^2$.

The intensity distribution across the beam in the far field can be approximated by a Gaussian distribution. The energy is distributed in the main lobe (about 84%) and the side lobes (rest of energy).

15.3 Ultrasonic Generation

In the last years therapeutic ultrasound techniques have been followed with increasing interest and the future of this technique highly depends on the development of high performance transducers. They must fulfill a set of requirements, some of which are common for diagnostic transducers and others are related to the specific task of therapy, i.e. transducers capable of producing high-power waves for a long time.

15.3.1 Piezoelectric Material

The piezoelectric material most widely used in medical ultrasonic transducers is piezoelectric ceramics type PZT. Among the different kinds of piezoelectric ceramics, PZT-4 has been generally accepted in therapy applications. PZT-4 is suitable for driving high level signals.

Its main singular properties are the high resistance to depolarization, small dielectric losses under high electric fields, high electromechanical coupling, high resistance to depolarization under high mechanical stresses and great deformation ability.

The typical value for the dissipation factor (%) is 0.40, almost four times smaller than for PZT-5, commonly used in diagnosis transducers and mechanical Q =750.

Hyperthermia ultrasound applicators (planar waves) are basically in the diameter range of 3 cm^2 to 10 cm^2, with frequencies from 0.5 MHz to 5 MHz. Consequently, the treated region belongs to the nearby field.

Anomalous lateral lobes and ultrasonic field patterns having local energy peaks in some points inside and away the treated zone have been detected in single-aperture diathermy applicators. The origin of this behavior is related to the quasi-narrow band ultrasonic radiation associated with continuous-wave (CW) and long-burst pulsed modes usually adopted in the diathermy practice [2]. This behavior could strongly influence the treatment. However it is generally accepted that the proximity of maxima and minima results in a "homogenization" of the temperature distribution because of thermal conduction.

15.3.2 The Therapy Transducer

Lead zirconate titanate (or PZT) is the most widely used material. In addition, the mechanical structure of the transducer is very important. Figure 15.3 shows the basic transducer (single crystal probe) for therapy. The maximum stress wave is obtained when the thickness of the plate $d = \lambda/2$ or an odd multiple of $\lambda/2$.

Its acoustic and electrical properties are very well known and there are many references that have studied the basic problem of the planar piston transducer with the equations describing its behavior [1, 3]. Typical driving configurations of the transducer include a signal generator or an oscillator and an RF amplifier.

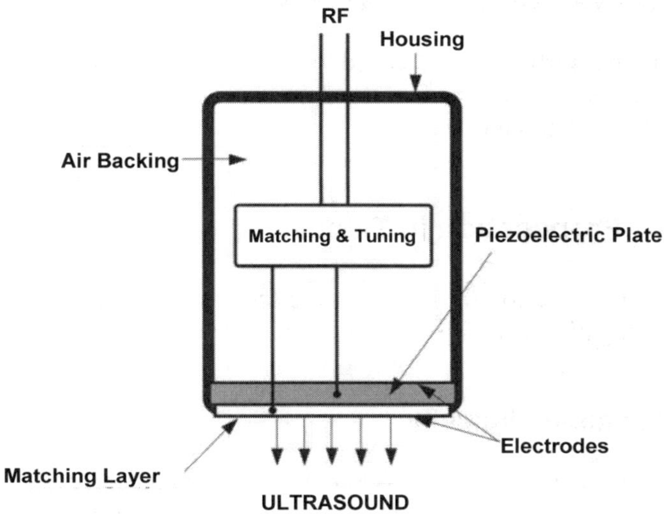

Fig. 15.3. The basic ultrasonic transducer for therapy

15.3.3 Additional Quality Indicators

We have seen that near the transducer surface the ultrasonic field has strong non-uniformity. Also, the behavior of the real transducer as an ideal planar piston is an approximation. In practice not the whole area of the transducer is effectively radiating waves. Then, it is convenient to perform the characterization of both phenomena.

15.3.4 Beam Non Uniformity Ratio

One further quality indicator is a value attributed to the Beam Non uniformity Ratio (BNR). This gives an indication of this nearby field interference. It describes numerically the ratio of the intensity peaks to the mean intensity, that is,

BNR = spatial peak intensity/spatial average intensity

where *spatial peak* means a "peak dosage at any part of the sound head", and *spatial average* the "average dosage over the entire sound head" ((0.1 to 4.0) W/cm^2 for common therapy units)".

A high BNR (>6) results in risk of "hot spot" causing periosteal pain and transient cavitation. Ratios between 2:1 and 6:1 seem to be clinically acceptable.

15.3.5 Effective Radiating Area (ERA)

It represents the portion of the transducer's surface area that actually produces US waves. It is defined as the area receiving at least 5% of maximum intensity at a depth of 5 mm.

15.4 Wave Propagation in Tissue

In ultrasonic propagation through a tissue there are two important characteristics: propagation velocity and wave attenuation.

15.4.1 Propagation Velocity

The acoustic properties of soft tissue have been studied extensively [4, 5]. Frequency dependence of the acoustic velocity may be neglected in soft tissues.

Typical values for air, water, various soft tissues, and bone are given in Table 15.1. Most soft tissues have an acoustic velocity within ±3% with an average value of 1540 m/sec. Fat is an exception with nearly 6% less.

Table 15.1. Acoustic velocities and impedances for several media.

Medium	Velocity (m/s)	Impedance (Mrayls)(Z values)
Air	330	0.00004
Water	1,480	0.148
Fat	1,450	0.138
Blood	1,570	0.161
Kidney	1,560	0.162
Liver	1,550	0.165
Muscle	1,580	0.170
Bone	4,080	0.780

15.4.2 Acoustic Impedance

When ultrasound travels through the medium it can cause changes in its acoustical properties. An important characteristic is acoustic impedance.

Acoustic impedance is defined as the complex ratio of acoustic pressure to the particle at a specific point of the acoustic field velocity $Z_{ac} = P/VS$.

For planar waves traveling in x direction, the general expression for the specific acoustic impedance is:

$$Z_{sp} = \sqrt{\rho_0 c}\left(\frac{P_+ e^{-jkx} + P_- e^{jkx}}{P_+ e^{-jkx} - P_- e^{jkx}}\right) \tag{15.5}$$

For a free planar wave traveling in one direction, $P_- = 0$ and the impedance is the characteristic impedance $(\rho_0 c)^{1/2}$.

Acoustic impedance is important in the determination of acoustic transmission and reflection at the boundary of two materials having different acoustic impedance and in the design of ultrasonic transducers. The reflected energy is the square of the difference divided by the sum of the acoustic impedances of the two materials.

15.4.3 Attenuation

It is well known that sound energy decreases with the distance traveled.

Attenuation is the decrease of sound intensity with distance. In ideal materials, sound pressure is only reduced by the scattering of the wave. Natural materials, however, produce attenuation of the sound pressure. The rate of energy flow through a unit area normal to the direction of the wave propagation is called acoustic intensity (I). If no wave distortion is assumed for a planar wave, intensity $I(x)$ at depth x is described by:

$$I(x) = I(0)e^{-2\mu x} \tag{15.6}$$

where $I(0)$ is the intensity at the surface and μ is the amplitude attenuation coefficient per unit length. Thus, wave propagation in tissue shows an attenuation of ultrasonic energy according to an exponential law.

For the amplitude attenuation coefficient we have $A_z = A_0 \exp(-\mu_a z)$, where A_z is a wave variable of the ultrasonic wave traveling in the z direction, with a starting value of A_0 (in z = 0), and $\mu_a = -(1/z) \ln (A_0/A_z)$ Nepers per centimeter [Npcm^{-1}].

This coefficient can also be expressed as $\alpha = 20\,\mu_a \log_{10} e = 8.686\,\mu_a$ decibels by centimeter, [dBcm^{-1}]. The attenuation of soft tissue is generally between (0.5 to 1) dB/(cm MHz).

For a purely elastic medium the energy in an ultrasonic field is kinetic or potential with the pressure wave in phase with the particle velocity. The viscous forces between the moving particles in a real medium play a very important role. They result in a delay between particle pressure and velocity (change in density) and, consequently, a loss of energy during each cycle.

Soft tissues have exponential attenuation coefficients that are proportional to frequency. The absorption in a viscoelastic medium should depend on the square of the frequency (f^2). In the case of tissues, absorption has shown to increase almost linearly as a function of frequency; $\alpha_c = \alpha_{c1}(f)^n$, where α_{c1} is the absorption coefficient at 1 MHz and f is the frequency in MHz, the values of α_{c1} and n depend on the tissue type and n (n= 1-1.2, for soft tissue [6]).

Absorption and attenuation are topics of great interest in order to develop efficient generators for heating tissues, and modeling this behavior. Values measured for different tissues are summarized in Table 15.2.

Table 15.2. Typical values for ultrasonic attenuation coefficients.

Medium	Attenuation (dB/cm at 1 MHz)
Air (STP)	12
Water	0.002
Fat	0.63
Blood	0.18
Kidney	1.0
Liver	0.94
Bone	15

The divergence between the measured absorption coefficient values and the estimated values based on the classical absorption theory shows that additional absorption mechanisms exist.

The classical absorption is the main effect on the heating process in therapy, but other sources of absorption are also important because of their influence on the ultrasonic energy useful in the target. Ultrasonic attenuation in tissues is the sum of the losses due to absorption and scattering.

In the scattering process, the elastic discontinuities within the tissue absorb the energy and then re-emit it in a different direction of propagation.

15.5 Ultrasonic Hyperthermia

Hyperthermia is used in the clinical management of cancer and benign diseases. High-temperature hyperthermia (>42°C) alone is being used for selective tissue destruction as an alternative to conventional invasive surgery.

Ultrasound technology has important advantages: a higher degree of dynamic and spatial heating control in comparison with other heating techniques commonly used. The control of the temperature increase, spatially and temporally as a result of hyperthermia therapy, is difficult due to the often heterogeneous and dynamic properties of tissues and the blood perfusion. This is considerably overcome by the advantages of ultrasonic hyperthermia, including a favorable range of energy penetration characteristics in soft tissue and the ability to shape the energy deposition patterns.

However, the focalized region for a simple focused ultrasound beam is too small (1-3 mm of full width half maximum (FWHM)) to heat a large tumor volume unless it is *scanned.*

In order to get a convenient temperature distribution induced during ultrasonic hyperthermia we need to consider several factors in the design of the therapeutic systems:

1. The transducer properties
2. The power deposition patterns
3. The pattern and speed of scanning
4. The output power
5. The measuring of temperature increase
6. Other factors to be considered:

- The thermal properties of the tissue treated (specific heat, thermal conduction and blood perfusion and flow). Thermal properties are both heterogeneous and dynamic [7]
- The ultrasonic properties of the tissue (absorption, density and speed of sound, nonlinear parameter and cavitation threshold).
- The geometrical characteristics of tumors, often irregularly shaped.
- The structure of tissues and tissue interfaces (gas, soft tissue-bone). The scattering and resulting reflection-refraction phenomena have an important effect on the amount of absorbed power and its distribution.

An ideal power source for hyperthermia should be able to customize power deposition to each individual tumor field.

15.6 Focusing Ultrasonic Transducers

Transducer focusing is a very important factor in ultrasonic therapy. The goal is to obtain highly focused transducers having an f-number (focal distance/active diameter relation) of the order of $f \leq 1$.

Focusing changes the shape of an ultrasound beam emitted by a transducer. This shape can be achieved by using self-focusing radiators, lenses or electrical focusing (i.e., transducer arrays that are driven by signals having the proper phase difference to obtain a common focal point).

The wavelength and the shape restrict the size of the focal region by the ratio between the apertures of the array to the wavelength.

15.6.1 Spherically Curved Transducers

At the focus in spherically curved radiators [8] the approximate half-intensity beam width (d_{xy}) can be obtained from

$$d_{xy} = 1.417\left(\frac{R}{2a}\right) \tag{15.7}$$

where R is the radius of curvature and $2a$ is the diameter of the transducer.

The axial length of the focus (d_z) is

$$d_z = 7.17\left(\frac{R}{2a}\right)^2 \tag{15.8}$$

Then, the shape of the focus is long and narrow with dimensions depending on the focusing properties of the transducer.

15.6.2 Ultrasonic Lenses

Acoustic lenses are made of materials in which the speed of sound (c_L) is different from that in the coupling medium (c_{zm}). Planar concave is a desirable shape for a lens where the generating curve of the concave surface is elliptical and $c_L > c_m$.

15.6.3 Electrical Focusing

The beam convergence can also be obtained by using phased array applicators. Using arrays of transducers the ultrasonic beams can be focused. Each element is driven by a signal of certain specific phase so that the waves generated by each element are in phase at the focal point. The elements must be small enough compared with wavelength in order to act as a point source

Figure 15.4 shows a one-dimensional array. The center-to-center spacing of the elements is a very important factor in the construction of a phased array (maximum size of the elements). This spacing must be as maximum one half of a wavelength in order to avoid grating lobes [9]. Using curve arrays make the element size larger (up to two wavelengths), [10-12].

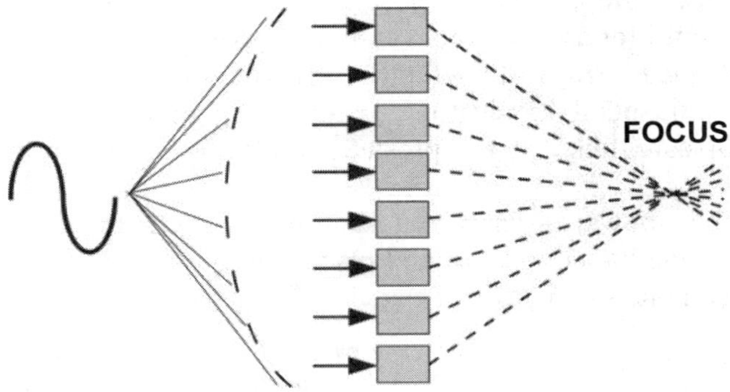

Fig. 15.4. Focusing by one-dimensional array

15.6.4 Transducer Arrays

The array theory requires that the spacing of the array elements be less than half a wavelength to avoid grating lobes, which for CW excitation can be as intense as the main lobe. In order to localize a heated volume at depth in attenuating tissue, large aperture arrays are necessary. In particular, two-dimensional arrays are desirable due to their ability to control a beam in three dimensions.

Complex beam-forming techniques can be applied to electronically focus and to drive the ultrasound energy in 3-D or to conform complex beam patterns. Versions of this technology include:

1. Spherical and cylindrical section arrays [10, 11].
2. Concentric ring arrays [13].
3. Tapered phased arrays [14].
4. Sector-vortex phased arrays [15].

These driving and control techniques have been developed and offer significant advantages over mechanically scanned and focused systems [10, 11].

15.6.5 Interstitial Transducers

The piezoelectric elements are manufactured in the shape of a cylinder with electrodes on its inner and outer surfaces. If an RF voltage is applied to the electrodes, the cylinder's wall thickness will expand and contract with the voltage. This generates a cylindrical ultrasound wave, which propagates radially outward.

To treat deep tumors that are close to a body cavity, intracavitary techniques are useful for applying conventional hyperthermia. Design and construction of linear arrays of PZT tubes was the initial development of intracavitary ultrasound applicators, [16-20], the most important for the treatment of carcinoma of the prostate and benign prostatic hyperplasia (BPH).

Figure 15.5 shows an intracavitary transducer mounted on a plastic structure that facilitated support and placement in the cavity where the tumor is only on one side [21].

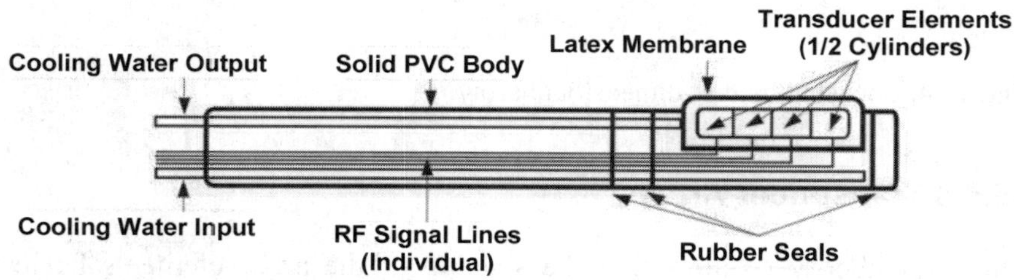

Fig. 15.5. Intracavitary applicator

The radial emission of heating energy from the length of each transducer segment, and the power applied along the length of the applicator is adjustable to the heating distribution.

Linear phased arrays of tubular sections (Fig. 15.6) have been considered to increase spatial control and depth of penetration [19, 20]. Interstitial methods are used for treating tumors or sites that are difficult to reach by external or intracavitary methods.

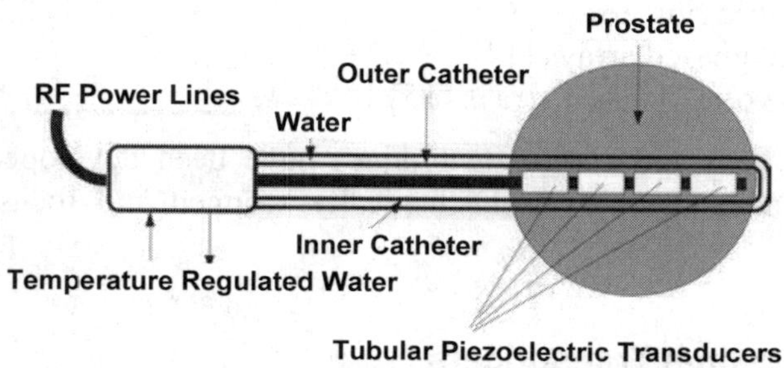

Fig. 15.6. Transrectal applicator, arrays of tubular section.

Despite the invasive nature of the techniques, the heating sources are implanted directly into the tumor, thereby localizing heat in the target volume. Several types of interstitial ultrasound applicators are currently being investigated, and can be classified as catheter-cooled and direct-coupled devices.

15.7 Trends

The technological advances in electronic and computer science would allow the potential development of improved and effective ultrasonic systems for new therapy applications. Efforts should be addressed to the determination and evaluation of computational tools and specific procedures [22] for: i) the precise analysis of the propagation in biological tissues and of their thermo mechanical effects, ii) to improve by electronic means the capacity to concentrate high-frequency ultrasonic energy in a controlled way and selective on the target of interest and iii) to minimize potential collateral damages in the neighbor tissues.

References

1. Kino GS (1987) Acoustic Waves: devices, imaging, and analog signal processing; Prentice Hall, Inc.
2. González G, Azbaid A, Leija L, Ramos A, Rami X, San Emeterio JL, Moreno E (2002) Experimental Evaluation of Some Narrow-Band Ultrasonic Transducers as Therapy Applicators, FORUM ACUSTICUM SEVILLA, Spain
3. Well PNT; Biomedical Ultrasound; Academic Press, 1977
4. Goos SA, Johnson RL, Dunn F (1978) Comprehensive compilation of empirical ultrasonic properties of mammalian tissues, J. Acoustical Soc. Am.; 64:423-457
5. Goos SA, Johnson RL, Dunn F (1980) Compilation of empirical ultrasonic properties of mammalian tissues. II, J. Acoust. Soc. Am. 68:93-108
6. Goos SA, Frizzell LA, Dunn F (1979) Ultrasonic absorption and attenuation of high frequency sound in mammalian tissues; Ultrasound Med. Biol. 5:181-186
7. Anhalt DP (1995) et al. Patterns of changes of tumor temperatures during clinical hyperthermia: implications for treatment planning, evaluation and control, Int. J. Hyperthermia, 11:425-36
8. Hunt JW (1987) Principle of ultrasound used for hyperthermia. In: Field S.B., Franconi C. (Eds.) Physics and technology of hyperthermia, NATO ASI Series E: No. 127, Martinus Nijhoff Publishers, Boston
9. Steinberg BD (1976), Principles of aperture and array system design, John Wiley, New York

10. Ebbini ES, Cain CA (1991a) Experimental evaluation of a prototype cylindrical section ultrasound hyperthermia phased-array applicator, IEEE Trans. Ultrasonics Ferroelectrics Frequency Control, 38:510-520
11. Ebbini ES, Cain CA (1991b) A spherical section ultrasound phased array applicator for deep localized hyperthermia, IEEE Trans. Ultrasonic Ferroelectrics Frequency Control, 38:634-643
12. Ebbini ES, Umemura S-I, Ibbini M, Cain C (1988) A cylindrical-section ultrasound phased array applicator for hyperthermia cancer therapy, IEEE Trans. Ultrasonic, Ferroelectrics and Frequency Control 35:561-572.
13. Ibbini MS, Cain CA (1990) The concentric-ring array for ultrasound hyperthermia: combined mechanical and electrical scanning, Int. J. Hyperthermia, 6:401-419
14. Benkeser PJ, Frizzel IA, Ocheltree KB, Cain CA (1987) A tapered phased array ultrasound transducer for hyperthermia treatment, IEEE. Trans. Ultrasonic, Ferroelectrics, and Frequency Control, 34:446-453
15. Umemura S, Cain CA (1989) The sector-vortex phased array: acoustic field synthesis for hyperthermia, IEEE, Trans. Ultrasonic. Ferroelectrics. Frequency Control, 36:249-257
16. Chapelon JY, et al (1993). The feasibility of tissue ablation using high intensity electronically focused ultrasound. Proc. IEEE Ultrason. Sympos., 2:1211-1214
17. Gelet A, et al (1996). Treatment of prostate cancer with transrectal focused ultrasound: early clinical experience. Eur. Urol., 29:174-183
18. Hutchinson EB, Hynynen K (1996) Intracavitary ultrasound paced arrays for noninvasive prostate surgery, IEEE. Trans. Ultrasonic, Ferroelectrics, and Frequency Control, 43:1032-1042
19. Hutchinson EB, et al. (1996)Design and optimization of an aperiodic ultrasound paced array for intracavitary prostate thermal therapies, Med. Phys., 23:767-776
20. Buchanan MT, Hynynen, K (1994) The design and evaluation of an intracavitary ultrasound phased array for hyperthermia; IEEE Trans Biomed Eng 41:1178-1187
21. Diederich C, Hynynen K (1991) The feasibility of using electrically focused ultrasound arrays to induce deep hyperthermia via body cavities, IEEE Trans Ultrasonic Ferroelectrics and Frequency Control; 38:207-219
22. Pulsets - Cyted international cooperative project, *Determination of Efficient Ultrasonic Patterns for Safe Therapy by Control of Distributed Pulsed Radiation* (2002-2005)

16 Data analysis and Interpretation in Bulk Acoustic Wave – Thickness Shear Mode Sensors

Yolanda Jiménez[1], Marcelo Otero[2] and Antonio Arnau[1]

[1]Departamento de Ingeniería Electrónica, Universidad Politécnica de Valencia.
[2]Facultad de Ciencias Exactas y Naturales, Universidad de Buenos Aires.

16.1. Introduction

The bulk acoustic wave - thickness shear mode resonator (BAW-TSM), whose major representative is the AT-cut quartz crystal, has been introduced in Chap. 1 as a microbalance sensor (QCM). In chapter 4, the basic concepts of modeling were introduced making the use of the resonator evident as a sensor device. Figure 4.1 showed the general schema of a quartz crystal resonator with a multilayer coating, which can be reduced to that of Fig. 16.1 by modeling a 3-layer compound resonator formed by the quartz crystal in contact with a finite viscoelastic layer contacting a semi-infinite viscoelastic medium. This reduced model is appropriate for representing a large number of applications. Changes in the physical properties of the coating are transferred to the electrical admittance or impedance of the resonator, through the acoustic load impedance Z_L (see Chap. 4 and Appendix 4.A), thus allowing its use as a sensor device.

Some typical applications of this acoustic wave sensor are the detection of special chemical species, the monitoring of electrochemical processes or the detection of biological molecules (see Chaps. 7, 9, 10 and 13).

In chapter 6, the three main steps involved in a TSM resonator sensor were introduced: 1) measurement of the appropriate experimental parameter values for a convenient representation of the resonant response of the sensor, including suitable electronic and cell interfaces (see Chap. 6); 2) extraction of the effective parameter values of the physical model used for representing the resonator sensor (2, 3 or n-layer model – see Chap. 4), and 3) final interpretation of the physical, chemical or biological phenomena responsible for the change in the effective equivalent parameters of the selected physical model.

The final interpretation indicated in the third point depends on the application, but it must be coherent with the changes in the effective physical properties of the coating, which have been used to define the physical model of the sensor. Thus, it is necessary an adequate previous interpretation of the changes in the physical properties associated with the changes in the experimental parameters measured. This is a common issue in the different applications; it is not possible to make a coherent interpretation of the chemical or biological phenomena, responsible for the changes in the experimental data, without a correct interpretation of the changes in the properties which define the model. An appropriate answer to this question is not an easy task and also depends on the data analysis. This chapter attempts to clarify this aspect and deals with the interpretation of the data in terms of changes in the effective parameter values of the physical model in Fig. 16.1.

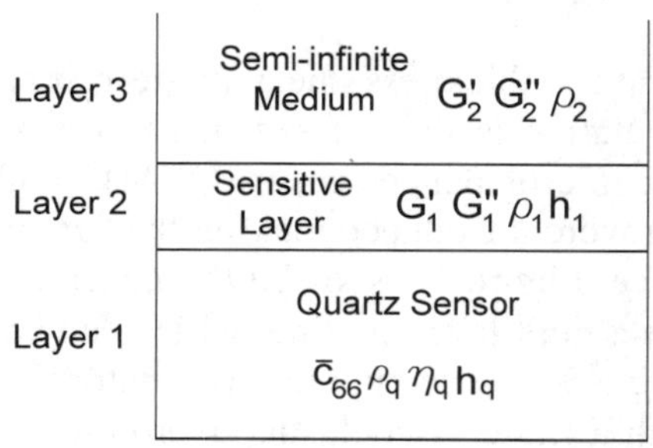

Fig. 16.1. Cross-section of a thickness shear mode resonator loaded by a two-layer viscoelastic medium

16.2. Transmission Line Model, Experimental Data and Interpretation

The most comprehensive model to characterize the electrical impedance of the compound resonator in Fig. 16.1 is the transmission line model, TLM, introduced in Chap. 4 (Sect. 4.A.3). The expression of this impedance (Eq. (4.A.12)) is written again for practical reasons as follows.

$$Z = \frac{1}{j\omega C_0}\left(1 - \frac{K^2}{\alpha}\frac{2\tan\frac{\alpha}{2} - j\frac{Z_L}{Z_{cq}}}{1 - j\frac{Z_L}{Z_{cq}}\cot\alpha}\right) \tag{16.1}$$

When parasitic capacitances C_p are taken into account, the complete admittance is:

$$Y = \frac{1}{Z} = j\omega C_0^* + \frac{1}{Z_m} \tag{16.2}$$

where $C_o^* = C_o + C_p$ and Z_m is given by the following expression (Eq. (4.A.21)):

$$Z_m = \frac{1}{\mathrm{j}\omega C_0}\left(\frac{\frac{\alpha}{K^2}}{2\tan\frac{\alpha}{2}} - 1\right) + \frac{1}{\omega C_0}\frac{\alpha}{4K^2}\frac{Z_L}{Z_{cq}}\frac{1}{1 - \frac{\mathrm{j}\frac{Z_L}{Z_{cq}}}{2\tan\frac{\alpha}{2}}} = Z_m^0 + Z_m^L \tag{16.3}$$

Z_L in the former equation is the acoustic load impedance acting at the surface of the quartz plate (Eq. 4.A.26), which corresponds with the following equation for the model in Fig. 16.1:

$$Z_L = Z_{c1}\frac{Z_{c2} + \mathrm{j}Z_{c1}\tan\left(\omega\sqrt{\frac{\rho_1}{G_1}}h_1\right)}{Z_{c1} + \mathrm{j}Z_{c2}\tan\left(\omega\sqrt{\frac{\rho_1}{G_1}}h_1\right)} \tag{16.4}$$

where Z_{c1} and Z_{c2} are the characteristic acoustic impedances of the layers, given by Eq. (4.A.25b).

Once the appropriate mathematical model for the physical model shown in Fig. 16.1 has been explicitly stated, it is important to define which magnitudes are considered as experimental and which as unknowns magnitudes to be extracted from it. The total admittance, Y, is the experimental data and the acoustic load impedance is the physical magnitude to be extracted and interpreted. Additional parameters of the sensor must be obtained with the appropriate calibration, such as: thickness h_q, viscosity η_q, static capacitance C_o of quartz and parasitic capacitance C_p. To a certain extent, they can also be considered as unknowns, especially C_o; this aspect will be discussed in more detail later on.

The complex acoustic load impedance, Z_L, can be directly obtained from experimental data. However, the extraction of the physical parameters associated to Z_L is a major problem [1-4]. Equation (16.4) depends on 6 unknowns: the density ρ_1, the thickness h_1 and the storage and loss shear moduli, G_1' and G_1'', of the first layer and the storage and loss shear moduli, G_2' and G_2'', of the medium (see Chap. 12). Thus, it is impossible, in principle, to extract 6 unknowns only from two data: the real and imaginary parts of Z_L evaluated from experimental data; even though the shear moduli of the second medium are known, four parameters remain. In principle, more than two data could be considered if the acoustic load impedance is evaluated at each frequency in a range around resonance, but the practical usefulness of these data will depend on the sensitivity of the acoustic load impedance change in the range of frequencies considered, and the magnitude of this change in relation to the unavoidable errors during experiment.

In chapter 13, the problem of extracting the physical parameter values from the typical experimental data (frequency shift Δf_s and resistance change ΔR_m – see Chap. 6) was presented. Alternative techniques, which could provide additional knowledge of the loading in order to reduce the number of unknowns, were introduced. However, there are cases where the TLM (Eq. 16.1) can be simplified, thus reducing the number of unknowns and making the interpretation of the data easier; in these cases it is necessary to know the limits of these approximations to minimize the error in the interpretation. In other applications the TLM should not be simplified and additional knowledge of the loading properties by alternative techniques [1-8], or a major definition of the experimental resonant behavior [9-11], for instance by the complete plot of the admittance around resonance [11, 12], in addition to appropriate data analysis, are required to provide an adequate interpretation in terms of the effective physical properties of the model. In order to clarify the problem it will be divided in different sections later on, but first the typical experimental parameters for sensor characterization and its relation to the acoustic load impedance will be analyzed, which will be very useful for an appropriate interpretation, at least in simple cases.

16.2.1. Experimental Parameters for Sensor Characterization

From Eq. (16.2), an evaluation of the total motional impedance as a function of the experimental data Y and C_o^* can be obtained as follows:

$$Z_m = Z_m^o + Z_m^L = \left.\frac{1}{Y - j\omega C_0^*}\right|_{EXP} = Z_m\big|_{EXP} \tag{16.5}$$

In the previous equation Z_m is the motional impedance evaluated from the TLM and $Z_L|_{EXP}$ is the motional impedance evaluated form the experimental data. If the sensor in the unperturbed state, or perturbed with a known acoustic load impedance (see Sect. 16.3.2), has been calibrated by an adequate selection of parameters h_q, C_o and η_q, the motional impedance in the unperturbed state can be evaluated by Eq. (16.3) with $Z_L = 0$. Thus, the part of the motional impedance associated to the load can be solved from Eq. (16.5) and results in

$$Z_m^L\Big|_{EXP} = Z_m\Big|_{EXP} - Z_m^o\Big|_{CAL} = \frac{1}{\omega C_0}\frac{\alpha}{4K^2}\frac{Z_L}{Z_{cq}}\frac{1}{1-\dfrac{\mathrm{j}\dfrac{Z_L}{Z_{cq}}}{2\tan\frac{\alpha}{2}}} \tag{16.6}$$

where the first member in the previous equation has been derived from experimental data and sensor calibration and the second member is the part of the TLM to be adjusted to experimental data.

Solving Z_L in Eq. (16.6) results in

$$Z_L = Z_{cq}\frac{4K^2\omega C_o}{\alpha} Z_m^L\Big|_{EXP}\frac{1}{1+j\dfrac{4K^2\omega C_o}{\alpha\, 2\tan\dfrac{\alpha}{2}}Z_m^L\Big|_{EXP}} \tag{16.7}$$

Thus, for an accurate evaluation of the acoustic load impedance, it is necessary to measure the admittance (both the conductance G and the susceptance B) of the sensor, at least at one frequency, the parasitic capacitance C_p and to make an appropriate calibration of the resonator (h_q, C_o and η_q). This can be "easily" made with a network or impedance analyzer or with special mini impedance analyzers specifically designed for this purpose [13]. However, the typical experimental measures for the sensor characterization, usually employed by researchers, are the resonant frequency and motional resistance shifts (see Chap. 6). What is the reason?

Let us analyze the term accompanying the unit in the denominator of Eq. (16.7). In order to evaluate its magnitude we will consider a typical AT-cut 10 MHz resonator with $C_o = 5$ pF and a quartz electromechanical coupling factor $K = 0.089$. Also a 1000 Ω modulus load impedance will be considered and a wave phase shift in the quartz $\alpha = 0.9\,\pi$ (near the resonant frequency the phase shift α tends to π and $\tan \alpha/2$ tends to ∞, which makes null the global term in a first approach). With the considered parameters we attempt to evaluate a more real approximation, which results in:

$$\frac{4K^2\omega C_o}{\alpha\, 2\tan\dfrac{\alpha}{2}}\left|Z_m^L\right|_{EXP} \approx 5.6\cdot 10^{-4} \tag{16.8}$$

Therefore, in many cases, the acoustic load impedance in Eq. (16.7) can be approximated to:

$$Z_L = Z_{cq} \frac{4K^2 \omega C_o}{\pi} Z_m^L \Big|_{EXP} = \frac{\pi Z_{cq}}{2\omega_o L_q} \Delta Z_m \Big|_{EXP} \tag{16.9}$$

It can be noticed that Eq. (16.9) also results from Eq. (16.6) for $Z_L << Z_{cq}$ which is the so-called "*small surface load impedance condition*".

In the previous equation it has been assumed the resonant frequency of the resonator $\omega = \omega_o$, and the ideal wave phase shift $\alpha = \pi$. Thus, Eq. (16.9) can be split in two, associated with the real and imaginary parts, as follows:

$$R_L = \frac{\pi Z_{cq}}{2\omega_o L_q} \Delta R_m \tag{16.10a}$$

$$X_L = \frac{\pi Z_{cq}}{2\omega_o L_q} \Delta X_m \tag{16.10b}$$

where R_L and X_L are the real and imaginary parts of the acoustic load impedance, Z_L, and ΔR_m and ΔX_m are the real and imaginary parts of the motional impedance shift evaluated from experimental data (from now on the subscript *EXP* will not be included for simplification). The motional resistance shift can be directly calculated as the shift in the reciprocal of the maximum conductance value with a very low error in most practical cases [14]. However, ΔX_m can be associated with the motional frequency shift if one assumes that the BVD circuit is appropriate for characterizing the loaded resonator (this is the main approximation which is not completely fulfilled in general, but is an optimal approach in a great deal of applications). It is also necessary to assume that the motional capacitance of the BVD circuit remains constant and that the load only increases the motional inductance (this condition is not fulfilled, for instance, near the film resonance). In these conditions, the shift in the imaginary part of the motional impedance is $\Delta X_m = L_L \omega$, which can be related to the motional frequency shift as follows [15]:

$$\Delta f_s \approx -\frac{f_o}{2} \frac{L_L}{L_q} = -\frac{f_o}{2} \frac{L_L \omega_o}{L_q \omega_o} \rightarrow \Delta X_m \approx -2 L_q \omega_o \frac{\Delta f_s}{f_o} \tag{16.11}$$

Thus, Eq. (16.10b) results in:

$$X_L = -\frac{\pi Z_{cq}}{f_o}\Delta f_s \qquad (16.12)$$

Equations (16.10a) and (16.12) correspond to the acoustic load approximation already introduced in Chap. 4 (Eqs. (4.A.31) and (4.A.32)). Thus, the reason why Δf_s and ΔR_m are the typical experimental data for the characterization of loaded sensors is because the acoustic load impedance can be directly related with them, with a high level of accuracy. These parameters can be directly obtained from the frequency corresponding to the maximum conductance and the magnitude of the conductance at this point, with regard to those values in the unperturbed state, as follows:

$$\Delta R_m = \frac{1}{G_{\max}} - \frac{1}{G^o_{\max}} \qquad (16.13a)$$

$$\Delta f_s = f_{G\max} - f^o_{G\max} \qquad (16.13b)$$

where the superscript "o" indicates the unperturbed state.

An important implication of these results, from a practical point of view, is that the conductance of the sensor is not affected by the parasitic capacitances in parallel.

Some researchers have used the value of the half power spectrum of resonance $W = f_{1/2}^+ - f_{1/2}^-$ (see Chap. 4) instead of ΔR_m. This provides coherent units for the experimental data, which makes the comparison between Δf_s and W easier; however, W can only be explicitly related to the motional resistance change in case of BVD circuits through the electrical quality factor (see Sect. 4.4); even in BVD circuits the change in the electrical quality factor is not only dependent on the motional resistance. Thus, the use of W can offer additional information only if it is used with the complete expression of the TLM, otherwise it gives a lack of accuracy in the evaluation of ΔR_m.

Well-designed oscillators may provide straight values for Δf_s and ΔR_m (see Chap. 6). Impulse or decay method techniques extract the resonant frequency f_o and the decay constant τ_m assuming the sensor characterized by a BVD circuit. As in the case of W, the decay constant, τ_m, can be related to the motional resistance through the electrical quality factor, thus leading to the same problem. Other techniques, such as the transfer function method used by Etchenique and Calvo [3], assume the sensor characterized by a BVD circuit and obtain the motional resistance and motional resonant frequency by fitting (see Chap. 6). These techniques are very simple and useful but pose a problem near the limits of application of the

BVD circuit for characterizing the sensor such as, for instance, near film resonance or for heavy loads. Apart from the errors associated to the characterization parameters dealt with in Chap. 6, simple approximate expressions can be obtained to relate the acoustic load impedance to experimental data. As an example, for the typical 10 MHz AT-cut resonator used to calculate Eq. (16.8) the following simple expressions are obtained from Eqs. (16.10a) and (16.12) and $Z_{cq} = 8{,}852{,}147$ N s m^{-3}:

$$R_L \approx 30\Delta R_m \tag{16.14a}$$

$$X_L \approx -3\,\Delta f_s \tag{16.14b}$$

These simple expressions permit an easy interpretation of the changes in the parameters of the physical model in Fig.16.1, but only in these simple cases, as we will see next.

16.2.2 Interpretation of Simple Cases

There are applications where the number of unknowns is reduced due to the nature of the load; some of these cases have already been presented in Chap. 4 as special cases. Next, these and other special cases will be discussed from the interpretation point of view.

The analysis made in the previous section has shown that, within those cases when the load acoustic approximation formalized through Eqs. (16.10a) and (16.12) is valid, the real and imaginary parts of the acoustic load impedance Z_L are directly related to the motional frequency and resistance shifts (in the simple cases analyzed in the next section, this approximation is valid). Thus, for data interpretation it is exactly the same to use Δf_s and ΔR_m or directly R_L and X_L. However, the use of R_L and X_L has the following important advantages: 1) they have a direct physical meaning regardless of the resonator, while the values of Δf_s and ΔR_m for the same values of R_L and X_L depend on the resonator; for instance, Eqs. (16.14a-b) establish that $R_L/X_L \approx -10\,\Delta f_s/\Delta R_m$, but this relation is only valid for that 10 MHz AT-cut quartz resonator. This is an important observation since the relative magnitudes between R_L and X_L have a relevant physical significance, as we will see next, 2) R_L and X_L, in contrast to Δf_s and ΔR_m, have coherent units and their magnitudes can be compared. This is another important observation because the direct use of Δf_s and ΔR_m, without performing the transformation into R_L and X_L, could make the evident relative weights between R_L and X_L unnoticed. This effect can be understood through Eqs. (16.14a-b) where, for a 10 MHz resonator, the magnitude of R_L is reduced 30 times when ΔR_m is used, while X_L is reduced only 3 times when using Δf_s, and 3) the use of R_L and X_L permits the representation of the data by displaying X_L vs. R_L, what clarifies the interpretation in terms of the physical properties of the model, at least in the simple cases analyzed next.

One Sauerbrey-Like-Behavior Layer

The acoustic load impedance of a thin rigid layer is (Eq. 4.A.35):

$$Z_L = \mathrm{j}\omega\rho h \tag{16.15}$$

In this case, $R_L = 0$, which is the condition to validate the Sauerbrey condition. It should be noted that a polar plot Δf_s vs. ΔR_m used to display the data for interpretation would reduce 10 times the relative sensitivity of R_L with regard to X_L for a 10 MHz AT-cut quartz. Thus, erroneously one could make valid the Sauerbrey condition.

In this case, the only experimental parameter needed to extract the property of interest is Δf_s which is directly related to the surface mass density ρ_s through Sauerbrey's equation (Eq. (4.A.37)). For a 10 MHz AT-cut quartz, combination of Eqs. (16.14b) and (16.15) gives the following simple expression:

$$\rho h = \rho_s = -\frac{3}{2\pi f_o}\Delta f_s \approx 4\Delta f_s \text{ pgr cm}^{-2}\text{Hz}^{-1} \tag{16.16}$$

Thus, when the Sauerbrey behavior of the coating has been validated, the interpretation of the experimental data is simple: *a decrease or increase in the resonant frequency shift, with regard to the resonant frequency in the unperturbed state, means a decrease or increase in the surface mass density, ρ_s, deposited on the sensor*. However, even in this simple case, the extraction of the two remaining parameters, density ρ and thickness h, starting from the surface mass density, requires the knowledge of one of them.

On the other hand, when the TLM is applied to these cases, the response of the sensor is due almost completely to the inertial contribution of the coating; the contribution of the viscoelastic properties of the load to the sensor response is so small that any minimal mismatch between the experiment and the TLM, or any error due to experiment, make the error in the extracted viscoelastic properties very important.

One Semi-Infinite Newtonian Liquid

The acoustic load impedance generated by a semi-infinite viscous (Newtonian) liquid is (Eq. (4.A.39)):

$$Z_L = \frac{(1+j)}{\sqrt{2}}\sqrt{\rho G''} \tag{16.17}$$

In this case $R_L = X_L$, which is the condition to validate the Newtonian behavior. The parametric plot X_L vs. R_L would show a straight line of slope equal to the unit. The values of R_L and X_L increase proportionally to $\sqrt{\rho G''}$.

Kanazawa's equation (Eq. (4.A.41)) gives an appropriate characterization of the medium from the experimental measure, Δf_s. Equations (16.14b) and (16.17) provide a simple expression of Kanazawa's equation for a 10 MHz AT-cut quartz as follows:

$$\sqrt{\rho G''} \approx 4\Delta f_s \tag{16.18}$$

Since $R_L = X_L$, only one experimental datum is necessary to extract $\sqrt{\rho G''}$, but it is necessary to measure both Δf_s and ΔR_m to validate the Newtonian condition which, from Eqs.(16.10a) and (16.12), is:

$$\frac{\Delta R_m}{2\omega_o L_q} = \frac{|\Delta f_s|}{f_o} \tag{16.19}$$

When the experimental data ΔR_m and Δf_s comply, with low mismatch, with the former expression, one can assume that the semi-infinite medium is Newtonian and an easy interpretation of the physical properties can be made: *an increase or decrease in Δf_s or ΔR_m implies an increase or decrease of $\sqrt{\rho G''}$*. The extraction of the remaining parameters ρ and G'', also requires the knowledge of one of them, typically density.

Equation 16.19 can be used from a different point of view. As it will be seen later, the appropriate calibration of the resonator is important to fit the TLM to experimental data. Thus, a proper calibration of h_q, η_q and C_o is necessary, specially C_o. Equation 16.19 can be used to calibrate the value of C_o when one knows that the semi-infinite medium is Newtonian. Hence, the value of C_o can be extracted from Eq. (16.19) with Eq. (4.A.30d) as follows:

$$C_o = \frac{|\Delta f_s|}{\Delta R_m}\frac{\pi}{8K_o^2 f_o^2} \tag{16.20}$$

The former equation can provide considerable errors when the experimental setup modifies, to a large extent, the response of the resonator. It can occur, for instance, when a galvanostat is connected to the sensor for EQCM applications. In this case, the complete TLM must be used to perform the calibration (see Sect. 16.3.2).

One Semi-Infinite Viscoelastic Medium

A semi-infinite viscoelastic medium generates a surface acoustic impedance:

$$Z_L = \sqrt{\rho G} \tag{16.21}$$

where $G = G' + jG'' = |G|(\cos(\delta) + j\sin(\delta))$ and $Q_L = \tan\delta = G''/G'$.

From Eq. (16.21) R_L and X_L can be solved as follows:

$$R_L = \sqrt{\frac{\rho G''}{2}}\sqrt{\sqrt{Q_L^2+1}+Q_L} \tag{16.22a}$$

$$X_L = \sqrt{\frac{\rho G''}{2}}\sqrt{\sqrt{Q_L^2+1}-Q_L} \tag{16.22b}$$

In this case, $R_L \neq X_L$ and both parameters increase with $\sqrt{\rho G''}$ for Q_L = cte, as shown in Fig.16.2a where the Newtonian case has also been included as limit for $Q_L = 0$. Figure 16.2b shows the variation of X_L and R_L as Q_L increases and $\rho G''$ remains constant, showing the transition between Newtonian to viscoelastic medium. On the other hand, the polar plot X_L vs. R_L in Fig. 16.3 shows straight lines only dependent on the value of Q_L.

The combination of Eqs. (16.22a) and (16.22b) provides an explicit expression of the quality factor Q_L as a function of the two magnitudes, as follows:

$$Q_L = \frac{1}{\tan\delta} = \frac{1-\left(\dfrac{X_L}{R_L}\right)^2}{2\dfrac{X_L}{R_L}} \tag{16.23}$$

The former expression permits a rapid interpretation of changes in Q_L from the ratio X_L/R_L and then from the experimental data $|\Delta f_s/\Delta R_m|$: *in semi-infinite viscoelastic media the quality factor Q_L decreases when the ratio X_L/R_L increases, and viceversa; when X_L/R_L is maintained constant, the quality factor remains constant as well.*

From Eqs. (16.22a-b), the properties of the medium are [16]:

$$G' = \frac{R_L^2 - X_L^2}{\rho}\ ;\ G'' = \frac{2R_L X_L}{\rho} \tag{16.24}$$

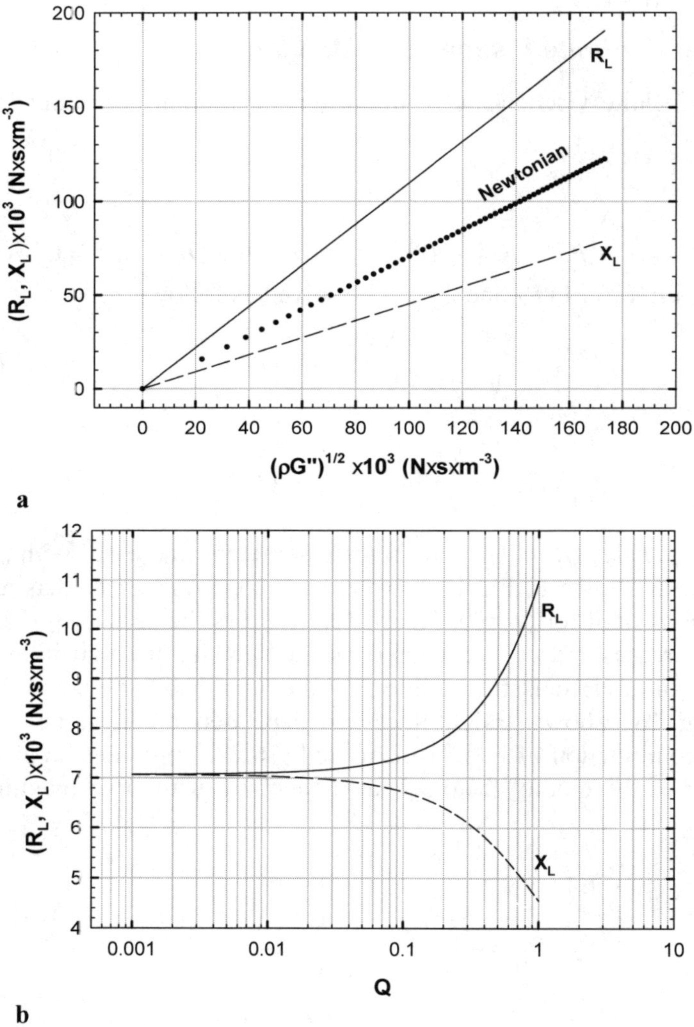

Fig. 16.2. Plots showing the variation of R_L and X_L: **a** as a function of $(\rho G'')^{1/2}$ for $Q_L = 1$, and **b** as a function of Q_L for $(\rho G'')^{1/2} = 10^4$ N s m^{-3}, where $\rho = 1$ Kg m^{-3} and $G'' = 10^5$ N m^{-2}

Notice that in this case X_L and R_L do not provide separately useful information of the medium properties. However, the parameter X_L/R_L is very useful for a simple interpretation of the relative changes between G' and G'' through the quality factor Q_L, as shown in Fig. 16.3.

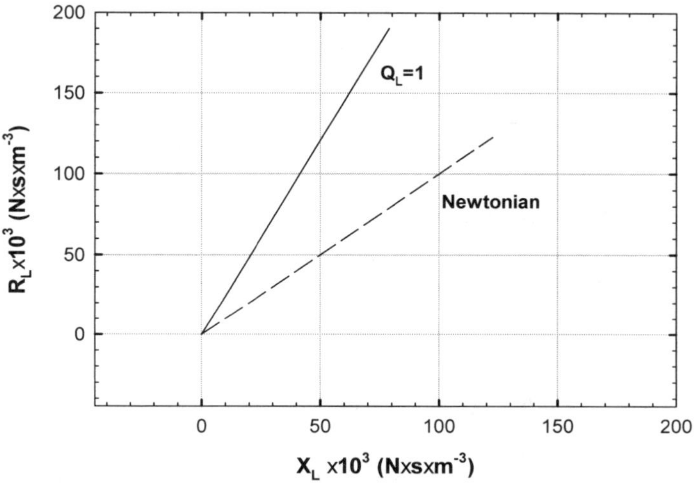

Fig. 16.3. Polar plot X_L vs. R_L for viscoelastic media with $Q_L = 1$

In this case, Kanazawa's extended equation is appropriate to describe the properties of the medium starting from the experimental data Δf_s and ΔR_m. The expression corresponding to an extension of Kanazawa's equation can be found elsewhere [14, 15] and is:

$$\Delta f_s = -\frac{f_o}{\pi Z_{cq}}\sqrt{\frac{\rho G''}{2}}\sqrt{\sqrt{Q_2^2+1}-Q_2} = \Delta f_N\sqrt{\sqrt{Q_2^2+1}-Q_2} \qquad (16.25)$$

where Δf_N is Kanazawa's frequency shift for the Newtonian liquids given.

The expression for ΔR_m is [14, 15]:

$$\Delta R_m = \frac{2\omega_o L_q}{\pi Z_{cq}}\sqrt{\frac{\rho G''}{2}}\sqrt{\sqrt{Q_2^2+1}+Q_2} = \Delta R_N\sqrt{\sqrt{Q_2^2+1}+Q_2} \qquad (16.26)$$

where ΔR_N is the resistance shift for a given Newtonian case.

For a 10 MHz AT-cut quartz a simple expression of X_L/R_L as a function of Δf_s and ΔR_m is $X_L/R_L = (\Delta f_s/\Delta R_m)/10$.

One Thin Rigid Layer Contacting a Semi-Infinite Medium

In this case, the acoustic load impedance results from Eq. (16.4) when the tangents can be approximated by their arguments:

$$Z_L = \frac{Z_{c2} + \mathrm{j}\rho_1\omega h_1}{1 + \mathrm{j}\frac{Z_{c2}}{G_1}\omega h_1} \tag{16.27}$$

When the second term in the denominator can be considered much lower than the unit, the acoustic load impedance corresponds to the addition of the impedance of the first layer and second medium, and the real and imaginary parts result in:

$$R_L = R_2 = \sqrt{\frac{\rho_2 G_2''}{2}}\sqrt{\sqrt{Q_2^2 + 1} + Q_2} \tag{16.28a}$$

$$X_L = X_2 + \rho_1\omega h_1 = \sqrt{\frac{\rho_2 G_2''}{2}}\sqrt{\sqrt{Q_2^2 + 1} - Q_2} + \rho_1\omega h_1 \tag{16.28b}$$

Thus, in this case, ΔR_m provides information only about the second medium, but it is not possible to distinguish if a change in R_L (ΔR_m) is due to a change in G_2'' or in Q_2. Even with the evaluation of both R_L and X_L it is impossible, in general, to differenciate the contribution of each property to the acoustic load impedance. Only when the second medium is Newtonian, Eqs. (16.28a-b) turn into Martin's equation, (Eqs. (4.A.44) and (4.A.42)), and it is possible to distinguish different contributions assuming the density of the medium as constant. Otherwise, only when the properties of the viscoelastic medium can be considered as constant, which can be validated through a second reference resonator in contact with the medium, it is possible an easy interpretation of the *differential frequency*, i.e., the difference between the frequency shift provided by the sensor and the frequency shift provided by the reference, as a function of the surface mass density of the sensitive layer; on the contrary, an interpretation of the changes in the properties cannot be made only by starting with the typical data, Δf_s and ΔR_m, and additional techniques are necessary to extract the parameters of interest.

As it can be noticed, even in this simplified model a coherent interpretation of the experimental data cannot be made without an additional reference sensor. In this case, the reference sensor, directly in contact with the viscoelastic medium, permits the evaluation of the changes in the viscoelastic properties of the medium and provides differential measurements of

Δf_s and ΔR_m, which permit the direct interpretation of the changes in the surface mass density of the sensitive layer.

Summary

Table 16.1 can be used as a guide for a rapid interpretation of experimental data in the simple cases discussed. The first column indicates the type of approximation; the second column indicates the relevant physical parameters in each approximation; the third column includes the experimental parameters, and the forth column indicates the interpretation of the variation of experimental parameter as a function of the physical relevant properties.

Table 16.1 Summary of the interpretation of experimental data in simplified cases

Case	Parameters	Experim. Data	Interpretation
One Sauerbrey layer	$\rho_S = \rho h$	Δf_s ; $\Delta R_m = 0$	$\uparrow\downarrow \Delta f_s \rightarrow \uparrow\downarrow \rho_S$
One Newtonian medium	$\sqrt{\rho_L \eta_L}$	Δf_s or ΔR_m and Eq. (16.4)	$\uparrow\downarrow \Delta f_s$ or $\uparrow\downarrow \Delta R \rightarrow$ $\rightarrow \uparrow\downarrow \sqrt{\rho G''}$
One semi-infinite[a]	ρ_L, G_L', G_L''	Δf_s and ΔR_m	$\uparrow\downarrow \left\lvert \frac{\Delta f_s}{\Delta R_m} \right\rvert \rightarrow \downarrow\uparrow Q$
One Sauerbrey + Newtonian[b]	$\rho_s, \sqrt{\rho_L \eta_L}$	Δf_s ; $\Delta R_m = \text{cte}$	$\uparrow\downarrow \Delta f_s \rightarrow \uparrow\downarrow \rho_S$
One Sauerbrey + viscoelastic[c]	$\rho_S, \rho_L, G_L', G_L''$	Δf_s^D; ΔR_m^D and $\Delta R_m^D = \text{cte}$	$\uparrow\downarrow \Delta f_s^D \rightarrow \uparrow\downarrow \rho_S$

[a]The interpretation in this case assumes the density of the medium as constant.
[b]In this case, $\Delta R_m \neq$ cte means that the properties of the medium are changing or that the sensitive layer cannot be considered as rigid and its viscoelastic properties should be considered.
[c]This case requires differential measures with a second reference sensor. In this case, $\Delta R^D \neq$ cte means that the first sensitive layer does not behave as rigid and its viscoelastic properties should be considered.

6.3.2. Limits of the Simple Cases

Loading approximations are useful for an easier interpretation of the experimental data. However, the greater the approximation the easier the interpretation and the lesser the information obtained from the data analysis. This means that when making approximations in order to reduce the number of unknowns, the contribution of these variables to the global response is also lost and, then, a loss of accuracy can occur in the interpretation as well. On the other hand, the advantage is that without these simplifications it would be very difficult, in some cases, to make any interpretation. Thus, it is important to be aware of the approximation limits; an example of these limits can be shown in piezoelectric biosensor applications. In these applications the typical simplification is to consider a 3-layer model formed by quartz, a thin rigid layer (sensitive layer) and a semi-infinite Newtonian medium. Martin's simplification can be applied in this case, in such a way that experimental Δf_s data are directly related to changes in the surface mass density of the first layer; the properties of the medium are considered, in general, as constant. However, more and more sensitivity is demanded in order to detect lower and lower concentrations of analytes in solution and considerable errors can be made if the limits of the reduced model have not been taken into account.

The Sauerbrey regime considers an ideal rigid film or, in other words, the film must behave rigid for the wave propagating through the layer. When the film is very thin, it is assumed that the out of phase of the acoustic wave through the layer is null, but it depends on the layer viscoelastic properties. Thus, a better approach could be to consider the film as very thin but with viscoelatic properties. The acoustic load impedance in this case fits to Eq. (16.27), but now the simplification in the second term in the denominator will not be made, giving the following expressions for R_L and X_L:

$$R_L = \frac{R_2|G_1|^2 + \omega h_1 G_1''|Z_{c2}|^2 + \rho_1\omega^2 h_1^2\left(X_2 G_1'' + R_2 G_1'\right)}{|G_1|^2 + 2\omega h_1\left(R_2 G_1'' - X_2 G_1'\right) + \omega^2 h_1^2|Z_{c2}|^2} \tag{16.28a}$$

$$X_L = \frac{X_2|G_1|^2 + \rho_1\omega h_1|G_1|^2 - \omega h_1 G_1'|Z_{c2}|^2 + \rho_1\omega^2 h_1^2\left(R_2 G_1'' - X_2 G_1'\right)}{|G_1|^2 + 2\omega h_1\left(R_2 G_1'' - X_2 G_1'\right) + \omega^2 h_1^2|Z_{c2}|^2} \tag{16.28b}$$

where R_2 and X_2 are the real and imaginary parts of the characteristic acoustic impedance of the medium $Z_{c2} = (\rho_2 G_2)^{1/2}$.

When the thickness of the layer is very thin, additional simplifications in Eqs. (16.28a-b) result in:

$$R_L = R_2 + \omega h_1|Z_{c2}|^2\frac{G_1''}{|G_1|^2} \tag{16.29a}$$

$$X_L = X_2 + \rho_1\omega h_1 - \omega h_1|Z_{c2}|^2\frac{G_1'}{|G_1|^2} = X_L\big|_2 + X_L\big|_{1/2} \tag{16.29b}$$

As it can be noticed, Eqs. (16.29) do not represent the load impedance as the additive contribution of the layer acoustic impedance and the medium impedance as it is the case in Martin's approximation.

Equation (16.29b) can be split into two terms: a first one that depends only on the properties of the medium and a second one which has two contributions, one depending only on the first layer and the other depending on the interaction between the layer and the medium. Expression $\Delta f_s|_{1/2}$ associated to the second term in Eq. (16.29) results from Eq. (16.12) in:

$$\Delta f_s\big|_{1/2} = -\frac{f_o}{\pi Z_{cq}}\omega\rho_1 h_1\left(1-\frac{|Z_{c2}|^2}{\rho_1}\frac{G_1'}{|G_1|^2}\right) \tag{16.30}$$

Considering the medium as Newtonian, $\omega \approx \omega_s$ near resonance, the relationship between the complex shear modulus G and the shear complex compliance J (Eq. (12.4)), and the relation between the resonant frequency f_o and Z_{cq}, Eq. (16.30) results in:

$$\frac{\Delta f_s\big|_{1/2}}{f_s} = -\frac{\rho_1 h_1}{\rho_q h_q}\left(1-\frac{\omega\eta_2\rho_2}{\rho_1}J_1'\right) \tag{16.31}$$

Equation 16.31 coincides with that obtained recently by Kankare [17] and Voinova [18]. These effects have also been considered in Chap. 4 as *small phase shift approximation* and it should be taken into account in biosensor applications since the contribution of the second term in Eq. (16.29b) can be a high percentage of X_L [18].

The previous analysis shows that the additive contribution of the acoustic load impedance of the first layer and second medium, in the context of Fig. 16.1, is a simplification which must be carefully considered. Figure 16.4a shows the deviation, as the thickness of the first layer increases, between the complete acoustic load impedance from Eq. (16.4) and the simplified acoustic load impedance corresponding to the additive contribution which is:

$$Z_L^{Add} = Z_{c2} + \mathrm{j}Z_{c1}\tan\left(\omega\sqrt{\frac{\rho_1}{G_1}}\,h_1\right) \tag{16.32}$$

Equation (16.4) is marked in Fig. 16.4a as "non additive" while Eq. (16.32) is marked as "additive". It can be noticed that for small thicknesses the deviation is very small while the greater the thickness the higher the deviation. When the thickness of the first layer reaches values higher than the penetration depth length, the error is equal to the impedance of the second medium.

The simple cases analyzed are extreme limits. When the thickness of a finite first viscoelastic layer grows from very small thicknesses, the behavior of the layer starts being Sauerbrey-like and finishes as semi-infinite. The parametric polar plot of Fig. 16.4b is very useful to describe the process. For small thicknesses, only inertial contribution associated with an increase of X_L can be observed; as the thickness increases both inertial and viscoelatic contributions are transferred in the plot as variations both in X_L and R_L. Finally, when the thickness can be considered as semi-infinite magnitudes X_L and R_L reach constant values.

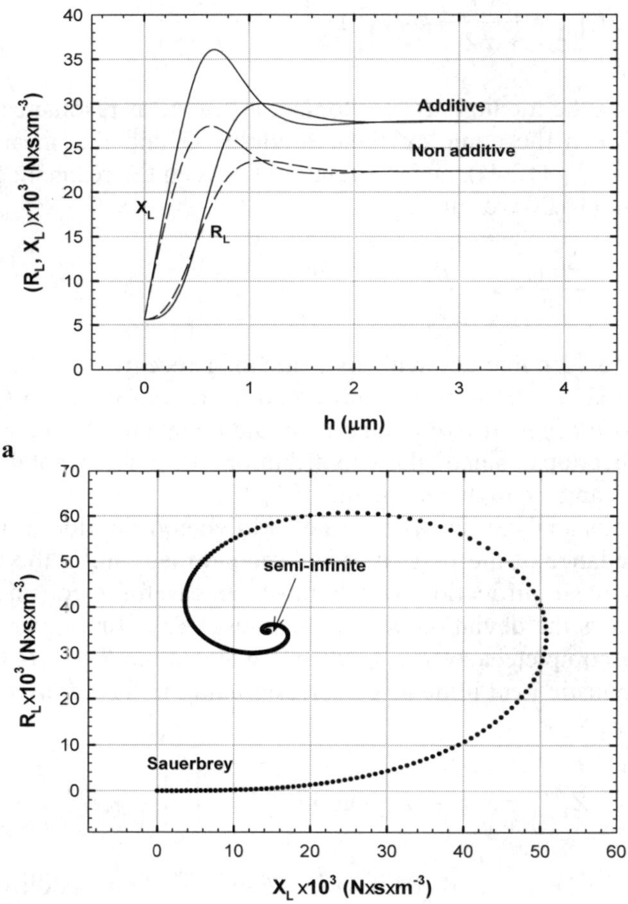

a

b

Fig.16. 4. a Plot showing the deviations between the complete acoustic impedance of a load like that described in Fig. 16.1 and the acoustic impedance simplified with the additive contribution approach; **b** polar plot associated with the thickness growth of a finite layer with viscoelastic properties $G = 10^6 + j10^6$ Nm^{-2}

In general, Sauerbrey condition is also a very restrictive one. When a finite first layer with viscoelastic properties is considered, the error in the evaluation of the surface mass density, from the experimental motional frequency shift through Sauerbrey's equation, increases with the thickness. This effect is called "*viscoelastic contribution*" [15, 19] and is explained in Fig. 16.5.

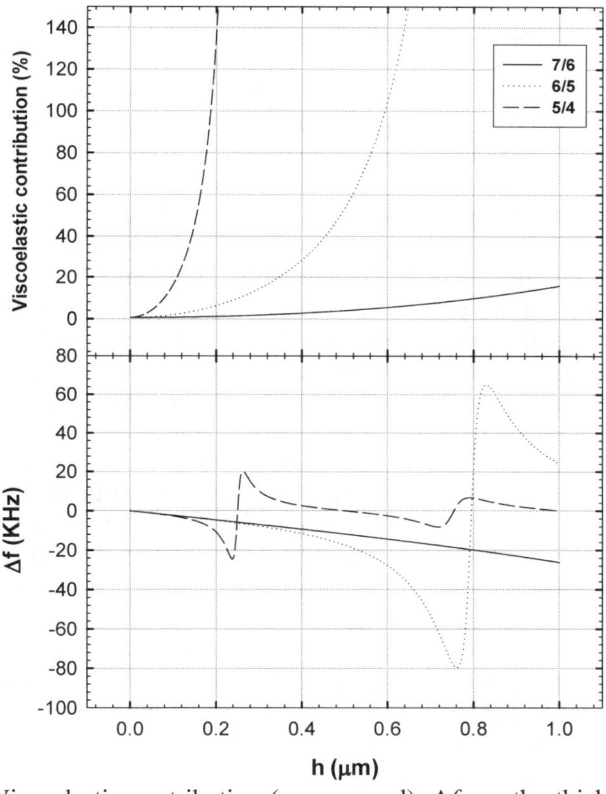

Fig. 16.5. Viscoelastic contribution (upper panel), Δf_s vs. the thickness h (lower panel); the properties of the medium are indicated as log (G'$_1$/Pa)/log (G''$_1$/Pa), in which the slash (/) represents a separator, not a quotient

Figure 16.5 plots the Δf_s associated with an increase in the thickness of the first viscoelastic layer. As it can be observed, the greater the thickness the greater the viscoelastic contribution and the more errors can be made when considering a direct relation between the thickness and Δf_s through Suerbrey's equation (or Martin's equation). The deviation between Sauerbrey's equation and the Δf_s obtained from the TLM can start from very low thicknesses depending on the viscoelastic properties of the layer. When there is a contacting semi-infinite medium, the deviations from Sauerbrey begin for lower thicknesses. This occurs in biosensors applications.

16.3 The General Case. The Problem of Data Analysis and Interpretation

In the context of Fig. 16.1, when no-approximation can be done about the first sensitive layer, i.e., when it is not reasonable to consider a Sauerbrey-like behavior of the first layer, or the "small surface load impedance condition" cannot be applied, the appropriate admittance model for characterizing the electrical response of the sensor, turns into the TLM (Eq. 16.2). The physical properties to be extracted and interpreted are concentrated in the acoustic load impedance Z_L (Eq. 16.7). Z_L depends on 6 unknowns which can be reduced to four: ρ_1, h_1, G_1' and G_1'' assuming the properties of the second medium G_2' and G_2'' known by, e.g., an additional reference sensor or during certain steps of a process where only the second medium is in contact with the sensor; it occurs in the first steps of EQCM or biosensor applications. The characteristic impedance of the second medium can also be obtained from theory, in case of Newtonian liquids, with low error.

The remaining four unknowns can be lumped in three: ρ_s, Q_1 and Z_{c1} [11]. Then, three unknowns is the minimum number of unknowns to which the TLM can be reduced without loading simplifications. Thus, the typical experimental measures considered until now Δf_s and ΔR_m, or the alternative magnitudes X_L and R_L, are not enough to extract without uncertainty the physical properties of the first layer. Infinite possible solutions provide the same couples of experimental data and it is impossible to make any coherent interpretation of experimental data in terms of the physical properties of the model (ρ_s, Q_1 and Z_{c1}) [1-4]. This is illustrated with a parametric plot in Fig.16.6. As it can be seen, infinite sets of properties match the same couple of X_L and R_L.

Therefore, it is necessary to restrict the number of possible solutions; this can be done in two ways: a) increasing the knowledge of the physical model by using alternative techniques which permit the evaluation of certain properties of the first layer, in general the thickness (see-Chap. 13) [1-3, 5], or by assuming the knowledge of some of them in order to reduce the set of possible solutions to a certain range, and b) increasing the knowledge of the admittance response in order to obtain more than two fitting conditions [9, 11, 12]. In this way, the range of possible solutions is reduced by fitting the models to experimental data.

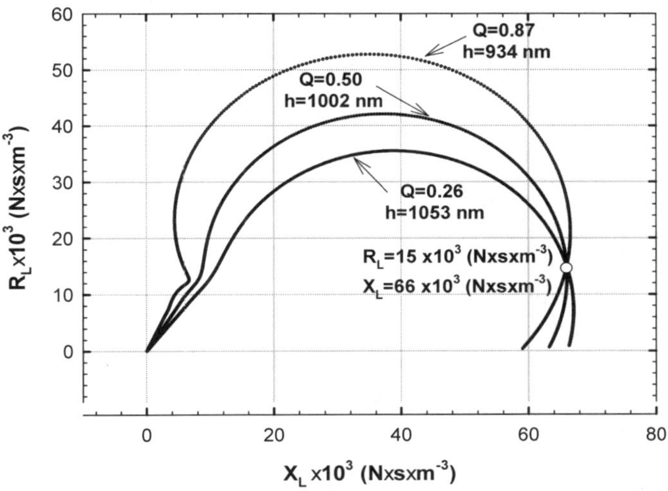

Fig. 16.6. Parametric plot showing the problem of indetermination in the property extraction starting only from one couple of R_L and X_L

16.3.1 Different Strategies to Face the Problem

Restricting the Solutions by Increasing the Knowledge about the Physical Model

The strategies used for increasing the knowledge of the physical model can be classified into three topics: 1) by measuring or evaluating the thickness by alternative techniques [1-3, 5], 2) by assuming certain values for some of the properties of the first layer, different from the thickness [3], and 3) by a controlled change of the properties of the second medium [5, 6].

One of the most typical strategies posed to carry out the extraction of the four coating properties is by measuring the thickness with an alternative technique, while the density is obtained from theoretical calculations or from the literature. Thus, the remaining unknowns are reduced to two which can be extracted with the typical measures Δf_s and ΔR_m.

Measuring the Thickness by Cooling

Behling and Lucklum [2] propose the evaluation of the thickness of a polymeric coating by cooling. In this situation, the viscoelastic properties of the coating change in such a way that the coating behaves like a rigid layer, and Sauerbrey's equation can be used to measure the surface mass density. By assuming or evaluating the polymer density, the thickness is obtained. It is necessary to be accurate in the measure of the thickness because the authors showed that multiple solutions with similar thicknesses and very different viscoelastic properties exist. They propose an alternative to Sauerbrey's equation for the calculation of the thickness [20]:

$$h_1 = \frac{Z_{cq}}{2\pi f_s \rho_1}\left(\frac{2}{\tan\left(\frac{\alpha(f_s)}{2}\right)} - \frac{4K_0^2}{\alpha(f_s)}\right) \tag{16.33}$$

Lucklum et al. stated that this technique is very accurate in comparison with other techniques such as ellipsometry, and estimated the error around 0.1% [21]. The main critical points to this technique are two: the density and the thickness of the coating can change with the change in temperature, and the continuous monitoring of the thickness is difficult to perform in certain applications.

Another important point to take into account is that the extraction of the remaining parameters G_1' and G_1'' is not a simple task despite the determination of the thickness. The TLM is not as simple as the simplified models described before; explicit expressions for G_1' and G_1'' cannot be obtained and the algorithm used to find the best fitting has to search for the best values of G_1' and G_1'' in a broad range of possible solutions, from 10^3 to 10^9; this makes the fitting more difficult because of the existence of local minima in the complete range of solutions. In general, this fitting is tedious and very time-consuming because of the necessity of evaluating Δf_s and ΔR_m obtained from the model for each tentative couple of parameters G_1' and G_1''. Additionally, the TLM does not offer explicit expressions for Δf_s and ΔR_m, which have to be calculated for each couple of viscoelastic properties with an additional algorithm [20]. One possibility to reduce this second algorithm is to use the LEM or the extended BVD model to obtain Δf_s and ΔR_m, but it would reduce the accuracy in the properties extracted, especially when the "small surface loading" condition cannot be applied; this occurs as the thickness comes near the film resonance [22]. Lucklum et al. propose a fast method for the extraction algorithm [21]. The main limitation of their fast method is that it is only valid for only one layer and in the range of thickness before the first film resonance; when a second medium exists or when the thickness is far from the first film resonance, the algorithm cannot be applied.

Measuring the Thickness by Ellipsometry

Film thickness, refractive index and extinction coefficient can be assessed with ellipsometry using the expressions derived from Fresnel equations, that relate the changes of elliptically polarized light parameters measured after incidence on the sample [23, 24].

Ellipsometry has been employed to study self-assembled multilayers with proteins [24-27]. Calvo et al. [28] have studied self-assembled multilayers composed of alternate layers of Glucose oxidase (GOx) and a redox polymer $Os(bpy)_2ClPyCH_2NH$-poly(allylamine) (PAH-Os), deposited on a thiolated modified gold surface under "ex-situ" and "in-situ" conditions by ellipsometry and QCM. Ellipsometric techniques have been used by other authors to measure the thickness of the coating in a QCM [9, 28, 29].

As an example, Fig. 16.7 shows in-situ acoustic mass - ellipsometric thickness plots obtained for $(PAH\text{-}Os)_{14}(GOx)_{14}$ self-assembled multilayer in the reduced state.

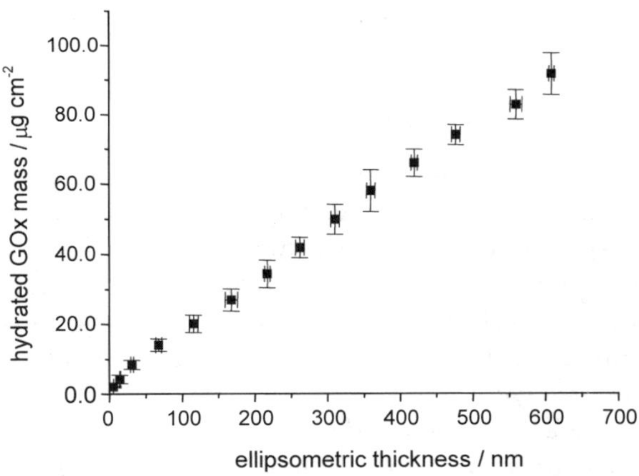

Fig. 16.7. Adsorbed enzyme acoustic mass vs. ellipsometric thickness for $(PAH\text{-}Os)_{14}/\ (GOx)_{14}$. Data taken after each enzyme adsorption and washing step under water or 10 mM Tris buffer pH = 6.4

The reduced polymer multilayer, with ellipsometric thickness ranging from 200 to 600 nm, behaves as an acoustically thin film. Therefore, Sauerbrey's equation can be used for the gravimetric evaluation of film growth and QCM and ellipsometry can be used to determine the thickness and the density of the film under the gravimetric regime.

During voltammetric oxidation of the film, a negative frequency shift Δf_s and an increase of the motional resistance ΔR_m is observed. These changes are not only due to changes in the thickness of the film but also due to changes in G' and G''. Therefore, the gravimetric regime is left and it goes into the viscoelastic regime. Under these conditions it is possible to estimate the values of G' and G'' from the values of Δf_s and ΔR_m and with the thicknesses evaluated by ellipsometry during the oxidation. As the refractive index of the film was constant during the electrochemistry perturbation, the density was supposed to be constant and equal to the density estimated for the reduced film.

Then, the viscoelastic contribution defines the limit in the thickness determination using Sauerbrey, while in terms of errors in the shear parameters one must expect extreme error sensitivity under thin film conditions. This defines one of the application limits in the shear parameters extraction.

The main problem associated with the use of ellipsometry to evaluate acoustic thickness is that the thicknesses measured by ellipsometry are different to those measured by QCM [9, 28, 30]. On the other hand, ellipsometry cannot be applied to opaque surfaces and the lack of information about the optical properties of the sample makes the use of ellipsometry inoperative [30].

Measuring the Thickness in EQCM Applications

In EQCM applications the measure of the charge by Faraday law is used to obtain an alternative and independent measure of the thickness. Bandey and Hillman [4] introduce and experiment where a layer of conductive polymer is generated by electro-polymerization. They estimated the density of the polymer layer and, by measuring the charge transferred during the process, estimated the deposited mass through Faraday law and then the thickness of the layer.

Faraday law states that *the number of moles of the substance deposited on the electrode equals the number of moles of electrons interchanged during the reaction.* If it is assumed that the efficiency is 100%, i.e., all the change of mass can be considered due to the total change of charge, and the density of the material deposited onto the quartz electrode is known, then it is possible to estimate the thickness of the layer as follows:

The number of moles of the substance deposited is:

$$N_1 = \frac{m}{M} \qquad (16.34)$$

where m is the mass and M is the molecular weight.

On the other hand, the number of moles of electrons interchanged is:

$$N_2 = \frac{Q}{nF} \tag{16.35}$$

where Q is the charge involved in the process, which can be obtained through the product of the current measured and the time that the process takes; n is the number of electrons interchanged in the reaction, which is known and $F = 96485$ C/mol is Faraday's constant.

Equating Eqs. (16.34) and (16.35) results in:

$$m = \frac{MQ}{nF} \tag{16.36}$$

and the thickness is obtained from Eq. (16.36) as follows:

$$h_1 = \frac{Q}{nFS}\frac{M}{\rho_1} \tag{16.37}$$

where S is the surface where the substance is deposited and ρ_1 is the density of the substance.

The thickness evaluated by an alternative technique can also be used to evaluate the density of the layer. If during the first steps in a process, the deposited mass on the quartz is small enough to assume that the behavior of thin layer can be considered as Sauerbrey, the surface mass obtained by the QCM can be plotted versus the evaluated thickness, then the slope corresponds to the density [1]; this can be applied when using ellipsometry and QCM simultaneously as described before. In electrochemical applications, the simultaneous measures of mass surface by Faraday Law and QCM can be used to estimate the efficiency or, alternatively, if the efficiency is assumed, they can be used to calibrate the effective surface S from Eq. (16.37) as follows:

$$S = \frac{Q}{nF}\frac{M}{\rho_s} \tag{16.38}$$

where ρ_s is the surface mass density obtained from Sauerbrey's equation.

Another strategy posed to carry out the extraction of the four coating properties is by assuming the knowledge of properties of the layer different from the thickness, while the density is obtained from theoretical calculations or from the literature. The remaining unknowns are reduced to two which can be extracted with the typical measures Δf_s and ΔR_m.

Assuming the Knowledge of Properties Different from the Thickness

Calvo et al. [3] introduce the characterization results of layers of Poly-(aniline) and Glucose oxidase deposited by electrostatic adsorption and submitted to cyclic voltammetry. To extract the parameters from the first layer they follow two strategies: in the first one they assume the shear storage modulus $G_1' << G_1''$, and determine the remaining parameters h_1 and G_1''. In the second one, they assume a constant Q_L and determine the remaining h_1 and $|G_1|$. In both cases the density is estimated by the literature. The main limitation of this method is that the assumption cannot be compared with an alternative technique. For example, the quality factor can change during a cyclic voltammetry [31].

By a Controlled Change of the Properties of the Second Medium

Another strategy to extract the properties of the first layer is by changing the acoustic load impedance of the second medium. In this case, and assuming the properties of the first layer unchanged when the second medium is altered, which is the main criticism to this method, the experiment provides two or more couples of Δf_s and ΔR_m, or the alternative magnitudes, X_L and R_L.

This strategy, introduced by Lucklum et al. [5], is based on the following method:

By solving the characteristic impedance of the first layer Z_{c1} from the Eq. (16.4), one obtains:

$$Z_{c1} = \frac{Z_1 Z_2 Z_L}{Z_2 + Z_1 - Z_L} \tag{16.39}$$

where Z_1 is the acoustic load impedance of the first layer without contacting the second medium, which can be measured during the experiment before depositing the second medium; Z_2 is the acoustic impedance of the second medium which can be measured with a second resonator in contact only with the second medium, and Z_L is the acoustic load impedance of the first layer in contact with the second medium.

Thus, by following the method an experimental value of the characteristic impedance of the first layer can be obtained, which permits a direct calculation of the shear moduli G_1' and G_1'' when the density of the first layer is known; the thickness is then calculated from the previous value of Z_1. Lucklum et al. applied this method using a second thin layer of gold whose thickness is measured during the deposition with a third QCM sensor.

Etchenique and Weisz use the same basis with a different method, in which they obtain additional information by changing the properties of the second semi-infinite medium [6]. It is, in fact, a generalization of Luclum's method.

Restricting the Solutions by Increasing the Knowledge about the Admittance Response

A different strategy is to increase the knowledge in the admittance response; i.e., in the electrical response of the sensor, by changing the frequency. It is the reverse case to the change of the physical properties of the second medium in order to increase the knowledge about the acoustic mechanical impedance at one frequency. Two different methods have been used:

By Measuring the Admittance Response of the Sensor to Different Harmonics

Johannsmann et al. introduced a sophisticated method to extract the properties of thin films, based on the measuring of the admittance response to different harmonics [8-10].

Starting from the complex resonant frequency shift, assuming very thin thickness and developing in Taylor series the tangent associated with the acoustic load impedance of the layer, they obtained the following approximated expressions for the frequency shift and the half power resonant spectrum W:

$$\frac{\Delta f}{f/n} = -\frac{2 f_s m_s}{\sqrt{c_{66} \rho_q}} \left[1 + J'(\omega) \frac{4\pi^2 m_s^2}{3\rho} f^2 \right] \quad (16.40a)$$

$$\frac{W}{f/n} = -\frac{16 f_s m_s^3 \pi^2}{3\rho \sqrt{c_{66} \rho_q}} J''(\omega) f^2 \quad (16.40b)$$

From the previous equations, the strategy consists in measuring the frequency shift Δf and W at different harmonics. Then, assuming properties J' and J'' (see Chap. 12) unchanged with the frequency, which is the main criticism to this method, they fit straight lines to the data and plot the data vs. f^2. From Eq. (16.40a) the intersection with ordinate axis provides the measure of the surface mass, and the slope provides property J'. From Eq. (16.40b) the slope provides property J''.

By Measuring the Admittance Response of the Sensor in the Range of Frequencies around Resonance

The limitation of the previous method can be avoided by measuring the admittance response in a range of frequencies around resonance. In this range it can be assumed that the viscoelastic properties remain unchanged.

Recently, Jimenez et al. have introduced an algorithm for the first layer properties extraction that uses the admittance plots around resonance (conductance and susceptance) to find the properties by fitting the TLM to these experimental data [11]. The algorithm can solve the problem of the first layer parameter extraction in ideal conditions; i.e., when the admittance plots obtained from the TLM starting from certain properties of the first layer are considered as experimental data, the algorithm is able to extract these properties starting from the conductance with very few errors in all the cases [11]. The main limitation of the method is that the acoustic load impedance does not change too much in the range of frequencies considered. Consequently, the applicability of the algorithm to real experimental data depends very much on the accuracy of the obtained experimental data. There are cases where the change in the acoustic load impedance is greater than the error due to the experimental setup. In these cases the algorithm allows the extraction of the properties and an estimation of the possible error made in the extraction.

16.3.2 Additional Considerations. Calibration

Throughout the whole chapter it has been assumed that the parameters of the unperturbed sensor (h_q, C_o and η_q) were perfectly known. However, this is not the case, which complicates even more the problem of parameter extraction. A calibration is necessary to determine the best set of parameters. The main problem is that these parameters change with the load. This change affects mainly the static capacitance. The change in the static capacitance is pronounced at the beginning, when the load increases from the unperturbed state, and tends to stabilize as the load increases. An error in the characterization of the static capacitance produces an important error in the parameter extraction for small loads; the effect of an error in the static capacitance has a lower effect as the load increases [5].

Therefore, a method is necessary to calibrate the resonator parameters. The normal method is to calibrate the sensor in the unperturbed state, but recently, Jimenez has shown that it is preferable to calibrate the parameters starting from the experimental data obtained with known Newtonian liquids as second medium, for instance, water. Thus, the TLM is applied with a known Z_L and the quartz parameters are calibrated to assure that the theoretical acoustic load impedance of the second medium is obtained from it. The larger the number of different Newtonian liquids with different loads the better the calibration in a concrete range of loads. As an example Figure 16.8 shows the effect of calibration.

Figure 16.8 shows different conductance plots. In both figures it is shown the experimental conductance plot of a resonator coated with a thin film. This resonator was calibrated, previously to be coated, in air and in water as second media. Figure 16.8 shows the conductance plots obtained from the TLM through the algorithm introduced by Jimenez et al. [11] in two cases: one with the sensor calibrated in air, and the other with the calibration made in water. As it can be noticed, a much better fit in the plots is obtained with the parameters calibrated in water - as shown by the corresponding fitting errors in Fig. 16.8.

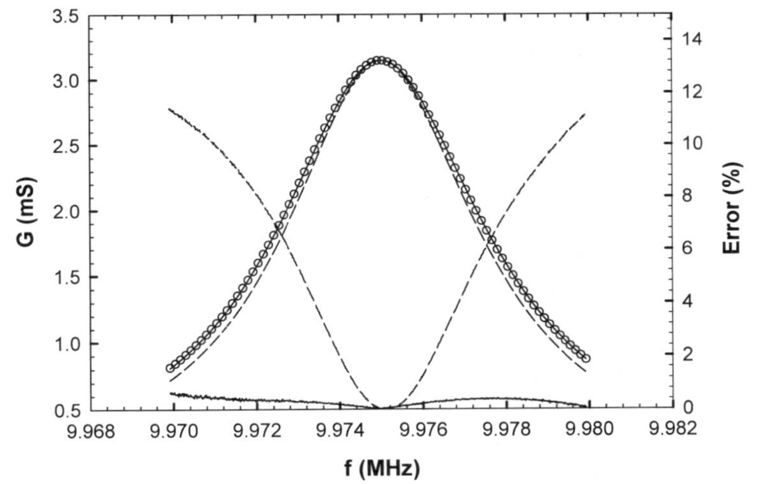

Fig. 16.8. Conductance plots obtained from the TLM and the corresponding fitting errors with regard to experimental data (open circles), for the parameters of the resonator calibrated in air (dashed line) and water (solid line)

16.3.3. Other Effects. The N-layer Model

The three layer model considered in Fig. 16.1 can be complicated or simplified by adding or subtracting some layers, in order to obtain a more comprehensive interpretation of some specific phenomena.

For example, the studies that deal with absorption processes include an additional layer whose thickness varies with time [29]. Another typical situation which requires an additional layer is rough quartz.

If the quartz surface is rough, part of the coating adjacent to the crystal will be trapped and will move synchronously with the crystal surface. This new layer acts as a rigid mass and may contribute to energy storage in a different way from the rest of the film; this implies that an extra frequency shift is going to appear. On the other hand, a lack of homogeneity within the coating can disturb the acoustic wave propagation and contribute to acoustic energy loss that will result in a new shift in the motional resistance. Another effect introduced to improve the physical model is the slip between the quartz surface and the first layer. The TLM used in the book, assumes that there is no slip between quartz and coating. This slip can be included in the model [32] and does not modify the equations in a relevant way. Thompson has studied the effect of the slip and has indicated that the slip is transferred into effective viscoelastic parameters in the context of the model of Fig. 16.1

The models can become very complex, although a simple model can provide an easy way to obtain some effective parameters that could be processed for a more comprehensive interpretation later.

References

1. Hillman AR, Jackson A and Martin, S.J (2001) "The Problem of Uniqueness of Fit for Viscoelastic Films on Thickness-Shear Mode Resonator Surfaces", Anal. Chem., vol. 73, nº 3, pp. 540-549
2. Behling C, Lucklum R and Hauptmann P (1997) "Possibilities and limitations in quantitative determination of polymer shear parameters by TSM resonators" Sensors and Actuators A, vol. 61, pp. 260-266
3. Calvo, EJ, Etchenique, R, Barlett, PN, Singhal, K and Santamaria, C (1997) "Quartz crystal impedance studies at 10MHz of viscoelastic liquids and films", Faraday Discuss., vol. 107, pp. 141-157
4. Bandey HL, Hillmann AR, Brown MJ and Martin S.J (1997) "Viscoelastic characterization of electroactive polymer films at electrode/solution interface", Faraday Discuss, vol. 107, pp. 105-121
5. Lucklum R and Hauptmann P (1997) "Determination of polymer shear modulus with quartz crystal resonators", Faraday Discuss., vol. 107, pp. 123-140
6. Etchenique R and Dan Weisz A (1999) "Simultaneous determination of mechanical moduli and mass of thin layers using non-additive quartz crystal acoustic impedance analysis", vol.86, nº4, pp. 1994-2000
7. Bund A, Schneider M (2002) "Characterization of the viscoelasticity an the surface roughness of electrochemically prepared conducting polymer films by impedance measurements at quartz crystals", Journal of Electrochemical Society, vol.149, nº9, pp.331-339

8. Wolff O, Seydel E and Johannsmann D (1997) “Viscoelastic properties of thin films studied with quartz crystal resonators”, Faraday Discuss, vol.107,pp.91-104
9. Domack A, Prucker O, Rühe J and Johannsmann D (1997) “Swelling of a polymer brush probed with a quartz crystal resonator”, The American physical society, vol.56, nº1, pp.680-689
10. Johannsmann D, Mathauer K, Wegner G and Knoll W (1992) “Viscoelastic properties of thin films probed with a quartz crystal resonator”, Physical Review B, vol.46, nº12, pp.7808-7815
11. Jiménez Y., Sogorb T, Arnau A, Otero M and Calvo E (2003) “Systematic Error Analysis in the Determination of Physical Parameters of the coating in Quartz Crystal Resonators Sensors”, AWSWS IV, Salbris, Francia
12. Granstaff VE and Martin SJ (1994) “Characterization of a Thickness Mode Quartz Resonator with Multiple Nonpiezoelectric Layers”, J. Appl. Phys., vol. 75, nº 3, pp. 1319-1329
13. Schröder J, Borngräber R, Eichelbaum F and Hauptmann P (2002) “Advanced interface electronics and methods for QCM”, Sensors and actuators A, vol.97-98, pp. 543-547
14. Arnau A, Sogorb T and Jiménez Y (2000) “Thickness-shear mode quartz crystal resonators in viscoelastic fluid media”, J. Appl. Phys., vol. 88, nº 8, pp. 4498-4506
15. Arnau A, Jiménez Y and Sogorb T (2001) “An Extended Butterworth-Van Dyke Model for Quartz Crystal Microbalance Applications in Viscoelastic Fluid Media”, IEEE Trans. Ultrason., Ferroelect. Freq. Contr., vol. 48, nº 5, pp. 1367-1382
16. Bund A and Schwitzgebel G (1998) “Viscoelastic properties of low-viscosity liquids studied with thickness-shear mode resonators”, Anal. Chem., vol.70, pp. 2584-2588
17. Kankare J (2002) “Sauerbrey Equation of Quartz Crystal Microbalance in Liquid Medium”, Langmuir, vol. 18, pp. 7092-7094
18. Voinova MV, Jonson M and Kasemo B (2002) “’Missing mass’ effect in biosensor’s QCM applications”, Biosensors and Bioelectronics, vol. 17, pp. 835-841
19. R. Lucklum C, Behling C and Hauptmann P (1999) “Role of mass accumulation and viscoelastic film properties for the response of acoustic-wave-based chemical sensors”, Anal. Chem., vol. 71, No. 13, pp. 2488-2496
20. Lucklum R, Behling C, Hauptmann P, Cernosek RW and Martin SJ (1998) “Error analysis of material parameter determination with quartz-crystal resonators”, Sensors and Actuators A, vol. 66, pp. 184-192
21. Behling C, Lucklum R and Hauptmann P (1999) “Fast Three-Step Method for Shear Moduli Calculation from Quartz Crystal Resonator Measurements”, IEEE Trans. Ultrason., Ferroelect. Freq. Contr., vol. 46, nº 6, pp. 1431-1438
22. Cernosek RW, Martin SJ, Hillman, AR and Bandey HL (1998) “Comparison of Lumped-Element and Transmision-Line Models for Thickness-Shear-Mode Quartz Resoantor Sensors”, IEEE Trans. Ultrason., Ferroelect., Freq. Contr., vol. 45, nº 5, pp. 1399-1407

23. Azzam RMA and Bashara NH (1997) in “Ellipsometry and polarized light”. North-Holland, Amsterdam
24. Arwin H (2000) “Ellipsometry on Thin Organic Layers of Biological Interest: Characterization and applications”.Thin Solid Films, vol.48, pp. 377-378
25. Sun S, Ho-Si PH and Harrison DJ (1991) “Preparation of Active Langmuir-Blodgett Films of Glucose Oxidase”. Langmuir, vol.7, pp.727-737
26. Spaeth K, Brecht A and Gauglitz G (1997) “Studies on the Biotin-Avidin Multilayer Adsorption by Spectroscopic Ellipsometry”. J. Coll. Interf. Sci., vol.196, pp.128 - 135
27. Sirkar K, Revzin A and Pishko MV (2000) “Glucose and Lactate Biosensors Based on Redox Polymer/Oxidoreductase Nanocomposite Thin Films”. Anal.Chem. vol.72, pp.2930-2936
28. Calvo EJ, Forzani E and Otero M (2002) “Study of layer-by-layer self-assembled viscoelastic films on thickness-shear mode resonator surfaces”. Analytical Chemistry, vol.74, pp.3281
29. Lee SW, Hinsberg D and Kanazawa K (2002) “Determination of the viscoelastic properties of polymer films using a compensated phase-locked oscillator circuit”, Anal. Chem, vol. 74, pp.125-131
30. Höök F, Vörös J, Rodahl M, Kurrat R, Böni P, Ramsden JJ, Textor M, Spencer ND, Tengvall P, Glod J and Kasemo B (2002) “A comparative study of protein adsorption on titanium oxide surfaces using in situ ellipsometry, optical waveguide lightmode spectroscopy, and quartz crystal microbalance/dissipation”, Colloids and Surfaces B: biointerfaces, vol. 24, pp. 155-179
31. Otero M (2003) “Construcción y caracterización de estructuras complejas de biomoléculas con aplicación en el diseño de biosensores”, Doctoral Thesis, Universidad de Buenos Aires
32. Ferrante F, Kipling AL and Thompson M (1994) “Molecular slip at the solid-liquid interface o fan acoustic wave sensor”, J. Appl. Phys., vol. 76, nº6, pp.3448-3462

Appendix A Fundamentals of Electrostatics

Antonio Arnau and Tomás Sogorb

Departamento de Ingeniería Electrónica, Universidad Politécnica de Valencia

A.1 Principles of Electrostatics

It is usually attributed to Thales of Miletus (600 B.C.) the knowledge of the property of amber (in Greek *elektron*) of attracting very light bodies, when rubbed. This phenomenon is called *electrification* and the cause of the phenomenon *electricity*. The electric manifestations of electrification phenomena on different materials can be opposite and, in these cases, it is said that both materials have acquired qualities of opposite sign. The quality that a material acquires when electrified is called *charge*, establishing arbitrarily the sign of this quality in relation to the phenomenon that it manifests. In addition, it could be experimentally demonstrated that the electrified bodies interact with each other, and generate repulsive forces when the charges are of the same sign and, of attraction when they are of opposite sign.

The fact that the charge is manifested in two opposite ways involves that if a body has as many positive charges as negative, its external electric manifestations are balanced, and we say that it is electrically *neutral*. Matter is mainly neutral and it is not common to find bodies whose net charge has a considerable value, therefore, the electrical interactions between bodies are in general quite weak. For this reason, for a material to manifest electric properties it is necessary to subject it to some type of action, for example mechanical, as in the case of amber. This external action produces a loss of balance between the positive and negative charges of the rod and allows the electric interaction with other bodies.

At this time, it is necessary to introduce the *principle of charge conservation* in an isolated system. This experimental principle states that in a system in which charge cannot enter or leave, the positive and negative charges can vary with time but its net charge (positive + negative) remains constant. Therefore, when the amber rod is rubbed with a cloth, the charges of the cloth "arbitrarily called negative" pass to the amber rod that now has an excess of negative charge. A redistribution of the charges has taken place between the cloth and the rod but the net charge of the system cloth-rod has not been altered.

If the rubbing action on the rod is vigorous, it can be noticed that the electric interactions are stronger. That is to say, the small bodies that are used to show the electric interaction are attracted with more strength. This means that the amber acquires a larger negative charge. However, this charge increase is always in integer multiples of a minimum amount. This minimum amount of electric charge is the negative charge of the electron or the positive charge of the proton. This hypothesis establishes the *principle of charge quantization* as well as the existence of a natural unit of electric charge.

A.2 The Electric Field

It is important to notice that it has been necessary to bring the electrified amber rod close to small bodies in order to see the electric interactions. That is, it has been necessary to introduce a test element to check that the electrified amber rod had modified the electric characteristics of its immediate environment in such a way that other bodies, when coming closer to this environment, were subjected to an external force. However, the electrified amber rod modifies the electric characteristics of its environment even if no light bodies, which allow seeing this interference, are placed nearby. For example, we can use an apple to show the gravitational force but this still exists even without the apple. This concept is very important because it indicates that, similarly to the gravitation example, regions of the space are created around the positive and negative charges where they can show electric interactions. If these regions of the space appear due to the presence of punctual charges or superficial or volumetric distributions of static charge, the electric field created is called electrostatic field and the forces on the charges located in this field are determined by Coulomb's law[1]. This electric field is of special importance in the study of piezoelectricity.

---

[1] Coulomb's law provides the force that acts on a charge *q'* due to the presence of another punctual and fixed charge *q*. Its mathematical formulation is:

$$F = \frac{1}{4\pi\varepsilon}\frac{qq'}{r^2}$$

where ε is the permittivity of the medium and r is the distance between charges. Consequently, the force is proportional to the product of the charges and inversely proportional to the square of the distance that separates them. The direction is the straight line that links the charges. The direction is repulsion for charges of the same sign and attraction for charges of opposite sign.

Intensity of the field, or simply field, at a point, is the force that acts on the unit of positive charge placed at this point. Consequently, a charge with value q located at a point, P, of this field would be subjected to a force given by:

$$F(P) = qE(P) \tag{A.1}$$

where $E(P)$ is the intensity of the electric field at point P. Similarly to the force, the field is a vector magnitude with a direction.

Generally, electric fields are represented by their so-called *lines of force*. These lines are a graphic representation of the trajectories that a positive charge would follow, subjected to the influence of the field, in a succession of elementary paths, starting every time from rest.

Let us imagine a surface immerse in an electric field. Infinite lines of force will cross this surface. At each point of the surface, there will be a line of force that, at that point, will be characterized by one value of field intensity and one direction. It is called flux Φ of an electric field through this surface:

$$\Phi = \int \vec{E} \circ d\vec{A} \tag{A.2}$$

where dA is a differential element of the surface under study.

The product indicated in the previous equation is a scalar product of the two vectors. The previous definition provides an intuitive idea of the electric field as the density of electric flux per area unit. The value of the intensity of electric field at a point can be graphically represented by the number of lines of force that cross the surface unit at this point. Thus, there will be more lines of force in those points where the field intensity is stronger and more lines of force will cross the surface increasing the flux.

A.3 The Electrostatic Potential

If the potential energy of a punctual charge inside an electrostatic field is defined as a dependent function of point $U(P)$ so that the difference between its values in the initial and final positions is similar to the work acting on the charge, by the force of the field. This potential energy would be formulated as:

$$W\Big|_1^2 = \int_1^2 q\vec{E} \circ d\vec{r} = U_1 - U_2 \tag{A.3}$$

and its differential expression would be:

$$dW = q\vec{E} \circ d\vec{r} = -dU \tag{A.4}$$

From the previous expression, results:

$$\vec{E} \circ d\vec{r} = -\frac{dU}{q} = -dV \tag{A.5}$$

where $dV=dU/q$ is called potential difference.

V is called *electrostatic potential*, a scalar function dependent on the considered point, and it represents the work carried out to transfer the charge unit from infinite (where it is considered that $V(\infty)=0$) to this point. From the previous expression it is deduced that the field is equal and with opposite sign to the gradient of the potential, that is to say:

$$\vec{E} = -grad\, V = -\frac{\partial V}{\partial x}\vec{i} - \frac{\partial V}{\partial y}\vec{j} - \frac{\partial V}{\partial x}\vec{k} \tag{A.6}$$

where $\vec{i}$, $\vec{j}$ and $\vec{k}$ are the unitary vectors in the directions of the coordinated axes.

A.4 Fundamental Equations of Electrostatics

The main problem in Electrostatics is the calculation of the fields produced by different charge distributions. Gauss' law, which provides the value of the flux of the field through a closed surface, is one of the fundamental vector relationships that allow solving part of the problem. Gauss' law states that *the flux of an electric field through a closed surface is equal to the sum of all the charges enclosed in this surface divided by the permittivity of the vacuum* ε_o. Its mathematical formulation is as follows:

$$\oint_A \vec{E} \circ d\vec{A} = \frac{Q}{\varepsilon_o} \tag{A.7}$$

where Q represents the total inner charge to surface A.

If the distribution of the charge were a volumetric distribution, defined by a charge density ρ_v, the previous equation would be written:

$$\oint_A \vec{E} \circ d\vec{A} = \frac{1}{\varepsilon_o} \oint_V \rho_v \cdot dV \tag{A.8}$$

The integral of the first member of the previous expression can be written as an integral of volume keeping in mind the theorem of divergence[2], thus Eq. (A.8) is:

$$\oint_V div\vec{E} \cdot d\vec{A} = \frac{1}{\varepsilon_o} \oint_V \rho_v \cdot dV \tag{A.9}$$

or its differential expression that is:

$$div\, \vec{E} = \frac{\rho_v}{\varepsilon_o} \tag{A.10}$$

Notice that if there is no charge, or there are as many positive charges as negative, inside the closed surface, the field in the inside is constant.

Substituting Eq. (A.6) in Eq. (A.10) one reaches Poisson's Equation, which is the fundamental equation of electrostatics:

$$div\, \vec{E} = div\; grad\, V = \frac{\partial^2 V}{\partial x^2} + \frac{\partial^2 V}{\partial y^2} + \frac{\partial^2 V}{\partial z^2} = -\frac{\rho_v(x,y,z)}{\varepsilon_o} \tag{A.12}$$

Notice that if the volumetric density of net charge is null, or the positive and negative charges are balanced, inside a volume, the potential inside this volume keeps a linear relationship with the distance.

A.5 The Electric Field in Matter. Polarization and Electric Displacement

From the electrostatic point of view, two types of charges can be considered in the substances: *free charges* which are likely to move from

[2] The theorem of divergence establishes that the flux of the vector field through a closed surface is equal to the integral of the field divergence extended to the inner volume of such surface. Its mathematical formulation is as follows:

$$\oint_A \vec{E} \circ d\vec{A} = \oint_V div\, \vec{E} \cdot dV$$

where

$$div\vec{E} = \frac{\partial E_x}{\partial x} + \frac{\partial E_y}{\partial y} + \frac{\partial E_z}{\partial z}$$

and E_x, E_y and E_z are the components of the vector $\vec{E}$

one place to another inside the material, and *linked charges* whose movement within the material is limited to small changes of position around its equilibrium positions. The substances with free charge are called conductors. Metals, for example, are substances where the charges move easily, that is to say, the electrons, at least one per atom, can move throughout the solid and they are not bound to their corresponding atom. On the contrary, in dielectrics, such as paraffin, the electrons are firmly bound to their positive ions and they cannot move freely.

From the definition of conductor it is deduced that in a conductor in static state, or in equilibrium, i.e., when all its charges are at rest, the electric field is always null in its inside. Since the charges can move freely, if the field in the inside were not null the charges would move and the conductor would not be in equilibrium. Indeed, when a finite conductor, with the same number of positive and negative charges, is introduced in an electric field, E, the negative charges of one side will move in the opposite sense to the electric field to the conductor's surface, since it is supposed that they cannot abandon it, leaving positively charged ions in the other end (Fig. A.1); this creates a superficial distribution of charge that produces an induced field E_i in the opposite direction to the external field applied, reducing the total field inside the conductor to zero.

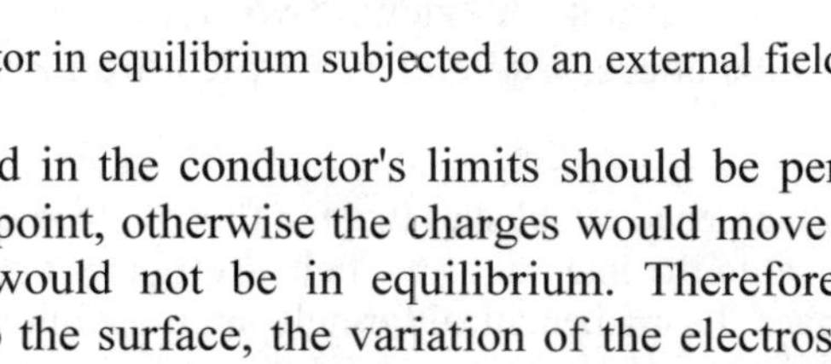

Fig. A.1. Conductor in equilibrium subjected to an external field

The total field in the conductor's limits should be perpendicular to its surface at each point, otherwise the charges would move tangentially, and the conductor would not be in equilibrium. Therefore, if the field is perpendicular to the surface, the variation of the electrostatic potential on the conductor's surface is null (see Eq. (A.5)) and this surface is an equipotential surface.

The distribution of charge induced in the conductor does not create any external field to the conductor since the conductor's overall charge is null, although there is a certain distribution of charge densities. Therefore, the application of Gauss' law to any external closed surface to the conductor would allow verifying this result. However, it creates an internal field to the conductor that compensates, at each inner point, the external field. In fact, the field created by the polarization of the charge at a nearby interior point to the conductor's surface can be calculated by applying Gauss' law to the closed surface shown in Fig. A.2. Since the conductor does not create any field outside and the flux through the lateral surfaces is null, there is only flux through the surface A_1. By applying Eq. (A.7) one obtains that the internal field generated is proportional to the superficial density of induced charge σ_i according to the following expression:

$$E_i = \frac{\sigma_i}{\varepsilon_o} \tag{A.13}$$

In the case of one conductor E_i=-E. That is, the charges move freely in the solid and generate the distribution of necessary charge so that this condition is fulfilled.

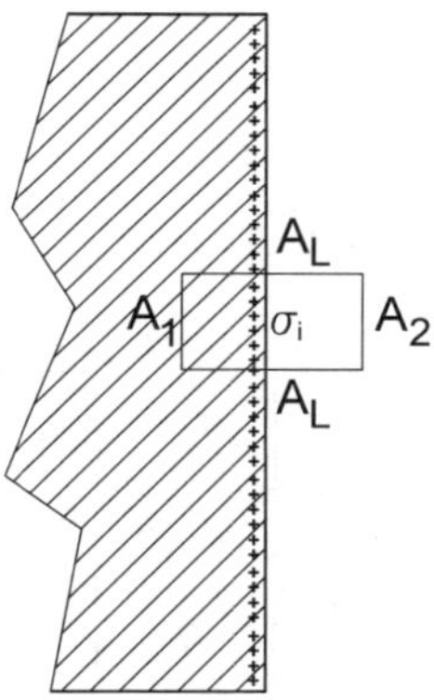

Fig. A.2. Application of Gauss' law to the closed surface of a conductor

When a dielectric material is in presence of an electric field, charges also appear on the surface of the dielectric. This phenomenon is called *polarization of the dielectric* and is similar to the polarization seen previously in the conductor, but its origin is completely different. In the conductor, polarization takes place by the migration of the charges, while in the dielectric, elementary dipoles are generated that try to align in the direction of the field, thus causing a bound charge density to appear on the surfaces of the dielectric. Polarization takes place for polar substances as well as for non-polar ones. In the polar substances, the molecules are true dipoles that are distributed at random making the dielectric, as a whole, to be discharged (Fig. A.3a). When subjecting it to a field, the dipoles are guided and the actions of the opposite close poles inside the dielectric are cancelled, thus a superficial density of charge appears in the sides of the dielectric (Fig. A.3b). In case of non-polar substances, the centers of gravity of the positive and negative charges in the molecules coincide (Fig. A.4a). However, in presence of an electric field, these centers of gravity separate and originate small dipoles that are guided in a similar way to the previous case, causing superficial densities of charge, similar to the previous case (Fig. A.4b).

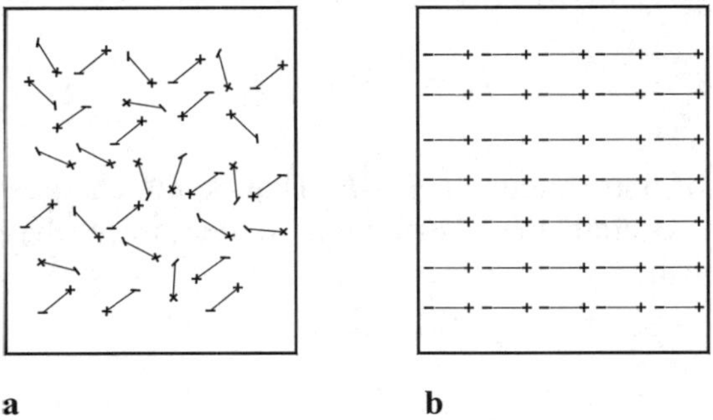

Fig. A.3. a Polar substance before being subjected to an electric field; **b** polar substance polarized by an electric field

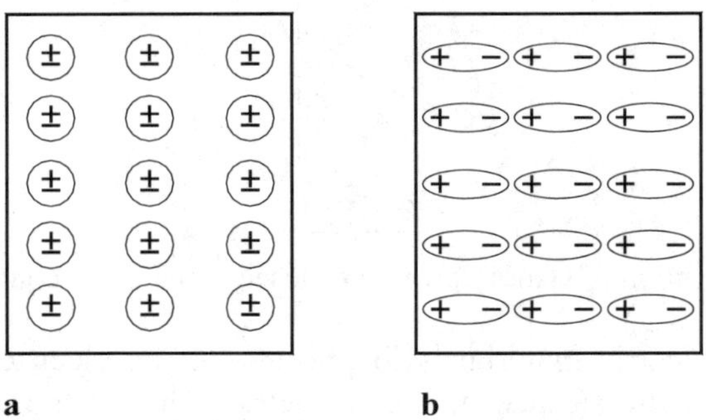

Fig. A.4. a Non polar substance before being subjected to an electric field; **b** non polar substance polarized by an electric field

As a result, when introducing a dielectric material in an electric field, for example between the plates of a charged and insulated capacitor, this material undergoes an initial electric field E_o (between the plates of the capacitor in absence of dielectric) (Fig. A.5) and the dielectric is polarized. This generates an electric field in the inside E_i that opposes to the initial field. The result of both is a new field E with the same direction as the initial field E_o (if the material is homogeneous and isotropic) and of smaller value. In fact, the alignment is never complete; the thermal agitation that increases with temperature is opposed to any type of alignment that tends to disarray the directions of the dipoles. Therefore, the dielectric properties depend strongly on the temperature. In any case, it is necessary to introduce some magnitude that allows measuring the state of polarization of the dielectric material when it is subjected to an electric field.

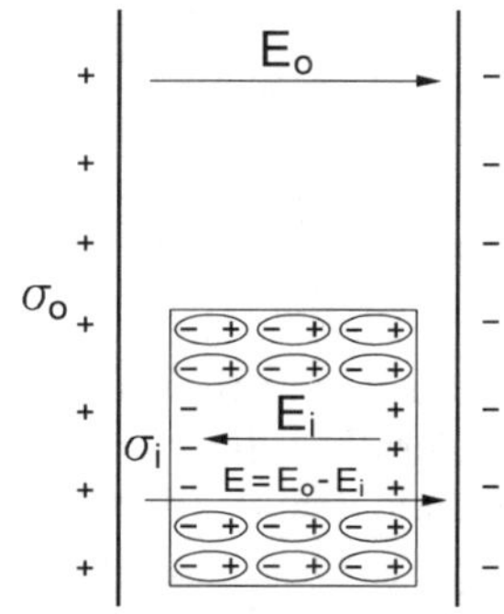

Fig. A.5. Dielectric material subjected to a field between the plates of a charged and isolated capacitor

The superficial density of charge induced on the surfaces of a dielectric by the action of an external field is called *density of dielectric polarization*. The vector whose module has the value of the surface density charge due to dielectric polarization and has a direction perpendicular to the surface considered is called *polarization vector P*. Consequently, if the polarization takes place on different surfaces, the polarization vector may have more than one component.

Polarization P can be expressed as a linear function of the field (the same conclusion can be reached experimentally for most of the dielectric materials where higher order effects can be neglected as it is the case of quartz). Therefore:

$$P = \chi E \tag{A.14}$$

where the constant χ is called *dielectric susceptibility* and it is, by definition, the polarization per unit of the existing electric field in the dielectric.

Let's call E_i the field nearby the dielectric due to the bound charges that have appeared in their surface. Gauss' law provides the value of this field according to the superficial density of charge and the permittivity in the vacuum as (Eq. A.13):

$$E_i = \frac{\sigma_i}{\varepsilon_o} = \frac{P}{\varepsilon_o} = \frac{\chi E}{\varepsilon_o} \tag{A.15}$$

Therefore, the existing field after the insertion of the dielectric will be:

$$E = E_o - E_i = E_o - \frac{\chi E}{\varepsilon_o} \tag{A.16}$$

That is to say,

$$E = \frac{\varepsilon_o}{\varepsilon_o + \chi} E_o = \frac{\varepsilon_o}{\varepsilon} E_o = \frac{E_o}{\varepsilon_r} \tag{A.17}$$

where $\varepsilon = \varepsilon_o + \chi$ is the dielectric permittivity of the material and ε_r the relative permittivity of the material to the vacuum.

Therefore, the field in the dielectric decreases with regard to the existing field in its absence.

Suppose that the dielectric material has been introduced between the plates of a capacitor that are subjected to a constant potential difference (Fig. A.6); under these circumstances, the field between the plates of the capacitor will be constant with or without the dielectric whenever it keeps the distance between plates. The charge density in the plates of the capacitor before introducing the dielectric will be $\sigma_o = \varepsilon_o E$, in this case E being the constant field between the plates of the capacitor.

When introducing the dielectric, its polarization causes an increase in the charge density in the plates due to the bound charge of the polarized dielectric. This increase of density corresponds to *polarization*. That is to say, there is an *electric displacement*, of free charges, from the conductors of the circuit towards the electrodes in order to compensate the effect of the polarization of the dielectric and to maintain the field inside the dielectric constant (Fig. A.6).

Maxwell, in order to facilitate the calculation in a great number of problems and allow for a more compact writing of certain expressions of electromagnetism, introduced the *displacement vector D*, and defined it as:

$$D = \varepsilon_o E + P \tag{A.18}$$

In the case of a homogeneous and isotropic dielectric, the polarization is $P = \chi E$ and the previous expression becomes:

$$D = \varepsilon E \tag{A.19}$$

In isotropic materials, the dielectric constant or permittivity is a characteristic of the medium. If the material is not homogeneous, it may be a function of the point; if it is anisotropic, it may depend on some direction, and if it is not linear, it may even depend on the field applied.

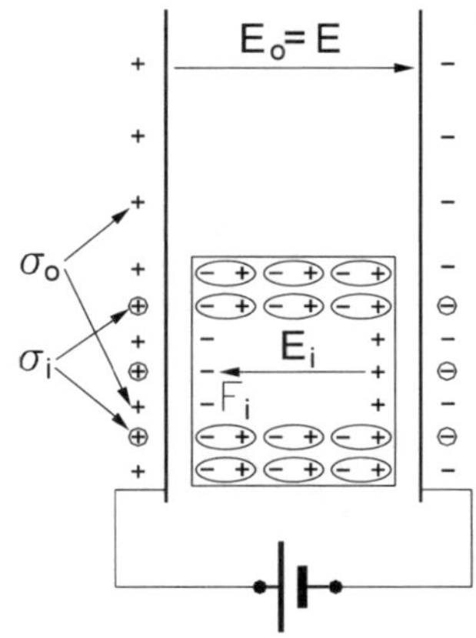

Fig. A.6. Dielectric material subjected to a constant field between the plates of a capacitor subjected to a constant potential difference

An important equation is obtained when the flux of the vector displacement is calculated in a closed surface *A*. That is:

$$\oint_A \vec{D} \circ dA = \varepsilon_o \oint_A \vec{E} \circ dA + \oint_A \vec{P} \circ dA \qquad \text{(A.20)}$$

The first term of the second member is the flux of the electric field, which is equal to the sum of all the charges inside the surface under study. That is to say:

$$\varepsilon_o \oint_A \vec{E} \circ dA = Q_f - Q_l \qquad \text{(A.21)}$$

where Q_f is the total free charge inside the surface *A*, and Q_l is the charge bound in *A*.

The second term is the flux of the polarization vector that is, in fact, the total bound charge inside A. Therefore, Eq. (A.20) results:

$$\oint_A \vec{D} \circ dA = Q_f - Q_l + Q_l = Q_f = \oint_V \rho_{vf} \cdot dV \qquad \text{(A.22)}$$

where ρ_{vf} is the volumetric density of free charge. Thus, keeping in mind the theorem of divergence, one obtains:

$$div\vec{D} = \rho_{vf} \tag{A.23}$$

The previous expression is known as first Maxwell Equation and it indicates that if there is no free charge inside a closed surface, the vector displacement in its interior is constant.

Appendix B Physical Properties of Crystals

Antonio Arnau, Yolanda Jiménez and Tomás Sogorb

Departamento de Ingeniería Electrónica, Universidad Politécnica de Valencia

B.1 Introduction

Crystals are, in general, anisotropic materials, that is, many of their physical properties depend on the direction in which they are considered. It is therefore not possible to understand their behavior and performance against external, mechanical or electrical effects or against changes in the environmental conditions such as changes in temperature, without due analysis of at least those properties most closely related to the object under study. The analysis of such properties provides the physical factors that relate a magnitude considered as the *cause* to another magnitude whose *effect* is the consequence of such magnitude. In the case under study, the physical factors include elastic, dielectric, and piezoelectric properties and thermal expansion coefficients.

B.2 Elastic Properties

The elastic properties characterise the behaviour of deformable solids studied in the theory of Elasticity. Any mathematical formulation for modelling physical phenomena is limited in the sense that it is necessary to develop some approaches to obtain a simple model that accurately represents the behavior of the phenomenon. One of the most common approaches is to consider physical phenomena as behaving linearly. In this case, linear elasticity studies the behavior of the elastic solid as a deformable, continuous, elastic, homogeneous and isotropic system of material points. In this theory, however, it is possible to include cases of heterogeneity corresponding to elastic constants dependent on point coordinates and anisotropy involving changes of the constants at a given point with respect to the direction.

The elastic constants are calculated by the constitutive equations that express the behavior of the material, i.e., they relate stresses and strains. These equations are phenomenological, their parameters are the elastic constants of the material, and they can only be obtained by experimental tests. However, their formulation requires knowing the physical constraints established by rational mechanics and by the equations of static equilibrium, which relate the acting forces with the static magnitudes called stresses, and by the equations of cinematic compatibility that represent the conditions between the displacement of the solid and the cinematic magnitudes called strains. In addition, the condition of linearity implies, on one hand, that the equations between strain and displacement be of first order, which means to neglect terms of a higher order; and on the other hand, that the material be elastic and Hookean, i.e., that the equations between stresses and strains be linear. Further, the elastic condition requires the solid to return to its original state without deformation when the forces acting on the material are removed.

B.2.1 Stresses and Strains

In order to obtain the equilibrium conditions at a generic interior point of the solid it is necessary to consider a rectangular contour with the sides parallel to the coordinate planes (Fig. B.1).

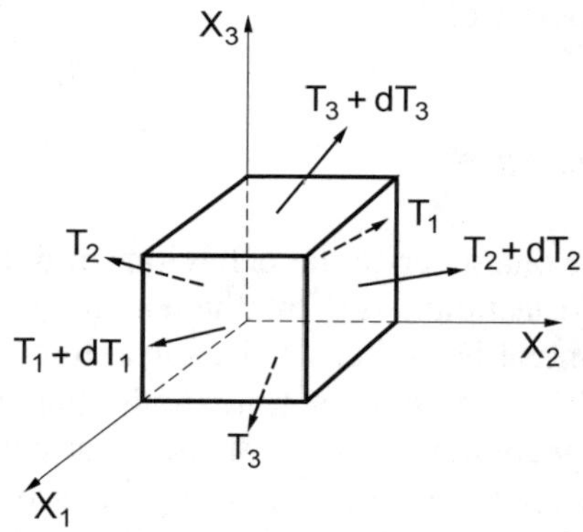

Fig. B.1. Generic point of a solid subject to mechanical stresses

On the sides of the elemental parallelepiped convergent with such planes, stresses T_1, T_2 and T_3 occur. Each stress can be broken down into its corresponding components depending on the direction of the coordinates as shown in Fig. B.2. Thus, stress T_i, will have as its components T_{i1} for direction X_i, T_{i2} for direction X_2 and T_{i3} for direction X_3. The stresses of each corresponding parallel plane will consider the changes in the components according to the direction of the coordinates. Therefore, the plane of the parallelepiped parallel to the coordinate plane on which stress T_1 is acting, will have as components $T_{11}+T_{11,1}\,dx_1$, $T_{12}+T_{12,1}\,dx_1$, $T_{13}+T_{13,1}\,dx_1$. The other components of the different planes are obtained in the same way.

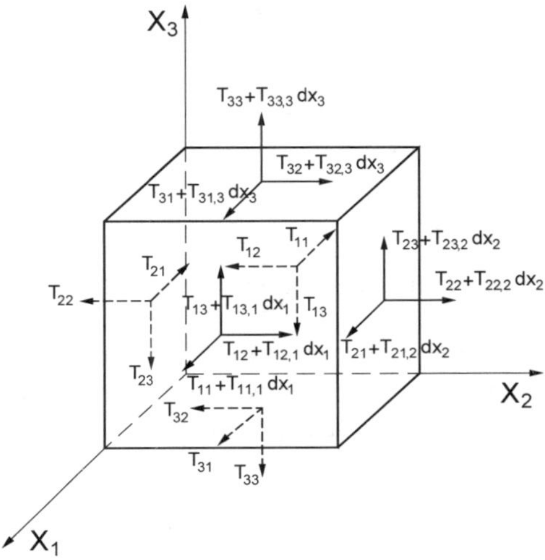

Fig. B.2. Decomposition of the stresses acting on a section of the solid.

The equilibrium of the forces acting on the volume of the elemental parallelepiped gives the following equations:

Resultant of null forces:

$$
\begin{aligned}
&F_1 - (T_{11,1} + T_{21,2} + T_{31,3})\,dx_1\,dx_2\,dx_3 = 0 \\
&F_2 - (T_{12,1} + T_{22,2} + T_{32,3})\,dx_1\,dx_2\,dx_3 = 0 \\
&F_3 - (T_{13,1} + T_{23,2} + T_{33,3})\,dx_1\,dx_2\,dx_3 = 0
\end{aligned}
\qquad \text{(B.1)}
$$

Null moment with respect to the gravity centre of the elemental volume:

$$
\begin{aligned}
&\sum M_1 = 0 \rightarrow T_{23} = T_{32} \\
&\sum M_2 = 0 \rightarrow T_{13} = T_{31} \\
&\sum M_3 = 0 \rightarrow T_{12} = T_{21}
\end{aligned}
\qquad \text{(B.2)}
$$

This result is obtained neglecting the terms of higher order. Therefore, from the static equilibrium conditions the stress tensor is considered to be symmetrical. In this way, any element of a material system may experience 6 types of stresses, three longitudinal depending on the directions of the coordinates, and three transversal or torsor type around the corresponding axes as shown in Fig. B.3.

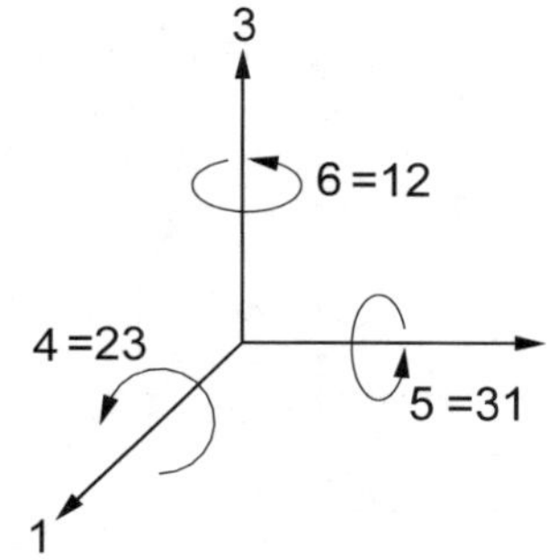

Fig. B.3. Different types of stresses or strains acting on a material

The stress components in the previous equations have been written following the typical tensor notation. However, other types of notation are often used whose equivalence is shown next

$$\begin{pmatrix} T_{11} & T_{12} & T_{13} \\ T_{12} & T_{22} & T_{23} \\ T_{31} & T_{23} & T_{33} \end{pmatrix} = \begin{pmatrix} T_{xx} & T_{xy} & T_{zx} \\ T_{xy} & T_{yy} & T_{yz} \\ T_{zx} & T_{yz} & T_{zz} \end{pmatrix} = \begin{pmatrix} X_x & X_y & Z_x \\ X_y & Y_y & Y_z \\ Z_x & Y_z & Z_z \end{pmatrix} = \begin{pmatrix} T_1 & T_6 & T_5 \\ T_6 & T_2 & T_4 \\ T_5 & T_4 & T_3 \end{pmatrix} \tag{B.3}$$

The first matrix follows the tensor notation, the second and third matrices show the directions of the stresses both longitudinal and torsors, whereas the fourth matrix is a tensor notation with reduced indices from 1 to 6 following the relationship previously established and considering Fig. B.3.

When a force is acting, all the points of an elastic solid experience a displacement. However, the stresses are not caused by the absolute movements of the points but by the separations and approximations between its particles, i.e., by the deformations.

To obtain the expression of the strains in terms of the displacements, let us consider a point P located at the origin of the coordinates, and a point Q with coordinates dx_1, dx_2, and dx_3 (Fig. B.4). During the deformation of the body, both points move to new positions P' and Q'. In order to estimate the strain, the difference in length before and after the deformation will be calculated, i.e., $P'Q' - PQ$.

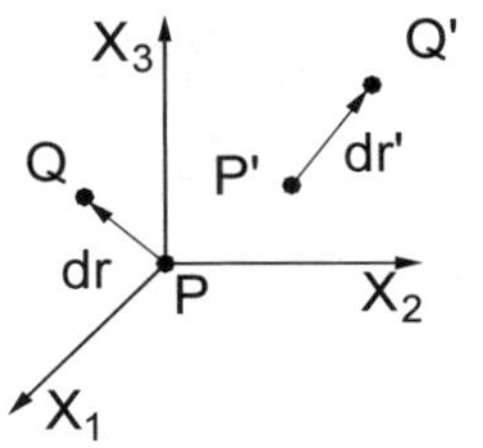

Fig. B.4. Deformation analysis of a segment *PQ*.

Let u_1, u_2 and u_3 be the displacements for axes X_1, X_2 and X_3 respectively. If ξ_1, η_1 and ζ_1 are the coordinates of point *P'*, the coordinates of *Q'* will be $dx_1+\xi_2$, $dx_2+\eta_2$ and $dx_3+\zeta_2$. Since the displacements are continuous functions of the coordinates, the relations between the displacements of each point will be given by:

$$\begin{aligned}
\xi_2 &= \xi_1 + \frac{\partial u_1}{\partial x_1}dx_1 + \frac{\partial u_1}{\partial x_2}dx_2 + \frac{\partial u_1}{\partial x_3}dx_3 \\
\eta_2 &= \eta_1 + \frac{\partial u_2}{\partial x_1}dx_1 + \frac{\partial u_2}{\partial x_2}dx_2 + \frac{\partial u_2}{\partial x_3}dx_3 \\
\zeta_2 &= \zeta_1 + \frac{\partial u_3}{\partial x_1}dx_1 + \frac{\partial u_3}{\partial x_2}dx_2 + \frac{\partial u_3}{\partial x_3}dx_3
\end{aligned} \tag{B.4}$$

As a consequence, the displaced differential vector *dr'* = *P'Q'* can be related to the differential vector previous to the displacement *dr* = *PQ* according to the following expression:

$$\overrightarrow{dr'} = \overrightarrow{dr} + J\,\overrightarrow{dr} \tag{B.5}$$

where *J* is the Jacobian matrix defined as:

$$J = \begin{pmatrix} \frac{\partial u_1}{\partial x_1} & \frac{\partial u_1}{\partial x_2} & \frac{\partial u_1}{\partial x_3} \\ \frac{\partial u_2}{\partial x_1} & \frac{\partial u_2}{\partial x_2} & \frac{\partial u_2}{\partial x_3} \\ \frac{\partial u_3}{\partial x_1} & \frac{\partial u_3}{\partial x_2} & \frac{\partial u_3}{\partial x_3} \end{pmatrix} = \begin{pmatrix} u_{1,1} & u_{1,2} & u_{1,3} \\ u_{2,1} & u_{2,2} & u_{2,3} \\ u_{3,1} & u_{3,2} & u_{3,3} \end{pmatrix} \tag{B.6}$$

which can be broken down into two matrices, one symmetrical matrix *S* and one asymmetrical matrix *A* defined as:

$$S=\begin{pmatrix} \frac{\partial u_1}{\partial x_1} & \frac{1}{2}\left(\frac{\partial u_1}{\partial x_2}+\frac{\partial u_2}{\partial x_1}\right) & \frac{1}{2}\left(\frac{\partial u_1}{\partial x_3}+\frac{\partial u_3}{\partial x_1}\right) \\ \frac{1}{2}\left(\frac{\partial u_1}{\partial x_2}+\frac{\partial u_2}{\partial x_1}\right) & \frac{\partial u_2}{\partial x_2} & \frac{1}{2}\left(\frac{\partial u_3}{\partial x_2}+\frac{\partial u_2}{\partial x_3}\right) \\ \frac{1}{2}\left(\frac{\partial u_1}{\partial x_3}+\frac{\partial u_3}{\partial x_1}\right) & \frac{1}{2}\left(\frac{\partial u_3}{\partial x_2}+\frac{\partial u_2}{\partial x_3}\right) & \frac{\partial u_3}{\partial x_3} \end{pmatrix}=\begin{pmatrix} S_{11} & S_{12} & S_{13} \\ S_{12} & S_{22} & S_{23} \\ S_{13} & S_{23} & S_{33} \end{pmatrix} \tag{B.7}$$

and

$$A=\begin{pmatrix} 0 & \frac{1}{2}\left(\frac{\partial u_1}{\partial x_2}-\frac{\partial u_2}{\partial x_1}\right) & \frac{1}{2}\left(\frac{\partial u_1}{\partial x_3}-\frac{\partial u_3}{\partial x_1}\right) \\ -\frac{1}{2}\left(\frac{\partial u_1}{\partial x_2}-\frac{\partial u_2}{\partial x_1}\right) & 0 & \frac{1}{2}\left(\frac{\partial u_2}{\partial x_3}-\frac{\partial u_3}{\partial x_2}\right) \\ -\frac{1}{2}\left(\frac{\partial u_1}{\partial x_3}-\frac{\partial u_3}{\partial x_1}\right) & -\frac{1}{2}\left(\frac{\partial u_2}{\partial x_3}-\frac{\partial u_3}{\partial x_2}\right) & 0 \end{pmatrix}=\begin{pmatrix} A_{11} & A_{12} & A_{13} \\ A_{12} & A_{22} & A_{23} \\ A_{13} & A_{23} & A_{33} \end{pmatrix} \tag{B.8}$$

Matrix S is symmetrical, since $S_{ij} = S_{ji}$. In addition, it possesses tensor properties and its components represent the different deformations experienced by the planes of the elemental parallelepiped when external forces act on it.

Figure B.5 shows the side of the elemental parallelepiped located on axes X_1 and X_2. Thus, the unitary longitudinal deformation in direction X_1 is defined as:

$$S_1 = \frac{O'A' - OA}{OA} = \frac{\partial u_1}{\partial x_1} = S_{11} \tag{B.9}$$

In this result, the effects of higher order corresponding to second partial derivatives have been neglected.

On the other hand, transversal strain in plane X_1X_2 is defined as the variation in the angle formed by two infinitely small segments when a solid experiences deformation, in both directions. Fig. B.5 shows that this variation, for small deformations, is:

$$S_6 = \langle BOA - \langle B'O'A' = \theta_1 + \theta_2 \approx \frac{\partial u_1}{\partial x_2} + \frac{\partial u_2}{\partial x_1} = 2S_{12} \tag{B.10}$$

Total transversal strain in plane X_1X_2 will be the sum of the tangential deformations or glidings according to the perpendicular directions and thus, $S_{12} = S_{21} = S_6/2$. The transversal deformations and the glidings or tangential deformations will be obtained in a similar way for the other planes.

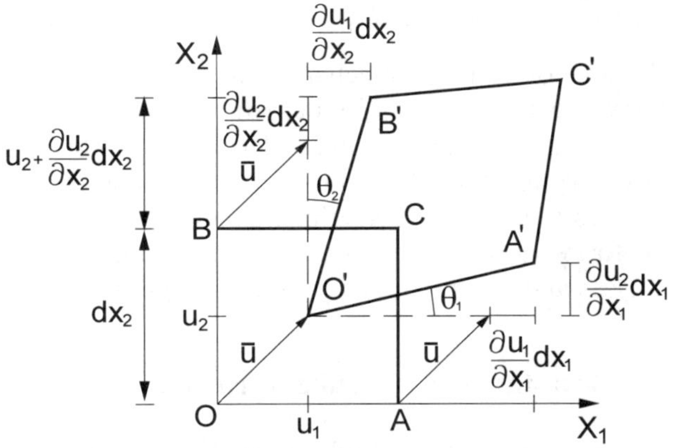

Fig. B.5. Deformation of the elemental parallelepiped

The previous notation corresponds to the reduced indices used for the stress tensor. Other common ways to express the strain tensor are:

$$\begin{pmatrix} S_{11} & S_{12} & S_{13} \\ S_{12} & S_{22} & S_{23} \\ S_{31} & S_{23} & S_{33} \end{pmatrix} = \begin{pmatrix} x_x & \frac{x_y}{2} & \frac{z_x}{2} \\ \frac{x_y}{2} & y_y & \frac{y_z}{2} \\ \frac{z_x}{2} & \frac{y_z}{2} & z_z \end{pmatrix} = \begin{pmatrix} S_1 & \frac{S_6}{2} & \frac{S_5}{2} \\ \frac{S_6}{2} & S_2 & \frac{S_4}{2} \\ \frac{S_5}{2} & \frac{S_4}{2} & S_3 \end{pmatrix} \tag{B.11}$$

On the other hand, each component of the antisymmetrical tensor ***A*** represents half turning of the bisectrix of each side of the elemental parallelepiped on the axis perpendicular to that side, the whole turn corresponding to the sum of the asymmetrical components. In this sense, when there is only deformation, the turnings are null.

The total internal energy stored in a general strain can be calculated as the sum of the energies caused by the different modes of distortion. In this way, the work performed in a longitudinal strain of the elemental parallelepiped in direction X_1 will be the product of the force in that direction by the displacement, that is, $T_{11}\, dx_2\, dx_3\, S_{11}\, dx_1$. Hence, the variation in the internal energy caused by the variation of the displacement in that direction is: $T_{11}\, dS_{11}\, dx_1\, dx_2\, dx_3$. The variation of internal energy due to the work performed by the transversal forces that cause an elementary shear deformation according to a turning around axis X_3 will be $T_{12}\, dx_2\, dx_3\, dS_{12}\, dx_1 + T_{21}\, dx_1\, dx_3\, dS_{21}\, dx_2 = 2\, T_{12}\, dS_{12}\, dx_1\, dx_2\, dx_3 = T_6\, dS_6\, dx_1\, dx_2\, dx_3$. Therefore, the internal energy stored for all the modes will be:

$$dU = \left[T_1 dS_1 + T_2 dS_2 + T_3 dS_3 + T_4 dS_4 + T_5 dS_5 + T_6 dS_6\right] dx_1\, dx_2\, dx_3 \qquad \text{(B.13)}$$

The application of rational mechanics has allowed establishing constraints that involve the static and cinematic behavior of an elastic solid subjected to the action of external forces. However, to characterise the material it is necessary to know the parameters that relate the stresses caused by external forces with the deformations generated. These parameters result from the formulation of constitutive or phenomenological equations that relate stresses and strains and are based on experimental results.

B.2.2 Elastic Constants. Generalized Hooke's Law

The experiments performed by Hooke show that for small displacements the more general strains in an elastic solid can be formulated through linear combinations of longitudinal and transversal deformations. Longitudinal deformations can occur in three directions parallel to the orthogonal axes. Similarly, any transversal deformation corresponds to the turnings around such axes (see Fig. B.3). In an isotropic material subjected to longitudinal stress, the unitary strain in the direction of the force for small displacements is proportional to the force per unit area or stress applied. This proportionality coefficient is referred to as *Young's modulus* and has the following expression:

$$Y = \frac{\frac{F}{A}}{\frac{\Delta l}{l}} \qquad \text{(B.14)}$$

where A is the transversal section on which force F acts, l is the original length of the material in the longitudinal direction of the force, and)l is the lengthening of the material in that direction.

The application of the law to transversal stresses relates shear stress with the resulting strain, defined as the shear angle, by the *transversal stiffness coefficient* or *transversal stiffness modulus* or *shear modulus*.

It is easy to understand that a stress in the longitudinal direction acting on an isotropic material may cause not only a longitudinal strain but also a deformation in the direction perpendicular to it. The relationship between longitudinal and lateral strains is referred to as *Poisson coefficient* and is dimensionless.

The situation in anisotropic materials is much more complex. In addition to the lateral and longitudinal strains, a longitudinal force stress can also cause transversal deformations. For instance, a stress acting on a thin bar of quartz cut in such a way that its length is parallel to axis *X*, not only lengthens it and makes it thinner (longitudinal and lateral deformation) but also tends to rotate it around that axis (transversal strain). Therefore, to thoroughly describe the relationship between stresses and deformations in crystalline materials, it is necessary to consider that a stress may produce any kind of deformation. Since there are 6 independent ways of causing stresses and of indicating deformation (Fig.B.3), 36 constants are needed to describe the general behavior of the solid. Usually there are two sets of equations depending on whether we want to express the stresses in relation to the deformations or vice versa. In the former case, the constants are called *stiffness coefficients or constants* and *elastic constants*, and in the latter *compliance coefficients or constants*. These equations are:

$$
\begin{aligned}
T_1 &= c_{11}S_1 + c_{12}S_2 + c_{13}S_3 + c_{14}S_4 + c_{15}S_5 + c_{16}S_6 \\
T_2 &= c_{21}S_1 + c_{22}S_2 + c_{23}S_3 + c_{24}S_4 + c_{25}S_5 + c_{26}S_6 \\
T_3 &= c_{31}S_1 + c_{32}S_2 + c_{33}S_3 + c_{34}S_4 + c_{35}S_5 + c_{36}S_6 \\
T_4 &= c_{41}S_1 + c_{42}S_2 + c_{43}S_3 + c_{44}S_4 + c_{45}S_5 + c_{46}S_6 \\
T_5 &= c_{51}S_1 + c_{52}S_2 + c_{53}S_3 + c_{54}S_4 + c_{55}S_5 + c_{56}S_6 \\
T_6 &= c_{61}S_1 + c_{62}S_2 + c_{63}S_3 + c_{64}S_4 + c_{65}S_5 + c_{66}S_6
\end{aligned}
\tag{B.15}
$$

where coefficient c_{ij} is an elastic constant that expresses the proportionality between deformation S_j and stress T_i. Using Einstein agreement the equations can be written as:

$$T_i = c_{ij}\,S_j \qquad i, j = 1 \;\; a \;\; 6 \tag{B.16}$$

The principle of energy conservation requires that $c_{ij} = c_{ji}$. From the equation that provides the variation of internal energy in terms of the strains it can be inferred that the stress component T_i is:

$$T_i = \frac{\partial U}{\partial S_i}$$

and then

$$c_{ij} = \frac{\partial T_i}{\partial S_i} = \frac{\partial^2 U}{\partial S_i \, \partial S_j} = \frac{\partial^2 U}{\partial S_j \, \partial S_i} = c_{ji}$$

since the order of differentiation does not vary the result. This reduces the number of independent elastic constants to 21.

The Generalized Hooke's Law can be formulated taking the stresses as independent variables. Then, we obtain:

$$\begin{aligned} S_1 &= s_{11}T_1 + s_{12}T_2 + s_{13}T_3 + s_{14}T_4 + s_{15}T_5 + s_{16}T_6 \\ S_2 &= s_{21}T_1 + s_{22}T_2 + s_{23}T_3 + s_{24}T_4 + s_{25}T_5 + s_{26}T_6 \\ S_3 &= s_{31}T_1 + s_{32}T_2 + s_{33}T_3 + s_{34}T_4 + s_{35}T_5 + s_{36}T_6 \\ S_4 &= s_{41}T_1 + s_{42}T_2 + s_{43}T_3 + s_{44}T_4 + s_{45}T_5 + s_{46}T_6 \\ S_5 &= s_{51}T_1 + s_{52}T_2 + s_{53}T_3 + s_{54}T_4 + s_{55}T_5 + s_{56}T_6 \\ S_6 &= s_{61}T_1 + s_{62}T_2 + s_{63}T_3 + s_{64}T_4 + s_{65}T_5 + s_{66}T_6 \end{aligned} \tag{B.17}$$

The values of coefficients s_{ij} can be obtained from the c_{ij} using the following relationship:

$$s_{ij} = \frac{(-1)^{i+j}\,\Delta^c_{ij}}{\Delta^c}$$

where)^c is the determinant corresponding to the matrix of coefficients c_{ij} of (B.17) and Δ^c_{ij} is the adjunct of term ij in the same matrix.

Constants c_{ij} and s_{ij} define the elastic behavior of the material. However, they do not possess tensor character since the deformation components S_1, S_2 ... S_6 do not have it.

The expressions used in the formulation of the physical properties of crystals are greatly simplified when the tensor notation is used. In addition, the calculation of the constants that characterise the properties for the orthogonal axes forming specific angles with the reference axes of the crystal becomes simpler due to the transformation laws of the tensors. On the other hand, the elastic and compliance constants provided by the researchers correspond to the expressions mentioned above. Therefore, it is necessary to establish the relationship between these constants and their corresponding coefficients in tensor notation.

The lack of tensor character of the strains defined in the expressions above is solved using as deformation tensor the symmetrical matrix $\boldsymbol{S}$ (Eq. (B.7)). The set of the nine elements that forms the matrix has a tensor character, like the symmetrical stress matrix (Eq. (B.3)). The relationship between both tensors of second order can be established using certain coefficients, depending on whether we want to express the stresses as a function of the strains or vice versa, which must form a tensor of fourth order. Then, the Generalized Hooke's Law can be written as:

$$T_{ij} = c_{ijkl}S_{kl} \tag{B.18}$$

or either its inverse form:

$$S_{ij} = s_{ijkl} T_{kl} \tag{B.19}$$

The elastic coefficients in tensor notation form a tensor of fourth order with 81 terms. However, as a result of the symmetry of the stress tensors T_{ij} and deformations S_{ij}, the tensor of elastic coefficients is also symmetrical. That is, its components fulfil the following equivalencies:

$$c_{ijkl} = c_{ijlk} = c_{jikl} = c_{jilk}$$

The relation between the tensor coefficients and the elastic constants is determined extending the tensor notation and comparing the equations obtained for each stress component T_{ij} with its corresponding one in Eq. (B.15) and considering that $S_{ij} = S_k/2$ when $i \neq j$ and $k = 4, 5$ *or* 6 (Eq. (B.11)). The equivalencies obtained can be summarised as follows:

$$c_{ijkl} = c_{\lambda\mu}$$

considering that subindex λ takes the values 1, 2 and 3 when the subindex pair *ij* takes the values 11, 22 and 33 respectively, and 4, 5 and 6 for 23, 31 and 12 and its permutations. The same happens with subindex μ in relation to the pair *kl*.

The equivalencies between the coefficients s_{ijkl}, of the equations that express the components of the deformation tensor, and the compliance constants in Eq. (B.17) are calculated in the same way and can be summarised as follows:

$$\begin{array}{lll} s_{\lambda\mu} = s_{ijkl} & \text{when} \quad i = j & \text{and} \quad k = l \\ s_{\lambda\mu} = 2s_{ijkl} & \text{when} \quad i \neq j & \text{or} \quad k \neq l \\ s_{\lambda\mu} = 4s_{ijkl} & \text{when} \quad i \neq j & \text{and}\, k \neq l \end{array}$$

with the same relations as for subindices λ and μ in relation to the pairs *ij* and *kl*.

Only 21 of the original 81 components are independent, like in crystals belonging to the triclinic system. However, this number decreases as crystal symmetry increases. The independent components correspond to half of the symmetrical matrix obtained from Eq. (B.15) (equation of the elastic constants) and are:

$$\begin{matrix} c_{1111} & c_{1122} & c_{1133} & c_{1123} & c_{1131} & c_{1112} \\ & c_{2222} & c_{2233} & c_{2223} & c_{2231} & c_{2212} \\ & & c_{3333} & c_{3323} & c_{3331} & c_{3312} \\ & & & c_{2323} & c_{2331} & c_{2312} \\ & & & & c_{3131} & c_{3112} \\ & & & & & c_{1212} \end{matrix} = \begin{matrix} c_{11} & c_{12} & c_{13} & c_{14} & c_{15} & c_{16} \\ & c_{22} & c_{23} & c_{24} & c_{25} & c_{26} \\ & & c_{33} & c_{34} & c_{35} & c_{36} \\ & & & c_{44} & c_{45} & c_{46} \\ & & & & c_{55} & c_{56} \\ & & & & & c_{66} \end{matrix}$$

For the case of the compliances the following relationship results:

$$\begin{matrix} s_{1111} & s_{1122} & s_{1133} & s_{1123} & s_{1131} & s_{1112} \\ & s_{2222} & s_{2233} & s_{2223} & s_{2231} & s_{2212} \\ & & s_{3333} & s_{3323} & s_{3331} & s_{3312} \\ & & & s_{2323} & s_{2331} & s_{2312} \\ & & & & s_{3131} & s_{3112} \\ & & & & & s_{1212} \end{matrix} = \begin{matrix} s_{11} & s_{12} & s_{13} & \frac{s_{14}}{2} & \frac{s_{15}}{2} & \frac{s_{16}}{2} \\ & s_{22} & s_{23} & \frac{s_{24}}{2} & \frac{s_{25}}{2} & \frac{s_{26}}{2} \\ & & s_{33} & \frac{s_{34}}{2} & \frac{s_{35}}{2} & \frac{s_{36}}{2} \\ & & & \frac{s_{44}}{4} & \frac{s_{45}}{4} & \frac{s_{46}}{4} \\ & & & & \frac{s_{55}}{4} & \frac{s_{56}}{4} \\ & & & & & \frac{s_{66}}{4} \end{matrix}$$

B.3 Dielectric Properties

In an anisotropic medium, the dielectric constant is actually a tensor that relates the components of the electric field with the vector of electric displacement according to the following expression:

$$\begin{aligned} D_1 &= \varepsilon_{11}E_1 + \varepsilon_{12}E_2 + \varepsilon_{13}E_3 \\ D_2 &= \varepsilon_{21}E_1 + \varepsilon_{22}E_2 + \varepsilon_{23}E_3 \\ D_3 &= \varepsilon_{31}E_1 + \varepsilon_{32}E_2 + \varepsilon_{33}E_3 \end{aligned} \tag{B.20}$$

where subscripts 1, 2 and 3 refer to axes X, Y and Z respectively.

Similarly, the susceptibility tensor results:

$$\begin{aligned} P_1 &= \chi_{11}E_1 + \chi_{12}E_2 + \chi_{13}E_3 \\ P_2 &= \chi_{21}E_1 + \chi_{22}E_2 + \chi_{23}E_3 \\ P_3 &= \chi_{31}E_1 + \chi_{32}E_2 + \chi_{33}E_3 \end{aligned} \tag{B.21}$$

These tensors represent the most common case. The transversal permittivities and susceptibilities (referring to the constants relating the displacement vector and the polarization vector in one direction with the field in one perpendicular direction) are only different from zero in the case of triclinic or monoclinic crystals. In the case of quartz, for example, the independent dielectric constants reduce to two.

B.4 Coefficients of Thermal Expansion

Another magnitude that depends on the direction is the strain generated by a change in temperature. In isotropic materials, thermal expansion is similar in all directions; however, in crystals, particularly in quartz, it depends on the direction. It is important to know the variation in length in a given direction due to changes in temperature since such variation will affect other physical properties, such as the temperature coefficients of the resonance frequencies. In addition, it may also affect the value of the constants that relate other magnitudes with the strain and may even make it difficult to determine the origin of certain electrical behaviors (e.g. piroelectricity).

By definition, the thermal expansion coefficients provide the deformation per temperature unit and can be expressed as the partial derivative of the strain with respect to temperature. It could be formulated as follows:

$$\alpha_i = \frac{\partial S_i}{\partial \theta} \tag{B.22}$$

B.5 Piezoelectric Properties

As we have already mentioned, in a deformable solid there are three fundamental directions according to axes *X, Y, Z* and six possible types of deformations and stresses, polarization in one direction may be due to the contributions of each strain or stress. In this way, a first approximation of the piezoelectric phenomenon can be made as a tensor of third order that relates polarization in one direction with each of the possible stresses or strains, that is:

$$\begin{aligned} P_i &= e_{ijk}\, S_{jk} \\ P_i &= d_{ijk}\, T_{jk} \end{aligned} \qquad i, j, j = 1, 2, 3 \tag{B.23}$$

The use of the first or second former expressions depends on whether we want to relate strains or stresses to polarizations.

Coefficients e_{ijk} and d_{ijk} represent the piezoelectric properties of the material and are called stress or strain piezoelectric coefficients or constants respectively. A reduced notation can be used, like in the case of the elastic constants, to simplify the mathematical formulation. Then, it can be written as:

$$\begin{aligned} P_1 &= d_{11}\,T_1 + d_{12}\,T_2 + d_{13}\,T_3 + d_{14}\,T_4 + d_{15}\,T_5 + d_{16}\,T_6 \\ P_2 &= d_{21}\,T_1 + d_{22}\,T_2 + d_{23}\,T_3 + d_{24}\,T_4 + d_{25}\,T_5 + d_{26}\,T_6 \\ P_3 &= d_{31}\,T_1 + d_{32}\,T_2 + d_{33}\,T_3 + d_{34}\,T_4 + d_{35}\,T_5 + d_{36}\,T_6 \end{aligned} \tag{B.24}$$

or

$$\begin{aligned} P_1 &= e_{11}\,S_1 + e_{12}\,S_2 + e_{13}\,S_3 + e_{14}\,S_4 + e_{15}\,S_5 + e_{16}\,S_6 \\ P_2 &= e_{21}\,S_1 + e_{22}\,S_2 + e_{23}\,S_3 + e_{24}\,S_4 + e_{25}\,S_5 + e_{26}\,S_6 \\ P_3 &= e_{31}\,S_1 + e_{32}\,S_2 + e_{33}\,S_3 + e_{34}\,S_4 + e_{35}\,S_5 + e_{36}\,S_6 \end{aligned} \tag{B.25}$$

Considering the symmetry of the stress and strain tensors and their relations with the components of the reduced indices, the following relationship between the piezoelectric coefficients in tensor and reduced notation is obtained:

$$e_{i\lambda} = e_{ijk} \quad \text{and} \quad \begin{array}{lll} d_{i\lambda} = d_{ijk} & \text{for} & j = k = \lambda = 1,2,3 \\ d_{i\lambda} = 2d_{ijk} & \text{for} & j \neq k \;;\; \lambda = 4,5,6 \end{array}$$

Equations (B.24) and (B.25) formulate the direct piezoelectric phenomenon. Lippmann predicted the existence of the inverse piezoelectric phenomenon by the application of the thermodynamic potentials to the piezoelectric effect. His reasoning was as follows: If a piezoelectric crystal is located in an electric field of intensity E and subjected to a mechanical stress T, the crystal is deformed by an amount S and a polarization P appears. If under these circumstances both the field and the stress experience a small variation in the values dE and dT respectively, the overall variation of the internal energy can be expressed as an exact differential $dU = PdE+SdT$. If the process is assumed to be reversible, then it can be written as:

$$\left(\frac{\partial P}{\partial T}\right)_E = \left(\frac{\partial S}{\partial E}\right)_T = d \tag{B.26}$$

This equation expresses the fact that the relationship between the polarization generated by a stress is equal to the strain caused by an electric field. This constant has been called d and corresponds to the strain piezoelectric coefficient previously mentioned.

The Curie brothers soon proved Lippmann's predictions, and since then, a number of studies on the topic appeared that generalised Lippmann's theory and used it to formulate their theories of piezoelectricity. Most of these studies have assumed that the phenomena are reversible and that all the relationships are linear. In addition, the theories based on thermodynamic potentials also allow the formulation of elastic, dielectric and piezoelectric phenomena as wells as their relationships. Thus, considering the relations for the dielectric and piezoelectric polarizations due to field and stress, the total electric displacement and total deformation in an anisotropic piezoelectric material can be written as:

$$\begin{aligned} S_i &= s_{ij}T_j + d_{mi}E_m \qquad i,j = 1 \ to\ 6 \\ D_m &= d_{mj}T_j + \varepsilon_{mn}E_n \qquad m,n = 1 \ to\ 3 \end{aligned} \qquad \text{(B.27}$$

The first equation corresponds to the reverse piezoelectric phenomenon and the second equation to the direct one. It is possible to obtain two additional formulations. By multiplying both equations by the elastic stiffness matrix c_{ij} and considering that this matrix is the inverse of the compliance matrix, we obtain:

$$c_{ij}S_j = c_{ij}s_{ji}T_i + c_{ij}d_{mj}E_m \qquad \begin{matrix} i,j = 1 \ to\ 6 \\ m = 1 \ to\ 3 \end{matrix} \qquad \text{(B.28)}$$

Then:

$$T_i = c_{ij}S_j - c_{ij}d_{mj}E_m \qquad \begin{matrix} i,j = 1 \ to\ 6 \\ m = 1 \ to\ 3 \end{matrix} \qquad \text{(B.29)}$$

Considering that the piezoelectric constant that relates the field to the stress is e, the stress can be written as:

$$T_i = c_{ij}S_j - e_{mi}E_m \qquad \begin{matrix} i,j = 1 \ a\ 6 \\ m = 1 \ a\ 3 \end{matrix} \qquad \text{(B.30)}$$

Therefore there is a relationship between the stiffness coefficients and the piezoelectric constants given by:

$$e_{mi} = c_{ij}\, d_{mj} = d_{mj}\, c_{ji} \qquad \text{(B.31)}$$

Multiplying the expression above by s_{ji} a new relationship is obtained:

$$s_{ji}e_{mi} = s_{ji}c_{ij}\, d_{mj} \quad \Rightarrow \quad d_{mj} = s_{ji}\, e_{mi} = e_{mi}\, s_{ij} \qquad \begin{matrix} i, j = 1\ a\ 6 \\ m = 1\ a\ 3 \end{matrix} \tag{B.32}$$

Replacing d_{mj} in the second equation of (B.27) and considering that $S_i = c_{ij}\, T_j$, a new expression is obtained for the direct piezoelectric effect:

$$D_m = e_{mi}S_i + \varepsilon_{mn}E_n \qquad \begin{matrix} i = 1\ a\ 6 \\ m, n = 1\ a\ 3 \end{matrix} \tag{B.33}$$

Thus, two equations are obtained for the direct effect and two equations for the reverse effect, shown next:

$$\left.\begin{matrix} D_m = d_{mj}T_j + \varepsilon_{mn}E_n \\ D_m = e_{mi}S_i + \varepsilon_{mn}E_n \end{matrix}\right\} \text{Direct Effect} \qquad \begin{matrix} i, j = 1\ a\ 6 \\ m, n = 1\ a\ 3 \end{matrix}$$

$$\left.\begin{matrix} T_i = c_{ij}S_j - e_{mi}E_m \\ S_i = s_{ij}T_j + d_{mi}E_m \end{matrix}\right\} \text{Inverse Effect}$$

$$e_{mi} = d_{mj}\, c_{ji}$$

$$d_{mi} = e_{mj}\, s_{ij}$$

Index

Printing: Strauss GmbH, Mörlenbach
Binding: Schäffer, Grünstadt